W9-BQV-169

DISCARD

COLLECTION MANAGEMENT

a chance for lasting survival

ECOLOGY AND BEHAVIOR OF WILD GIANT PANDAS

PAN WENSHI, LÜ ZHI, ZHU XIAOJIAN, WANG DAJUN,
WANG HAO, LONG YU, FU DALI, AND ZHOU XIN

edited by
William J. McShea, Richard B. Harris, David L. Garshelis, and Wang Dajun

translated by
Richard B. Harris

A Smithsonian Contribution to Knowledge

Smithsonian Institution
Scholarly Press
WASHINGTON, D.C.
2014

Published by
SMITHSONIAN INSTITUTION SCHOLARLY PRESS
P.O. Box 37012, MRC 957
Washington, D.C. 20013-7012
www.scholarlypress.si.edu

Cover images: Xiwang, at the age of 3 in February 1996 (top);
the upper montane region of Qinling, from 1,350 m to the highest peaks
at 3,700 m (bottom). Both photos: Pan Wenshi.

Library of Congress Cataloging-in-Publication Data

Pan, Wenshi.
  [Ji xu sheng cun de ji hui. English]
  A chance for lasting survival : ecology and behavior of wild giant pandas / Pan Wenshi,
Lu Zhi, Zhu Xiaojian, Wang Dajun, Wang Hao, Long Yu, Fu Dali, and Zhou Xin ; edited
by William J. McShea, Richard B. Harris, David Garshelis, and Wang Dajun ; translated by
Richard B. Harris.
    pages cm. — (A Smithsonian contribution to knowledge)
  Includes bibliographical references and index.
    ISBN 978-1-935623-17-5 (cloth : alk. paper) — ISBN 978-1-935623-30-4 (ebook)   1. Giant
panda—China—Qinling Mountains.   I. Title.
  QL737.C27P342613 2014
  599.789—dc23   2013009579

ISBN-13 (print): 978-1-935623-17-5
ISBN-13 (ebook): 978-1-935623-30-4

Printed in the United States of America

∞ The paper used in this publication meets the minimum requirements of the American
National Standard for Permanence of Paper for Printed Library Materials Z39.48–1992.

Contents

*The original chapter 10, "Genetic Diversity of Giant Panda Populations," has been omitted from this translation. See Introduction (p. 18).

Foreword to the English Translation

In 2001, soon after the original Chinese edition of *A Chance for Lasting Survival* was published, Kathryn S. Fuller, J.D., who was president of World Wildlife Fund (WWF) United States at the time, visited Peking University and told me that she wanted to translate the book into English. But I chose not to follow her suggestion because it was not yet possible to judge the effects that our actions in Qinling—halting logging and establishing a nature reserve in this montane-subalpine region—would have on the area's natural landscape and human society.

In April of 2011, 15 years after bidding farewell to Qinling, I returned to Xinglongling, where I had focused my giant panda research efforts. On that visit, I coincidentally ran into Dr. William McShea, who was investigating the area's biodiversity. This time, when Dr. McShea proposed an English edition of *A Chance for Lasting Survival*, not only did I immediately agree, I asked him to plan the project and do everything in his power to make it happen.

The timing could not be better for the publication of this English edition. Logging has been halted for a full 17 years, and Xinglongling's natural environment has had this period of time to recover. We can now see what has become of Xinglongling's inhabitants, including its humans, over the last 17 years, and we are finally able to more objectively evaluate the success of this conservation endeavor.

On a snowy day in 1984, I led a scientific team from Peking University into Xinglongling, the hinterland located between the western section of northern Qinling's southern slope and the western section of middle Qinling's northern slope. From the beginning, we viewed the protection of Qinling's giant pandas and all of the broadleaf forest, mixed broadleaf-conifer forest, and conifer forest ecosystems within this montane-subalpine region as an integral part of the global environmental protection effort. Over the last 28 years, we have continuously monitored developments in Qinling. On the worldwide stage of nature conservation, I believe that Qinling presents an important case for reference. With this conviction in mind, I have divided the recent natural history of Qinling into three phases.

Phase 1: Halting the Destruction of Xinglongling's Biodiversity and Establishing the Changqing National Nature Reserve

Early on, we made a crucial discovery: at an altitude of 3,071 m above sea level on the southern slope of Xinglongling lies the most biodiverse area in the Qinling mountain range, providing a refuge for many unique species, including the giant panda (*Ailuropoda melanoleuca*), takin (*Budorcas taxicolor*), golden snub-nosed monkey (*Rhinopithecus roxellanae*), and crested ibis (*Nipponia nippon*). These creatures depend on their intrinsic reproductive abilities and the natural environment of Xinglongling for their continued survival. In the last decade of the last century, however, this area's natural environment fell into critical danger. In 1991, the local Changqing Forestry Bureau extended a logging road up to an altitude of 2,800 m. This road opened the way for logging a pure primary-growth forest of Taibai larch (*Larix chinensis*), a rare tree species that grows endemically in Qinling's subalpine zone.

In October of 1993, I personally witnessed the wanton decimation of this forest that had grown since the end of the last ice age. We realized that because mankind's activities had created this ecological crisis, only mankind had the power to end it. After the failure of repeated attempts to stop the logging, we finally appealed to China's central government for authoritative political and financial support to protect this land. Losing little time, the central government halted the logging of Xinglongling's forests in July of 1994 and established the Changqing National Nature Reserve in 1995. The creation of this reserve was the most crucial step in the struggle to preserve Qinling's natural environment.

However, the total land area of Changqing National Nature Reserve was only 300 km²; even including the adjoining Foping Nature Reserve, the combined land area totaled only 600 km². Furthermore, these reserves were a mere speck surrounded on all sides by extensive swathes of logging and agricultural cultivation.

From the perspective of species conservation, such a minuscule "ecological island" cannot ensure the survival of its inhabitants. Logging continued unabated in other portions of Qinling, creating a geographically fragmented habitat. Because of the obstruction of healthy genetic flow, species in such a habitat inexorably head down the path toward extinction.

Phase 2: Xinglongling's Recovery and Significant Increases in Protected Land Area

In fall of 1995, although logging had, in fact, ceased in Xinglongling, the mountainsides I observed were scenes of utter devastation. I could not imagine how this seemingly ruined piece of land would be able to pull its scarred, weary body into a phase of postdisaster recovery.

However, the events that followed were much more fortuitous than I had imagined. In the years following the establishment of the Changqing National Nature Reserve, Qinling received serendipitous boosts from the Chinese government's environmental undertakings, including the Natural Forest Protection Program and the Conversion of Cropland into Forest Policy. One after another, new reserves were established in the area, expanding the protected panda habitat outward from Xinglongling to 2,000–3,000 km^2 of nearby land. It was during this time that Xinglongling's biodiversity entered a recovery phase.

In 2008, Dr. William McShea of the Smithsonian Conservation Biology Institute and Dr. Wang Dajun of Peking University, using unmanned cameras with infrared triggers, recorded the presence of many animals representative of ancient growth forests: giant pandas, golden snub-nosed monkeys, Asian leopards (*Panthera pardus fontanieri*), takin, serow (*Capricornis milneedwardsii*), tufted deer (*Elaphodus cephalophus*), yellow-throated martens (*Martes flavigula*), Asiatic black bears (*Ursus thibetanus*), porcupines (*Hystrix hodgsoni*), and wild boar (*Sus scrofa*). These animals, among others, had already returned in droves and were again starting to birth and raise their young in this peaceful montane-subalpine region. This was a real signal that the recovery of Changqing National Nature Reserve's ecosystems had begun.

Phase 3: Rebuilding Xinglongling's Vast Wilderness Area—Relying upon the Combined Forces of Nature and Mankind

Even after suffering the calamity wrought by logging, Xinglongling still possessed the vivacity required to resuscitate itself. Before logging had occurred, a belt of broadleaf deciduous forest grew in Qinling's warm temperate zone from the agricultural altitude limit at 1,350 m up to 1,800 m; in the temperate zone, from 1,800 to 2,600 m, there flourished a mixed conifer-broadleaf forest, and above that, in the cold temperate zone, a conifer forest previously reached up to 3,200 m.

We discovered that although the forests had been practically wiped clean off of Qinling's mountainsides by 1999, these previously forested areas were still covered by a thick layer of humus sheltering an abundance of dormant seeds. We believe that considering the soil conditions and given Xinglongling's favorable temperatures and precipitation, it will only be a matter of time before these forests reestablish themselves with their innate vitality.

Key to the healthy rehabilitation of Xinglongling's ecosystems will be a rationally designed network of wildlife corridors connecting the area's reserves. Because many corridors will have to extend through agricultural areas lower than 1,350 m, comprehensive, high-level scientific investigation is needed to sort out the complicated relationships among natural habitats, agriculture, industry, transportation, the human population, and wildlife. Even more importantly, there must be a concerted national effort to ensure that all of the necessary political, economic, and administrative considerations are in line. Otherwise, this essential advancement in conservation will not enjoy successful implementation.

There is hope that Xinglongling's vast wilderness, an area free of mankind's interference, will be reestablished within 20–30 years. Here humanity's reckless trespass will be prohibited, with only occasional visitors quickly passing through.

Although the content of this book focuses on Qinling's nonhuman species and habitats, any discussion of Qinling's recovery must not neglect the critical role of the local human population. In 1996, I passed through the town of Huayang at an altitude of 1,100 m. This is a place with 1,600 years of history, located at the confluence of Xinglongling's wilderness and the agricultural community below. At that time, the sudden loss of the logging industry and its accompanying commerce had dragged this little mountain town into an economic depression. Scenes of decline and destitution were everywhere. Upon seeing the sorry state of Huayang, my emotions became too complicated to describe, my mind examining and reexamining this question: had my advocacy for the logging ban and nature reserve been a grave error?

Fifteen years quickly passed, and I returned to Huayang in April of 2011, just in time to catch the warmth of spring and the flowers blooming. The scenes of prosperity that greeted me made my heart leap with joy. A healthy human community, bursting with vigor, glittered before me. Family after family had found considerable success, either by operating rural bed-and-breakfasts and other ecotourism businesses or by cultivating local agricultural specialties. On that visit, I finally understood the real significance of protecting Xinglongling's forests.

What we gained is not merely a natural refuge for giant pandas. Indeed, it is even more than a Garden of Eden sheltering countless living beings. Above all, Xinglongling's wilderness is the true wellspring of Huayang's happiness and prosperity. It is the land of milk and honey that will nurture and sustain Huayang's people for generations to come.

I express my sincere gratitude to Adrian Lu for translating this foreword into English.

Pan Wenshi
February 29, 2012

Foreword to the Chinese Edition

Nature conservation got a late start in China because for a long time, people failed to recognize the importance of protecting the environment and they did not know the best ways to protect nature and biodiversity. Although the concept that we should conserve giant pandas had been raised as early as 1946 in an article in the newspaper *Da Gong Bao,* pointing out that "at present there is a great deal of hunting. . . . Should this continue, pandas face the prospect of extinction," by 1949, when the Chinese government completely prohibited the export of giant pandas, no fewer than 73 pandas had already been sent abroad. Although the State Council issued an order in 1962 for all provinces to forbid hunting giant pandas, periodic cases of poaching persisted.

During the period from 1963, when the first two giant panda nature reserves were established in Sichuan Province, to 1978, China established 12 more reserves, but these covered only 20% of the known area of panda distribution. Logging activity continued on the remaining 80% of panda habitat. Since 1990, however, support of all kinds from the Chinese government has allowed the number of nature reserves for giant pandas to increase to 33. In October 1998, Sichuan promulgated a ban on logging within all of the province's natural forests, creating an unprecedented opportunity for the recovery of the giant panda's natural habitat. However, understanding the threats to the species' survival and helping it to persist in its natural environment remained unsolved problems.

In 1976, a large-scale flowering of the bamboo species *Fargesia denudata* followed landslides that attended the large earthquake in the northern Min Shan region. This flowering was the backdrop for the large panda mortality event that attracted so much attention. In 1980, deeply concerned with the possibility that we might lose the giant panda entirely, government bureaus charged with conservation within China came together with the World Wildlife Fund (WWF) to jointly establish a collaborative research project focusing on the giant panda. In 1985, this project published the first ecological study in the history of research on the giant panda, *The Giant Pandas of Wolong*. Among the findings reported in that book were that "the giant panda retains the simple stomach and intestines of a carnivore while leading the life of a specialized herbivore. Without physical and physiological specializations to process cellulose and other indigestible plant materials, the panda has in evolution tied itself to bamboo; it has, in fact, become so fettered by the bamboo's passive power that its fate now depends on that of the plant. . . . How is the giant panda adapted to bamboo? This is the main scientific question our report attempts to answer." The book was unable to address the original starting point of the project: "The primary purpose of the project is to study those aspects of the panda's life history which are most important to conservation. For example, to preserve the panda it is essential to determine the minimum population size needed for the species to persist in the small, isolated areas that now form its habitat. Such a determination requires information on birth and death rates, movement patterns, and carrying capacity."

In July 1983, when we researchers on the cooperative project met at Peking University's Shaoyuan Guesthouse to write and edit the above-mentioned book, we realized that the mountains in Wolong contained two species of bamboo (*Fargesia spathacea* and *Sinarundinaria fangiana*) and that both were abundant. When one species flowers, the other one provides sufficient food for pandas to buffer the effect. But in autumn 1983, when there was widespread flowering of *S. fangiana* within the Wolong Nature Reserve, media coverage led some people to overemphasize supposed difficulties pandas have with reproduction, which in turn led to an overly pessimistic view of prospects for the giant panda. From that time on, the question on many people's minds was, "Might the giant panda have already entered an evolutionary dead end and what we are now witnessing is merely the slow wait for extinction that could come at any time?" Or, alternatively, were pandas more like other species with hope for a brighter future?

Shortly afterward, in autumn 1984, we came for the first time to the Qinling Mountains to study pandas. While trudging through the vast tracts of uninhabited forests searching for evidence of the giant panda to document, we were surprised one day to discover, at an elevation of 1,800 m, relics of past humans nestled within the bamboo. Inscriptions on several steles in the area made clear that perhaps 1,000 years earlier, the site had been home to a large human population with a flourishing economy. This made plain that vegetation in these montane and subalpine regions was capable of recovering from agricultural development after people had abandoned their fields, given some time. This then suggested that pandas could also recover, reclaiming, as it were, those habitats that had once been denied them by humans. This realization provided us with a ray of light in what was otherwise a rather depressing picture as we considered pandas and how to conserve them.

x   Foreword to the Chinese Edition

Thus energized, we began organizing a team that could investigate in a deep way the chance that pandas had for persistence in the Qinling. Specifically, we set as our objectives understanding the effects of human activity on the long-term persistence of pandas and exploring effective methods to prevent their extinction. To achieve these goals, we required a thorough understanding of pandas' unique behaviors, as well as their complex relationships with their habitat, including humans.

Our research team, which began its work in spring 1985, was a multidisciplinary one, composed primarily of master's and doctoral students. We began with the difficult but crucial task of attempting to clarify the Qinling population's abundance and distribution. We realized that the most direct approach was to capture free-ranging pandas, equip them with radio collars, and monitor them over long time periods, documenting their activities, migrations, and patterns of habitat use. From June 1986 to March 1999, we collected some 10,000 data points from radio-collared pandas, which provided information on their locations, activity budgets, seasonal migrations, dispersal routes, dens, and home ranges, as well as behaviors during estrus, mating, parturition, and feeding. Beginning in February 1994, we began using a specially designed miniature closed-circuit television video recorder equipped with an infrared sensor to document behavioral relationships involving mothers and their newborn cubs. We collected hair found on tree branches or bamboo to supplement blood samples from captured animals for analysis of genetic diversity of the panda population. We collected panda feces, comparing samples with those from the species of living bamboo consumed; in addition to chemical analysis, we employed both optical and electron microscopy to assess how pandas used bamboo cytoplasm and cell wall contents. We used extensive ground observations, supplemented with remote sensing imagery, to analyze the spatial structure of panda habitat. We estimated the abundance and density of pandas within the 92 km² study area. On the basis of 15 years' data, we constructed a model of the Qinling panda population's dynamics and assessed its long-term viability. We documented a high level of species richness in the Qinling habitats that support giant pandas; many of the plant and animal species currently present in the Qinling formerly lived over a much larger geographic area.

The giant panda is the flagship species for this biological community; ensuring its survival requires also protecting thousands of other species that have persisted over evolutionary time. We documented that many aspects of this ecosystem are in flux; conserving the giant panda therefore requires constant monitoring because prediction alone will not suffice. Integrating our 15 years of multifaceted research effort on panda ecology and behavior, although we cannot claim to have full knowledge of every aspect of evolutionary pressure on the Qinling giant panda population, a general picture of how this local population can persist into the future has gradually emerged.

Thus, by publishing the results of our 15 years of research, we believe we can deepen understanding of the giant panda's biology and the threats it faces; ultimately, we trust this will assist us in selecting the most appropriate strategies for its conservation.

This introduction details the scientific content of this book, which has been 3 years in the making as we have organized field data, undertaken laboratory procedures, and conducted statistical analyses. Doing so has increased our realizations not only that free-ranging pandas living today have managed to maintain their populations in the face of fluctuating environments over the past two centuries or so but also that their behavior we view today is a manifestation and consequence of long-term evolutionary processes. This is the context within which we understand giant panda biology and the animal's relationships with the larger natural world, as well as with mankind.

When it comes to giant panda conservation, we strongly support applied research. The value of this type of research should be very clear: to facilitate existing populations to continue flourishing within their natural habitats rather than to see these animals as experimental subjects useful for addressing some purely scientific question. As in any other field of science, we find among panda researchers a variety of ideas and contrary points of view. This situation often arises when raw field data are scarce, although it can also be caused by inappropriate methods or interpretations. For example, what do we understand to be the cause of the panda's current endangered status? What is the best approach to conserving modern giant pandas? As these questions arise in the book, we point out the origins of differing opinions on these and other issues, as well as present our own viewpoints.

As for conservation of giant pandas in the Qinling, although some details remain unclear, a general consensus has now emerged. The giant pandas of Qinling are not, in fact, headed into an evolutionary dead end. However, at this point, their ultimate survival depends not so much on their own natural ability as it does on mankind's skill in conservation and management. Because logging ceased in the core distribution area of the Qinling pandas as of July 1, 1994, and logging was banned throughout the entire Qinling in the summer of 1999, we have reason to be truly optimistic regarding the future of giant pandas in the Qinling. In considering the potential of this population of giant pandas, we believe it is no different from other species of wildlife that can be expected to flourish and have a bright future.

Pan Wenshi
August 28, 2000

Preface to the Chinese Edition

In October 1984, in the midst of a national fervor to "save the giant panda" after bamboo had flowered and died off, I wrote a report to the Central Committee explaining my disagreement both with a plan to set up 13 panda feeding centers in the core of the panda's distribution area and with plans to continue taking wild pandas into captivity. I persisted in my view that the best way to preserve this species was to strengthen research on wild populations and find ways to allow them to continue to survive in their natural habitat.

In November of that same year, when I first had the opportunity to visit the Qinling, I was struck by how the remaining habitat for giant pandas was so clearly associated with elevation: the higher the elevation was, the more panda habitat there was; the lower in elevation one looked, the more that agricultural development had already taken hold. Once one got to about 1,350 m in elevation, agriculture ceased and panda habitat began. This critical fact laid the foundation for my decision to look further into the chance of lasting survival for pandas in the Qinling.

This was the first time that I had independently conducted research on pandas since leaving the Wuyipeng research station at the Wolong Nature Reserve, Sichuan Province. Although articles and reports of scholarly meetings on conservation of giant pandas had by now become as numerous as the hairs on an ox, my initial proposals for studying pandas and their habitats were met with indifference. It was made clear that my applications for research funding would be approved only if they addressed the issue of "how bamboo flowering had threatened the panda with extinction." Without question, this was a preconceived conclusion that my conscience as a scientist would not allow me to accept. In any case, I had already conducted 3 years of research in Wolong and had seen with my own eyes that the flowering of bamboo posed no threat to the pandas' survival.

In March 1985, I led two graduate students and an undergraduate student, all carrying heavy backpacks, into the snow-covered southern slopes of the Qinling. Only 39 days after we entered the mountains, one of the graduate students, 21-year-old Zeng Zhou, was tragically killed when he fell from a cliff while searching for panda sign. It was with profound grief and under enormous emotional stress that we buried our companion's ashes and returned to our research, which at that point had barely even begun. Three months later, the undergraduate student matriculated and was assigned work elsewhere, leaving only me and a young female graduate student, who had just turned 20, to carry forward our work, born as it was in adversity.

This female student, who had grown up in a big city and graduated from the Department of Biochemistry and Molecular Biology at Peking University, endured untold suffering during her 7 years of field research. The severe conditions at the high-elevation research area resulted in her suffering from arthritis for several years, and at its most severe she was nearly unable to walk. But she never gave up. Her firm belief and strong will drove her to complete devotion to her work. Late one night, while in a small mountain village after rescuing a brown-and-white colored panda named Dandan, she was overcome by kerosene gas in a small mountain village. The following year she once again suffered gas poisoning at a mountain guesthouse. Once she accidentally slid into a crevasse in the ice; only because of timely rescue was her life saved in each instance.

In 1992, she temporarily left the Qinling for the United States to conduct postdoctoral research at the Molecular Genetics Laboratory at the National Institutes of Health (NIH). First, she conducted outstanding analyses of the genetic structure of some of the planet's most endangered mammals, including the tiger and orangutan. Following up on this work, she discovered that the genetic diversity of living pandas was no lower than other mammals, pointing out that, contrary to popular belief, pandas had not come to an evolutionary dead end. In 1996, she turned down an offer from NIH to continue working at the molecular genetics laboratory there. Her mission to devote herself to scientific research and preserve biodiversity drove her to return immediately to her alma mater and immerse herself in our new conservation work.

To this day, it seems to me that many fortuitous factors led to the success of this research project. That said, I must admit that the project would have had died an untimely death if not for the participation of the 15 graduate students who took part from 1984 to 1999. Sixteen busy years have flown by, but I am still overcome with admiration every time I think of this crew of young people, full of energy but lacking experience, remaining united and giving their all in the face of natural hardships and dealing with the inevitable difficulties of human interactions.

Because our funding was limited, we had to economize daily. To get to the research site from Beijing was an arduous journey for the graduate students, first requiring a 35-hour train ride to Hanzhong (on which they had to sit on hard seats the entire way), followed by an 8-hour bus ride (without taking time to overnight in town), and then further travel in rough forestry bureau vehicles to get to the research station. Our young male graduate students had extra work to do: following each day's fieldwork, they had to push their already tired bodies toward the task of gathering firewood for our station's stove. To minimize expenses, we eschewed hiring workers, chopping firewood and cooking for ourselves.

In the Qinling, giant pandas live at 2,000- to 3,000-m elevation, where winters are so cold that even forestry workers descend seasonally. However, to follow the animals, we stayed in these mountains all winter. One winter, I spent Chinese New Year in the mountains together with a graduate student. To save time and firewood, we boiled up a hash of potatoes and rice in a big pot and ate nothing else for several days until we moved to a new work site, whereupon we boiled up another pot of it. This was our "gourmet field cuisine."

Conditions in the field were rough. In retrospect, our biggest problems were lack of water and food; at times it seemed as if we would starve we were so hungry. We endured these conditions for 8 years. During this time, we spent nights in drafty, abandoned timber sheds; we dared not even start a fire and could only burrow into our down sleeping bags and use one hand to hold a candle for light and the other to record our field notes. Upon awaking, we would find our washbasins occupied by a large ice block; during the coldest period of winter I took to washing my face only once every 2 weeks to preserve my skin and avoid getting frostbite. One of our female graduate students, who had delicate skin and washed daily, suffered a frost-bitten face.

Starting in August 1989 when Jiaojiao, our first radio-collared female, gave birth to her first cub, we tracked them daily from dawn to dusk to study their behavior. Later, in August 1992, when Jiaojiao produced her first daughter, we observed her continuously for 3-day bouts every 4 days, studying mother-cub interactions and movements. To keep up with them, my graduate students needed to trudge up and down the mountains carrying tents, instruments, and even a small generator; they continued this practice until the cub was 5 months old. In autumn 1994, we began using a small closed-circuit observation system installed in Jiaojiao's natal den for her third cub. There, we conducted round-the-clock observations for three straight days with only a single day off. During this time, we ourselves lived in a cave some 50 m away, recording all their sounds and behaviors. Only when the cub had grown to the age of 5 months and had left the den to begin its arboreal life did our own "den life" come to an end. We repeated this lifestyle again in autumn 1996, when we welcomed the long winter from our temporary abode high in the mountains where we were following Jiaojiao's life with her fourth cub. We spent 5 months in the cold and snow of the forest in our own little "den."

In 1992, we admitted to our team a female graduate from the cell biology program at Peking University after waiving the usual examination. I sent this young woman, who had lived in the city since childhood and had never lived in a room by herself, to do solo fieldwork. She worked without complaint for 8 straight months in the Qinling, during which time she established a rigorous and standardized recording system for field study. Once, in order to preserve specimens obtained with great effort (as well as to save money), she shouldered a 20 kg liquid nitrogen tank herself, trudging the 13 km back to our field station rather than hiring workers to do it for her. In 1995, she visited the Smithsonian Institution as a visiting scholar to analyze her data on mother-cub relationships. Because her work was so outstanding, an American professor invited her to pursue doctoral studies at the University of Maryland, prepared a research project, research funds, and a full scholarship for her, and even went so far as to contact someone with a PhD in computer science to assist her in her work. But in 1995, after completing her research in the United States, she announced that "there is still much more work to be done on the Qinling's giant pandas" and so returned to Peking University. Within only a few days, she was back in the Qinling, staying there until she completed her doctoral thesis in 1999. Her ground-breaking work reported for the first time the specific characteristics of giant panda mating systems, as well as previously unknown reproductive behaviors. This research showed for the first time that the giant panda had similar reproductive capability to other bears.

At that time, graduate students were only paid ¥331/month, barely enough to live on; on top of that, they had to endure months of harsh field research. After two of our PhD students married each other, I told them, "Now you are the poorest people in all of Beijing."

To my surprise, they responded, smiling, with "True, we don't have much money, but it's enough."

"But if you worked for a joint venture firm, you could earn several thousand per each month," I interjected. "Maybe you should consider having one of you go abroad to work, perhaps as a postdoctoral fellow?"

"We enjoy the research work we're doing now," they said. "Although we haven't much money, we're more interested in our lives being meaningful."

I often reflected on my own daughter, who never accompanied me to the mountains but was able to find a high-paying position in Beijing. Because of their academic achievements, my graduate students had the opportunity to follow me into the high mountains to study pandas but lost the chance to earn much money. I often worried about their lives of hardship, but when I heard their answer, I felt better.

Reporters have often asked us, "You could undoubtedly have lived a comfortable life in Beijing or abroad; why did you give that up to live such a hard life in the field? It seems crazy." In response, I have no simple explanation. I can only say to them, "If you try to understand us from the standpoint of religious devotion, then perhaps you can get a sense of our motivation."

The more we understand the difficulties giant pandas face, natural, social, economic, political, scientific, and cultural, the harder it is to ignore their plight, and the more we feel a sense of responsibility toward them. It is this almost sacred sense of mission that impelled us to trudge through the Qinling's towering mountains and deep valleys, day after day, year after year, in order not only to investigate if pandas could persist, but, in fact, to work for their security.

In 1988, we discovered a 1,670 km² region of the Qinling that may have been on its way to becoming the last remaining habitat for the approximately 250 giant pandas there. In 1992, we confirmed that Qinling giant pandas were reproductively normal when we learned that our radio-marked females were consistently producing cubs. This made clear to us that Qinling giant pandas had potential to persist. But in 1993, we discovered that not only had habitat for pandas not improved from 8 years earlier, it had deteriorated yet further because of uncontrolled deforestation. At the end of the day, did we want wood or did we want pandas?

Faced with the imminent loss of the last remaining old-growth larch forests on the main ridge of the Qinling at 2,800 m, forests that had formed some 12,000 years ago at the end of the last glacial epoch and were still maturing, and faced with the realization that the devastated southern Qinling slopes might therefore fail in their role as the last refuge of these pandas, we began appealing to the appropriate government agencies at various levels to reduce logging, as this appeared critical to giving pandas a chance. Alas, without exception, these efforts failed.

In August 1993, our research group penned a letter to our country's leaders. In September of that year, we joined with 29 other Chinese and foreign scientists at an international conference to write to the State Council, describing what we considered an "ecological crisis occurring in the Qinling and suggestions for resolving it." We had originally hoped to transmit it through various channels to the nation's leaders, but several efforts all ended in failure. At that time, a well-meaning friend advised us, "Just forget it; don't waste your efforts. Even if it gets passed on up, it's of no use. Better to write additional scientific reports; that course of action will be more beneficial to you personally." And to be honest, publishing scientific articles is valuable not only for the scientific community but for the authors' careers as well.

But were we to write a hundred books and a thousand scientific articles, what use would they be if the Qinling were to lose its forests and its human residents were to walk toward a deeper abyss of poverty? Protecting the Qinling's giant pandas required that we protect their forest habitats, and protecting these forests meant that we would be working for our

A Chance for Lasting Survival   xvii

own well-being. Thus, we redoubled our efforts to stop the logging in the Qinling.

In October 1993, we managed to get both of the above-mentioned letters transmitted to the Environmental Committee of the National People's Congress through the office of the Overseas Chinese Liaison. From there, they were submitted to the State Council and were answered with a memorandum by then Vice-Premier Zhu Rongji: "Stop the logging immediately and arrange alternative productive activities for the workers, and establish a new nature reserve." In late June 1994, Academician Song Jian, then a member of the State Council and the National Science Council, arrived in the Qinling to investigate for himself. Within days (on July 1, 1994), the Changqing Forestry Bureau ceased their logging operations on over 300 km². This was a result we had only dreamed of and the most critical step in protecting the Qinling's giant pandas.

By spring 1999, when we returned to our research area after 5 years with no logging, we saw that many of the roads that had been used to transport timber over the years were now overgrown with trees and grass and seemed full of life. Jiaojiao, who was then 15 years old, was leading her fifth cub out from a deep valley. Meanwhile, her four older offspring (two males and two females) as well as their own offspring still lived in the nearby pine and bamboo forests. Could this signify that the Qinling's giant pandas had managed to make it through the 12,000 difficult years in which agricultural advancement threatened them, all the way to the present? Could this signify that they had already welcomed a new dawn that would allow them to thrive? Although it is still too early to be sure, we now have great hope for the future of this population.

In autumn 1999, I was asked by a foreign reporter, "What is your greatest achievement from so many years of research on the giant panda?" Speaking honestly, I had never considered this question. At the time, I could only say, "Aside from increasing our understanding of the giant panda's biological characteristics and facilitating the creation of a new protected area in the most critical area of the Qinling, most important has been, through these difficult years, training and developing a group of young scientists who will continue the efforts and sacrifices to protect biological diversity."

Pan Wenshi
August 28, 2000

Acknowledgments in the Chinese Edition

Looking back over the entire history of this research project, we extend our special appreciation to former State Council member and National Science Council Chairman Academician Song Jian. His timely support at several key moments ensured that this long-term project could continue. In addition, we were fortunate to receive the continuous help and encouragement of Gao Zhengsheng, former director of the Changqing Forestry Bureau; Zhong Gaoshi, former director of the Shaanxi Forestry Department; and both Jiang Hong and Liu Jianhua, former director and vice-director of the Department of Fauna and Flora Conservation at the Ministry of Forestry (MOF) in Beijing. We extend our most heartfelt thanks to all these individuals.

We especially thank the former Changqing Forestry Bureau for their consistent support. It was only through the bureau's cooperative spirit and the excellent work conditions they provided us that we were able to successfully complete our 15-year-long research project. In addition, MOF (now the State Forestry Administration) and the Shaanxi Department of Forestry assisted us many times during the course of our work. We express our heartfelt thanks to these offices.

From the beginning of our project, we received support, encouragement, and direction from administrators of the Peking University, including also the Institute of Life Sciences and the Department of Natural Sciences. These individuals were Professor Gu Xiaocheng (former department head of the Biology Department), Professor Zhou Zengquan (Dean, Institute of Life Sciences), Professors Qiang Di, Shen Zhong, and Shi Shouxu (former directors of the Natural Sciences Department), Professors Wu Shuqing and Chen Jia'er (former presidents of Peking University), and Professor Ren Yanshen (former Party Secretary of Peking University).

In our scientific work we benefited from many teachers, colleagues, and friends. Among them are the late Professor Li Ruqi, a well-known geneticist at Peking University; the late Professor Chen Yuezeng, a prominent biologist at Peking University; Professor Chen Changdu, a well-known ecologist; and Professor Shao Jiao, a prominent behavioral psychologist. We also thank Professors Wang Ping, Yang Anfeng, Wang Jinwu, Cui Haiting, Dai Zhuohua, Chen Fengxiang, Cao Zhuo, Liu Xuzhuo, Ma Lailing, Xia Zhengkai, Zhang Miaodi, and Li Songgang, Senior Engineer Meng Guangli, and Senior Engineer Chen Maosheng. Several adjunct professors at Peking University also gave us a great deal of assistance, including the famous zoologist George Schaller of the New York Zoological Society, the prominent molecular geneticist Stephen O'Brien of the United States' National Institutes of Health, and the famous ethologist Devra G. Kleiman of the Smithsonian Institution. Additional experts outside Peking University who provided us with advanced technology and valuable data included Academician Jin Jianming.

Introduction

The Qinling Panda Research Project: Its Context and Importance

Richard B. Harris, David L. Garshelis, William J. McShea, and Wang Dajun

Reserve staff dressed in a panda suit feed a young panda. An appeal is made for help in planting bamboo in the mountains of Sichuan. New record numbers of births in captivity, with squadrons of cuddly panda cubs lined up with their caretakers. Scientific breakthroughs are reported as Chinese scientists work to save pandas by reintroducing some of these captive-born individuals back into their native habitats. A select group of foreign "Pambassadors" is announced, and many of their individual life stories are recounted on television and the internet. China's most famous basketball player expresses his concern for the balance of humans and nature while at a ceremony associated with the release of panda cubs into a new, larger, captive facility.

These are the images that would likely come to the mind of a wildlife enthusiast when considering the present state of giant panda conservation. The popular media continue to emphasize not only the perils inherent in the small numbers of giant pandas, but even more so, how direct human intervention is the best remedy. Popular websites in 2012 urged those devoted to the conservation of giant pandas to contribute to campaigns that would provide additional technical equipment to captive breeding centers or to rebuild infrastructure destroyed by the Wenchuan earthquake of 2008. Donations to the damaged breeding centers have far exceeded the extent of the damage brought about by the quake.

Many with an appreciation of conservation, giant pandas, or Chinese wildlife in general, be they scientists, conservation professionals, or simply wildlife advocates, view these images as distractions, the unfortunate if unavoidable trappings of a popular culture where attention and funds are directed to the latest crisis or most poignant personal story. In contrast, many in the conservation community consider that a species' conservation will not be achieved simply by producing individuals in zoos, by small-scale reintroductions lacking programs to address the underlying habitat needs, or by public ceremonies. Instead, the future of wild giant pandas depends largely on society's ability to provide this species with habitat of sufficient area, quality, and spatial configuration. It now seems that the decline in the geographic range and numbers of free-ranging pandas, ongoing for the past few centuries, may have finally been turned around, largely in response to the 1998 cessation of commercial logging in panda habitat and possibly lower incidental mortalities from snares set for other species. Nevertheless, pandas remain, in the current lexicon, "conservation reliant"; they are still the only species of bear given the status of "endangered" on the International Union for Conservation of Nature (IUCN) Red List (Lü et al., 2008).

Pandas exist in only a small fraction of their historical range. That historic range once stretched across a broad swath of lowlands in eastern China and southward to northern Vietnam, with the western limits in the rugged mountainous region where they exist today (Plate 1). They live today in this remote region because it could not be farmed, but it is likely that these mountain ranges represented the most marginal habitat within their former range, an area where they now must "make do." Although much of their current distribution is now formally protected under China's nature reserve system, these nature reserves remain poorly funded and managed (Yu and Deng, 2004; Yu, 2006). Moreover, the majority of these nature reserves are individually small, and many are separated from one another by unprotected habitats in which natural features, human-altered habitat, or both make demographic or genetic linkage of pandas living in well-protected areas difficult (Reid and Gong, 1999; Loucks et al., 2001; Dinerstein et al., 2004).

Our knowledge of what giant pandas need has evolved over the past three decades (Swaisgood et al., 2010, 2011; Wei et al., 2011). We now understand that giant pandas as a species are not, and indeed probably never were, threatened by periodic blooms of bamboo, their staple food. Although individual animals have been affected, populations have always had the resilience to withstand these natural events by switching to alternative bamboo species and/or by moving to another area where the bamboo had not flowered. We also now better understand their reproductive rates, which had once been a cause for concern largely because of previous difficulties of pandas reproducing in captivity coupled with assumptions that the same was true in the wild. Whereas reproductive rates of giant pandas are lower than most other mammals, they are not particularly low by the standards of their ursid relatives (Garshelis, 2004), nor are they in any way abnormal or indicative of some sort of evolutionary or genetic syndrome (Zhang et al., 2007). In the absence of poaching, which China has successfully stemmed over the past few decades, we now understand that the naturally high survival rates of giant pandas enable them to sustain or even grow their populations, despite their inherently slow replacement rates.

We also recognize that the challenges to long-term persistence of the species remain formidable. Although the threat from poaching may have

2   Introduction

eased and China's forestry practices have improved greatly during the past two decades, China's hunger for economic growth continues to exert great pressures on giant panda habitat. Within reserves, managers continue to be pushed to generate income, often in ways inconsistent with their fundamental conservation mandate. Tourism-related development, including improved roads, plush hotels, paved walking trails, and karaoke bars, draws more and more people into the "wild areas" set aside for pandas. Outside reserves, transport infrastructure and the demands for energy and materials generated by China's rapidly expanding ecological footprint make it increasingly difficult to provide the habitat connectivity that each small population will ultimately need (Hu and Wei, 2004; Lü and Liu, 2004; Zhu and Ouyang, 2004; Hu et al., 2010a, 2010b; Zhu et al., 2010b, 2011b). The low-hanging fruits of the conservation tree, reduction of poaching and increased production of captive pandas, have already been plucked. The more difficult and arduous work, that of adjusting society's demands on these lands in ways that are economically and socially just but still environmentally sustainable, has only recently begun (Lindburg and Baragona, 2004; Liu et al., 2004).

A host of experiences from the burgeoning field of conservation biology leads us to expect that solutions to the panda situation will be long, difficult, and messy. Conservation biologists have learned that species are adapted to particular environments, that populations naturally fluctuate, but that in the long term, they can tolerate only limited alteration of their essential habitat. We have become leery of claims that human intervention to "save" threatened species is a top priority, given that it so often competes with the ever-continuing quest to make life better and easier for humans. It now seems abundantly clear that the everyday activities of humans often compromise the future of those species that we profess to care about most.

Such awareness did not, however, characterize either the Chinese public or the scientific community within China when Dr. Pan Wenshi and his team of intrepid young graduate students began their work on giant pandas in the Qinling Mountains in the mid-1980s. At that time, despite the panda research that had just been completed in Sichuan's Wolong Nature Reserve (Schaller et al., 1985), the overriding concerns for pandas within China continued to be that bamboo was disappearing, free-ranging pandas were starving, and captive pandas did not sufficiently understand the concept of the birds and the bees to produce new cubs without human intervention. Countering these prevailing concerns were the results from the study of George Schaller, Hu Jinchu, and their team at Wolong, but the Chinese government and people were not listening. It took the work of an entirely Chinese team to open the eyes of the public and ultimately the government to the notion that habitat preservation and integrity were key requirements for saving pandas.

The Context: Chinese Wildlife Conservation in the Mid-1980s

China has changed so rapidly over the past three decades, both materially and intellectually, that it is all too easy for us to forget the backdrop within which Dr. Pan and his students began their panda studies in the Qinling. Although politically China long since had moved on from the Cultural Revolution (1966–1976), legacies of that era were still evident in the mid-1980s. China's university system had been essentially closed down during the first 5 years of the Cultural Revolution. Most universities reopened in the early 1970s, but they remained oriented toward educating only those with the "correct" background (i.e., workers, peasants, and soldiers), and academic standards were weak. As a result, a large gap was created within the cohort of scientists, as well as well-educated civil groups, who might be expected to advocate for conservation and rational use of natural resources. This gap was never entirely rectified, simply stepping up in age as the years passed. (Indeed, this gap is reflected in the age difference between Dr. Pan and even his oldest graduate students.)

The production, dissemination, and sharing of new information, which the academic community now takes for granted, was hampered in China during the mid-1980s, by both technological and cultural impediments. A university department might typically have had only a single telephone line, and even then, few scientists thought to call each other to share ideas. Libraries remained disheveled, collections of international scientific publications incomplete and disorganized, and systems to access the limited available information were archaic. The mentoring system was still the dominant education tool within China: all new knowledge and drive emanating from young scientists had to pass through the lens of the senior scientist. Senior scientists that survived the earlier decades were not always quick to embrace new ideas that directly contradicted colleagues or institutions. It was a rare scientist who allowed their graduate students academic freedom. Thus, a young Chinese wildlife biologist wishing to benefit from the latest technical and conceptual advances would have been faced with a daunting challenge in breaking beyond what could be imparted directly from his or her mentor.

Additionally, the Confucian tradition of focusing on the improvement of mankind as a central theme lent itself easily to focusing on how mankind could "help" nature (Santangelo, 1998; Sterckx, 2004; Harris, 2008). As China gradually became aware of the excesses of the Mao period (Shapiro, 2001), consciousness of its environmental problems increased. But it would be fair to characterize these early stages of Chinese environmentalism as superficial. Sentiments regarding the importance of preserving species and their habitat were expressed, both in public discourse and within laws and regulations, but these were not accompanied by realistic measures by which they could be implemented, given the confounding desire to produce more material goods and thereby raise people's living standards (Harris, 2008).

Thus, when the Chinese government and public became aware that only China had giant pandas (bolstering a sense of national pride), that pandas had declined to dangerously low levels (causing national embarrassment), and that individual pandas were starving as a result of bamboo flowering (eliciting universal sentiment for action), the stage was set: Organize society to "save" pandas and mobilize China's considerable human ingenuity to produce more of them through captive breeding. Serious scientists in China even discussed cloning pandas to increase their numbers.

Meanwhile, the field of wildlife management, which had a rich tradition in Europe and had been reinvigorated through various movements in North America during the early decades of the 20th century, remained embryonic in China until the 1990s. Government officials working in the wildlife field during the 1980s were few and far between, and generally worked under Forestry or Agriculture Departments (as most still do), and almost all received their education at a single institution (the Northeast Forestry Institute in Harbin, Heilongjiang, later renamed Northeast Forestry University). Their knowledge of panda ecology and useful conservation actions generally came from higher government officials rather than from direct observation or the scientific literature.

Research into giant pandas had begun by the early 1980s, but with the exception of the field study in the Wolong Nature Reserve, most of it focused on physiology, taxonomy, and captive breeding. Research topics were mainly limited by what one could learn from studying a few captive individuals. In their review of research conducted during the 1980s, Hu and Wei (1990) categorized by subject a total of 538 published papers or reports (mostly in Chinese), of which the plurality (172 papers) were concerned with captive breeding and an additional 232 focused on giant panda behavior, physiology, cytology, genetics, or "practical use" of pandas (most of which depended on captive animals for data). An additional 43 treated taxonomy or biogeography, which similarly did not depend on data from free-ranging animals. Many of the remaining "field" papers were short notes, descriptions of the presence of pandas, or anecdotes. (A similar emphasis was reflected in an independent accounting of Chinese scientific papers published about bears during the 1979–1991 period; Ma et al., 1992). Hu and Wei (1990) reflected the previous decade's emphasis in the English version of their abstract, which after summarizing the fields of research, provided statistics on the number of captive births and pandas living in captivity but said nothing of the conservation status of wild pandas. In summarizing recent panda research within China, Hu et al. (1990) illustrated the gradual shift in focus of Chinese researchers, with increased attention to ecological aspects potentially important to conservation (almost all of which came from the Wolong

Nature Reserve), but they also continued to feature purely academic studies of captive individuals. That is not to say that captive studies of pandas have not contributed substantially to our understanding of panda biology and behavior; certainly, they have (see Lindburg and Baragona, 2004, and Wildt et al., 2006, for reviews). But understanding how pandas live in the wild requires a study in the wild.

The ground-breaking ecological study of wild giant pandas in Wolong, China's first protected area dedicated primarily to conserving pandas, began in earnest in 1980. The project was a joint collaborative effort of the Chinese Ministry of Forestry and the Worldwide Fund for Nature (WWF) and built upon earlier work conducted by Hu Jinchu and his colleagues from Nanchong Normal University in Sichuan. Led by George Schaller, this study was the first to capture and radio track free-ranging pandas and arguably produced more ecological information about the species than had all previous studies combined.

The Wolong study provided a new wealth of material, particularly on the relationships between pandas and their staple food source, bamboo. Schaller et al. (1985) studied foraging, movements, activities, and aspects of social and reproductive behavior. They had to extract some information from captive pandas but, for the first time, were also able to obtain important insights into the lives of wild pandas through the use of radio telemetry. Although not previously a proponent of telemetry studies, Schaller found this technique essential in this situation. The Wolong team captured and collared six wild pandas and by triangulation of multiple radio bearings obtained roughly 1,500 locations. They concluded that pandas occupied relatively small home ranges and spent about 60% of their time foraging. Importantly, Schaller's team discovered that although pandas were bamboo specialists, they could persist through a mass-flowering event as long as an alternate bamboo species was available.

As the Wolong research project was wrapping up, it pointed to the need for continued information to support conservation efforts on giant pandas. Dr. Pan was in a position to know, having been transformed from a laboratory scientist to an accomplished field ecologist through his participation as one of the lead scientists in the Wolong work (Schaller, 1994). The intellectual backdrop within China at the time favored solving complex socio-ecological problems with technological solutions. Dr. Pan's embrace of the more holistic, interdisciplinary approach favored by Western conservation biologists was, in this context, both innovative and courageous. As his team members themselves have characterized it, their research was a venture in applied science. Rather than endeavoring to describe what made the giant panda unique, they sought to find what was needed to ensure its persistence. To be sure, they also pursued avenues of "pure science," but with the goal of potentially yielding additional conservation implications. His team truly took a conservation biology perspective toward their research, focusing on specific aspects of panda biology that mattered long term. Importantly, they also placed their biological findings within the context of human use of the panda's current and historic range within the Qinling. In contrast to many Chinese wildlife studies that avoided issues of human activities and how they might be modified to conserve a species, Pan and team took special interest in how pandas were affected by logging and road building. They went so far as to plumb historical records and attempt to interpret past events within this landscape in order to hypothesize how humans and pandas had coexisted during previous centuries.

This open-minded and impartial view toward research ultimately led Pan's team to a most ironic conclusion pertinent to the panda's future persistence: this unique, endemic, and emblematic icon of Chinese conservation was not so different from a great many other conservation-reliant species (Harris, 2004). Giant pandas were not gastronomic oddities or reproductive maladroits, nor were they vestiges of a dying line of evolutionary failures, destined to fizzle out even without human-imposed threats. Rather, giant pandas were highly specialized bears that long ago had found a way to live off an abundant but low-quality resource: bamboo. Whereas their overall reproductive rate was low, they certainly had no problem reproducing (and were not timid about it either). To be sure, their dependence on bamboo produced a sensitivity to ecological change, as would be the case for any dietary specialist. The risks to their future, however, had little to do with inefficient digestion or lack of libido, but rather the increasing destruction and fragmentation of their habitats by humans, their putative caregivers. The extinction of wild pandas, Pan and his team learned, was not a foregone conclusion of biology but a blunder that could be averted, not by technology focused on a few individuals but by habitat conservation on a large scale.

The Qinling Panda Study: Import and Importance

The significance of Pan's study of giant pandas in the Qinling was to refocus conservation emphasis on the innate ability of these animals to survive (and thrive) given adequate habitat and protection from human exploitation, to debunk the prevailing notion that pandas had a "reproductive problem," to relegate captive rearing to a more supportive than primary role in conservation, and to highlight what we recognize today as core conservation biology concepts. Some important core concepts include the following: (1) species have a functional role within ecosystems and do not exist in isolation, (2) field data are essential for assessing biological and demographic attributes of species, (3) population fragmentation poses serious risks to evolutionary processes and population viability, (4) each species is evolutionarily unique in some way (so species are an appropriate focus of conservation), yet they share many attributes (so insights can be gained from studies of similar species), and (5) systems vary in their resilience to human disturbances and modifications.

Unfortunately, until now, relatively little of Pan's work was known to those lacking access to the Chinese language. For example, Swaisgood et al. (2010) characterized some Chinese panda research as "flying below our radar" (where "our" refers to Western scientists). Times have changed, and scientific interchange between conservation biologists in China and other countries has blossomed; panda research has taken off, in part as a result of this international interchange of ideas and advanced technology.

As expected for any study that is already almost 20 years old, some of the specific results from Pan's work have become dated. Some of their results spurred further investigations that either confirmed or modified Pan's conclusions, and some were supplanted by better analytical tools, but several remain to this day as the most definitive findings on the particular subject. Below we have selected what seemed to be the most significant scientific contributions of this study, whether upheld through later work or not. We note that the project to translate this book was undertaken not only to make the scientific findings available to a wider audience but also to capture the historical setting and personal characters that infused the study and their process of interpreting their accumulating data. The exuberance and dedication of the team spills over in their writing as they describe the difficulties of data collection yet piece together how they were able to uncover intriguing and important aspects of the life history and ecology of this species. This is not just a scientific document but a quest of discovery, carried out mainly in isolation by one professor and his students. The limited number of previously published scientific papers that they cite is a reflection of the paucity of resources that they had access to; often, it was clear that one particular paper had a disproportionate effect on the types of analyses conducted or interpretation of results. This is not a criticism, but the reality of the constraints that flavored the study results. Any field study of this sort is subject to varying interpretations that are in part influenced by the people who took part and the context in which it was carried out. In most scientific papers, that context is typically not apparent; here it is uniquely preserved.

Most Significant Scientific Contributions

1. Prior to the Qinling team's investigation, it was well known that the pandas in the Qinling Range, at the northern extremity of their present distribution, were geographically isolated from pandas in Sichuan and Gansu Provinces (Plate 1). Pan's team also discovered geographic isolation at small scales within the Qinling. They identified four "mountain islands," which although all nominally within the Qinling panda range, were separated from one another by low-elevation habitats that no longer supported pandas

(chapter 2). They surmised, however, that occasional interchange of individuals among three of the four subpopulations might yet occur in the future because the constraint on their movement was due only to human-caused alterations of the habitat, and that habitat could recover if protected (as was the case following the 1998 logging ban). However, one of these "mountain islands," in the Tianhua Range, had become more permanently isolated by the construction of a highway and was dangerously small. The description of how panda populations may become threatened by small population size on a small scale, despite appearing to be a part of a reasonably large population when viewed on a larger scale, has resonated in panda conservation ever since.

2. When the Qinling team began their research, most of the area had yet to be officially protected. Indeed, their research station was the unused base camp for the recently departed logging operation. That they situated their study where human impacts had occurred, and to some extent continued, enabled them to investigate the effects on pandas of various degrees of forest disturbance. They concluded that although wholesale forest destruction and clear-cutting were inimical to pandas, bamboo could recover from certain types of selective logging and, with time, panda habitat could recover (chapter 2). Similarly, old, abandoned transportation routes, overtaken by forests, were not an impediment to panda use. However, with increased human use, roads and highways became significant obstacles to panda travel and functioned to fragment panda habitat and divide small pockets of individuals from one another (chapter 3).

3. Pandas in the Qinling were found to be seasonal migrants, switching between the two major bamboo species (each of which occupied unique elevation bands) from summer to winter (chapter 5). This pattern had been observed earlier in Wolong, but the extent and duration of migrations among Qinling pandas were greater because of a more extensive array of palatable bamboo than at Wolong. By following a larger number of pandas (over a longer time period) than had been possible in Wolong, the Qinling team added considerable detail to our understanding of panda home range use and movements, particularly the extent of variation among individuals (chapters 4 and 5).

4. One of the most striking findings from the Pan team, which may not have been fully recognized by the authors at the time for its uniqueness, was that although adult male pandas generally occupied larger home ranges than females, long-range dispersal of subadults was more likely for females (chapter 4). Pan et al. concluded that "among pandas, male offspring are philopatric, remaining near their natal area. They inherit a portion of their mother's home range, establishing their own home ranges very close to their mother. In contrast, the young female pandas disperse when they are ready to do so, establishing their own home ranges in a new location." This finding was

similar to that observed for other species of bears except that the sexes were reversed (McLellan and Hovey, 2001; Kojola et al., 2003; Ishibashi and Saitoh, 2004; Costello, 2010). As their results were based on a slender sample of only two subadult females, Pan's team remained rather cautious about overinterpreting their observations, especially (we presume) because it was counter to results of other bear studies that they knew about. Interestingly, however, two later panda studies in the Min Shan (Zhan et al., 2007) and the Liangshan Mountain (Hu et al., 2010b) ranges of Sichuan, using DNA extracted from feces, discerned corroborating evidence that dispersal was more common among females than males.

5. The Wolong study had offered some tantalizing evidence that males actively competed among themselves for mating privileges and at times an estrous female would become the attention of multiple males. The Qinling study further expanded our knowledge of the social dynamics within panda society during the mating period, providing evidence that at least in the portion of the Qinling where panda densities were high, males aggregated during the breeding season and contested for and possibly were selected by estrous females (chapters 4 and 8). The study team was able to identify specific geographic locations where such aggregations occurred and could go there to observe pandas. The combination of their large number of radio-marked individuals, identification of some unmarked individuals, and their ability to make observations at close range enabled the team to identify which pandas were involved in mating events. They documented that in addition to individual adult males mating with multiple adult females, some adult females were associated with several males within a single season, suggesting that both sexes bred with multiple partners. This sort of polygamous (or promiscuous) mating system is now recognized as being consistent with other species of bears (Joshi et al., 1999; Kovach and Powell, 2003; Steyaert et al., 2012), whereas 90% of other mammal species are polygynous (only males have multiple mates).

6. The Qinling team was able to generate a level of habituation to their presence among some of their study animals. This enabled them to obtain much more detailed information than had previously been known about the way in which free-ranging adult females care for their newborn young (chapter 9). The detailed observations of interactions (including many hours of videotape) within and adjacent to den sites were unprecedented. The Qinling team's findings that reproductive rates and offspring survival were sufficient to compensate for natural deaths of older animals were central to debunking the idea that wild pandas were inherently doomed.

7. One of the most basic but difficult questions in any field study of wild animals is how many there are. The study team was challenged with estimating the number of giant pandas within their study site and then also within the entire Qinling Range. Furthermore, they sought to develop a method

that could be used to estimate and monitor the total number of pandas in all six of the mountain ranges that they occupy. Pan's team pioneered the use of a mark-resight approach for estimating local panda abundance (chapter 6). They found that their 15 km^2 study area had about 30 pandas, or an unexpectedly high density of 2 pandas/km^2. Remarkably, this is even higher than the highest recorded density for American black bears (*Ursus americanus*; 1.6/km^2; Peacock et al., 2011); however, the authors noted that panda density changed seasonally and other parts of the Qinling likely had lower densities. Pan's team also made significant advances in developing a method to estimate panda numbers over large-scale landscapes (chapter 6). Building on existing knowledge that the lengths of bamboo fragments in droppings were somewhat individualistic, the team combined attributes of the droppings with distances between them. Using their telemetry data to quantify how far pandas moved per day, they developed a means of differentiating distinct individuals from spoor found along transects with a certain degree of confidence. This method became the basis for the subsequent national survey of pandas across their range.

8. Estimating the parameters needed to fully understand population dynamics is a daunting challenge for any wildlife study, particularly so for a long-lived species with delayed reproduction inhabiting a dense forest that obstructs visual observations. Sample sizes are often too small for biologists to ascertain even the most basic question of whether a population is increasing or decreasing. The Qinling team captured and monitored more than three times as many pandas as were studied in Wolong; even so, the data were barely sufficient to make some preliminary assessments of population trends (chapter 7). Needed for such an analysis are good estimates of reproduction and survival. In a decade of research, the team obtained reproductive data on five adult females, including 10 reproductive cycles, but variability among them was rather large. Estimating survival was even more problematic: whereas a number of male pandas died of natural causes during the study, none of the adult females died, so an estimate based on these data would suggest that females lived forever, and the population would thus continually grow. A number of assumptions and analytical "fixes" were necessary, some of which were appropriate and, in hindsight, some were not. In this volume we included only those that the scientific community is likely to find useful.

9. A particular strength of the earlier Wolong study had been the information on panda food habits, including the specific parts of bamboo ingested, panda strategies for maximizing nutritional intake, and their ability to obtain their dietary needs from such an unlikely seeming source (chapter 11). The Qinling team added considerable detail and nuance, while largely corroborating this information. A series of fascinating images, some obtained using electronic microscopy, helped elucidate that pandas were able

to digest some portion of the hemicellulose making up bamboo cell walls and thus increase their absorption of minerals and energy over that from merely the cell contents.

10. Similar to what had been found in Wolong, the Qinling team's calculation of bamboo feeding rates and digestion, combined with their estimates of bamboo availability, indicated that the existing panda population was well below carrying capacity (chapter 11). They also noted that the presence of two bamboo species would help buffer pandas from another mass-flowering event. We note that Pan and his team did not consider that some bamboo, on very steep slopes, was inaccessible to pandas and therefore should not have been used to calculate carrying capacity and that periodic flowering and die-offs do reduce carry capacity, but neither omission significantly changes their conclusions of abundant palatable bamboo for relatively few pandas.

Subsequent Developments Related to the Team's Findings

As identified early on by the Qinling team, habitat fragmentation and the resulting isolation of panda groups from one another continue to be among the most pressing conservation concerns (Loucks et al., 2001, 2003). The panda populations at the opposite (southwest) end of the existing geographic distribution from the northeastern Qinling Mountains, in the Da (large) and Xiao (small) Xiangling Mountain ranges (Plate 1), have aroused particular concern among conservationists because of their small size, history of rapid reduction due to an explosion of the human population, and isolation from other panda populations (Hu et al., 2010a; Zhu et al., 2010b, 2011b).

Much of our understanding today about the extent and effects of fragmentation and isolation of panda populations is attributable to rapid advances in genetic techniques (Wei et al., 2012). For example, recent genetic and morphological studies suggested that pandas in the Qinling may be sufficiently distinct to deserve recognition as a unique subspecies (Wan et al., 2003, 2005); although this position has not been widely adopted, it is intriguing to note that genetic analyses indicate that Qinling pandas have been separated from the larger distribution of pandas to the south for some 300,000 years, and during this time they apparently became locally adapted, through modification of a certain taste receptor gene, to greater tolerance of bitter bamboo leaves in their diet (Zhao et al., 2013). Sophisticated genetic analysis also revealed that Qinling pandas had undergone a very large historical population decline, apparently due to human-caused deforestation and other activities (including a road built over 2,000 years ago) (Zhao et al., 2013).

A host of recent panda studies have examined aspects of their ecology from the distribution and genetic analysis of their spoor (including differentiation of sex and individuals). Pandas defecate frequently, and their fibrous feces are long lasting and so can yield far larger samples of panda locations than a telemetry-based study like Pan's (and with much less effort). The new genetic techniques combined with enhanced techniques for ascertaining habitat preferences from presence data have enabled an avalanche of information to be extracted from panda scats. For example, a number of recent studies, based mainly on locations of panda scats, have followed up on Pan's concerns that forest cutting, roads, and other development pose the most serious constraints on panda populations. These studies found that pandas generally shy away from villages, that males are more likely than females to come near roads with vehicular traffic, whereas females readily use abandoned logging roads, and that the spatial distribution of roads with respect to habitat and human settlements largely affects whether pandas live nearby (Feng et al., 2009; Wang et al., 2009; Fan et al., 2011; Qi et al., 2011, 2012).

Advances in remote sensing using satellite imagery have also been a boon to conservation-related research on pandas, enabling investigation of habitat fragmentation at the landscape scale (Xu et al., 2006; Shen et al., 2008; Wang et al., 2010). Investigators have delineated habitat "blocks" where pandas survive; there is an active effort to match these blocks with protected areas and to link blocks with corridors so that pandas can move between them. It is now widely accepted that isolated blocks, harboring small populations of pandas, are (as Pan's team postulated) in jeopardy of extirpation, even if the larger-scale population is relatively secure. Recent spatial data from the Qinling, again based on locations of panda scats, confirmed Pan's suspicions that this population is divided into subpopulations, with limited interchange, and one subpopulation appears to be in dire jeopardy (Gong et al., 2010).

An important recent discovery is that pandas seem to be associated more with old-growth forest than previously thought (Z. Zhang et al., 2011). The data set used in this recent study by Zhang and colleagues (locations of feces found during the Third National Giant Panda Survey, 1999–2003) represented 70% of the entire panda range, so this finding, which challenged conventional wisdom, may have been related to scale. These authors speculated that large trees may be important for maintaining the nutritional quality of bamboo growing in the understory and/or provide important denning structures, which might be more limiting than previously considered. These results would seem to contradict the Qinling study team's finding that although old-age forest is optimal, pandas were able to achieve very high densities even where historical logging practices had largely eliminated large, old trees. Interestingly, however, Z. Zhang et al. (2011) only used samples from Sichuan Province, so it is uncertain whether their results can be extended to the Qinling, where the national survey found the highest density of pandas. Feng et al. (2009) and Gong and Song (2011) performed similar analyses with feces locations within the Qinling, concluding that panda presence was closely associated with topography (elevation, slope, and aspect), but they did not examine effects of forest age.

Thus, there seems to be no doubt about the ultimate importance of mature forest for pandas, there is still a degree of uncertainty about the relative value of regenerating forests that were previously logged. The answer to this question may impact future forest management strategies in China. One promising development is that staffs of the various nature reserves in the Qinling have begun conducting regular monitoring and patrolling, not only of pandas themselves, but of their habitats and of the effects of newer developments, such as road building (Wang et al, 2010).

In May 2008, the enormous Wenchuan earthquake, centered in the panda habitat of Sichuan Province, set off alarms within the conservation community. Besides the human fatalities and destruction of roads and buildings, many feared that panda populations would be significantly affected. Indeed, it was discerned, from satellite images, that a large swath of panda habitat was destroyed, and much of the remaining habitat was dissected by landslides (Xu et al., 2009). Appeals were made to use this catastrophe as a reason to give more serious considerations to panda habitat quality and connectivity during the restoration process (Wang et al., 2008), including restrictions on tourism and relocation of local human residents (Xu et al., 2009). Even greater damage to habitat was found when investigators examined vegetation on the ground, but it was learned that most of the devastation occurred in areas that were generally too steep for pandas and that pandas continued to use areas near earthquake-damaged areas (J. Zhang et al., 2011). Pan's findings that pandas could adapt spatially to an environmental disturbance (i.e., shift their home ranges accordingly) would indicate disturbances at the scale created by landslides do not severely limit panda populations. We predict that the ongoing Fourth National Survey will not find reduced panda population numbers in the region affected by the 2008 earthquake.

Habitat disturbance and fragmentation have become paramount concerns in panda conservation because it is still unclear what sorts of natural or anthropogenic barriers disrupt genetic interchange (Zhu et al., 2010a). As noted, genetic evidence has begun to corroborate the Qinling team's tentative findings that unlike most other bears, subadult dispersal in pandas is more common among females than among males (Zhan et al., 2007; Hu et al., 2010a). The implications of this phenomenon for population genetics and connectivity are currently a topic of interest among researchers (Wei et al., 2011). Genetic structure is also influenced by mating system and social structure, and the Qinling team's observation that males may aggregate and aggressively compete for the attention of an estrous female remains important. (Subsequently, Yong et al., 2004, observed as many as four adult males and one subadult male attracted to the vicinity of a single estrous female in Foping Reserve in the Qinling). Although considerable advances have since been made on endocrine (Nie et al., 2012b), chemical (Swaisgood et al., 2004; Nie et al., 2012a), and vocal cues (Charlton et al., 2012; Xu et al., 2012)

14   Introduction

related to reproductive activity, relatively little new information has been produced on female mate choice, individual reproductive success, or the distribution of paternities among adult males since Pan's team's work. These characteristics of panda social structure are all relevant to conservation genetics and thus to the long-term future of the species.

Many authors since the field studies in Wolong and Qingling have converged on the idea that panda life history and demographics are tightly bound to the low energy provided by their food source. Hence, a great deal of work has gone into studying the nature and composition of bamboo and how that varies across sites and seasons. We will not review that extensive literature here. However, particularly interesting is the recent discovery of how pandas are able to digest cellulose and hemicelluloses (Zhu et al., 2011a). Pan's group speculated that pandas might have microbial symbionts in their gut to aid in fermentation of this material but were perplexed how this could occur in the panda's short digestive tract. A later study using highly sophisticated genetic techniques to examine panda scats found that they contain gene segments indicative of cellulose-digesting microbes that are unique to pandas (Zhu et al., 2011a). That is, not only has this species of bear developed a unique pseudo thumb to hold the bamboo and extra powerful jaw muscles and bulbous teeth to grind it up, but it also evolved a distinct suite of microbes to help digest it. Nevertheless, it remains a mystery why pandas, once an omnivore like other bears, switched from eating nutritious foods like meat and fruits to concentrate almost exclusively on bamboo, a food for which they can utilize less than 20% of what they ingest (Zhao et al., 2010; Jin et al., 2011).

Pan's team recognized that the future of pandas was inextricably tied to access to plentiful supplies of bamboo, and their population projection modeling suggested that if humans provided pandas enough forested habitat, their "chance for lasting survival" was high. Indeed, two national reforestation programs, initiated in the late 1990s, have significantly increased forest cover throughout the panda's range (although much less so in Qinling; Li et al., 2013), which should generally bode well for their future. A looming question now concerns the potential impacts of a slowly warming climate, which is already driving some animals northward or to higher elevations. Pan's data suggest that pandas also could adapt and move, but will they have a place to move to? One recent study projected that less than half the current panda range will remain suitable habitat by the year 2080; other areas, currently outside the range, will become suitable, but some may be too far away for pandas to reach (on their own), and much of this new potentially habitable area is outside the current reserve system (Songer et al., 2012). A model specific to the effects of climate change on bamboo projected a much more dire scenario for Qinling pandas, involving a massive shrinkage in range of their primary foods by the end of this century (Tuanmu et al., 2012).

Given that Pan and his team were studying a seemingly relict species that subsisted on a marginal food source in degraded (logged) habitat at the extreme western edge of their former range and that the overall population size was quite small, it was surprising to them that their estimates of panda densities could be so high. The big question, however, was how representative was their selected study area within the greater Qinling range? Extrapolating densities from small study areas to larger landscapes is a nagging problem in wildlife studies. The significant advancement made during Pan's study was in developing an empirically based method for utilizing bite sizes of bamboo in scats to estimate population size. This method, employed during the Third National Giant Panda Survey (1998–2002) yielded a range-wide estimate of about 1,600 pandas. The technique, however, suffers from some bothersome assumptions; in dense populations these assumptions become untenable because there is likely to be large overlap in bite sizes of closely spaced individuals, so their feces become indistinguishable (Garshelis et al., 2008). Perhaps this is why Pan's group did not attempt to compare the bite-size-based estimate to the mark-resight estimate of their dense study population. Even if the technique yielded a reasonable range-wide estimate, it might not be sufficiently sensitive to detect a population whose density was increasing, an obvious flaw. The apparent answer to this problem came when scientists began to be able to reliably extract DNA from fecal material (which was initially difficult because it degrades through time), so this could replace bamboo bite sizes as the means for distinguishing individuals. Obtaining DNA remotely from scats was first attempted in the Qinling by Fang et al. (1996).

Mark-recapture estimates based on remotely sampled DNA may yet prove to be indispensable in improving estimates of panda abundance, either in selected reserves (Zhan et al., 2006) or across broader landscapes, as in the next national survey. But, as pointed out by Swaisgood et al. (2011), this may not be feasible or appropriate in all circumstances. Moreover, this method also entails certain assumptions, which, if violated, may produce biased estimates of panda abundance (Garshelis et al., 2008). Presumably, future studies will investigate and compare the bite-size- and DNA-based techniques more closely, so results from upcoming DNA-based surveys can be compared to the past (Zhan et al., 2006, 2009).

Like nearly all studies, although relevant in the present day, the significance of the Qinling work will fade through time as investigators use increasingly sophisticated techniques to obtain larger quantities of more detailed data and as more scientists are drawn into addressing the intriguing questions that persist about this fascinating animal. One thing we found striking, by discovering for ourselves the value of the information contained within the volume, is how infrequently it has been cited by other Chinese scientists. In contrast, the earlier Wolong work (Schaller et al., 1985) is still often cited for findings that were more thoroughly investigated in Qinling. Possibly, the legacy of Pan's top-down approach to conservation action (by bypassing local officials and writing letters to the highest level of government and international nongovernmental organizations) had taken its toll on the acceptance of the science work. Once the Changqing Nature Reserve was established, the new administration quickly controlled all activity at Pan's research site. Government officials now had more direct oversight of the project. Animal welfare concerns about capture and radio tracking of pandas led in 1998 to an unofficial but well-known ban on direct handling of animals for research. Although the ban was specific to Pan's project, it had a chilling effect on all wild panda studies in China. Only after 2007 were any pandas in China radio-collared specifically for research (at the Foping and Wolong Nature Reserves). We do not fully know why Pan's efforts led to so many kudos by the international conservation community (see postscript) but not within the Chinese scientific community. We hope this translated volume will prompt all giant panda researchers to reexamine their novel work.

One key aspect of the work that has been adopted was the use of a field station to collect long-term data on a marked population of animals. Pan's team expanded the successful model in Wolong by bringing diverse scientists and students to a single site for almost a decade. Each member of the team built on the discoveries of those before them and took advantage of the habituated animals and observations from previous years. In the Qinling, this long-term collaborative environment has been continued in Foping Nature Reserve by Wei Fuwen and Ron Swaisgood, and in Sichuan Province at Wolong Nature Reserve by Jack Liu and Ouyang Zhiyun and, more recently, in Pingwu County by Wang Dajun. The mounting new discoveries from these sites are a testament to the power of this research model.

Aims and Process of Translating This Book

We have approached the translation and editing of this volume as an opportunity to bring valuable natural history data to a broader audience and to place the work in a context suitable for this broader audience. Our first step was for one of us (RBH) to conduct the task of translating the entire volume so it could be reviewed by the coeditors and outside reviewers. We then reviewed each chapter for clarity and worked with the original chapter authors to understand their data processes and deductions. All chapters were reduced in scope to retain only information that linked directly to the central themes of the original volume (panda biology and conservation), contained direct observations that would be relevant to today's ecologists and conservation biologists, or directly resulted in important subsequent programs (such as the origination of the bite-size technique to estimate population size). There are multiple scientific names used in the volume, and we did not update any nomenclature that has changed since the volume's publication. However, there is potential confusion over names for bamboo species because of both frequent revision of nomenclature and the difficulty in identifying species in the field. We have provided an appendix with the scientific names used in the text, the synonyms or revisions for these names, and their English translation (Table I.A.1). We did not translate or reprint the original chapter 10 because it has rightfully found its way into an international journal (Lü et al., 2001). We regret the necessity of having to delete fascinating passages on other subjects. The original authors have been tremendously generous in allowing their texts to be edited and abbreviated to better suit the broader audience. Nevertheless, if some of the English appears imperfect or if explanations seem a bit "foreign," they were intended to capture the flavor of the original writing style and the process of discovery that so pervades the original book. To help readers, we have translated into English the Chinese names that were given to each individual study subject (Table I.A.2).

We worked closely with Zhang Lu, a young scientist at Peking University, as she assisted with author correspondence, figure and table translation, and revisions. Zhang Lu was an important bridge between the past and the present, and we needed her dedication to make this possible. Wang Fang (also of Peking University) graciously provided a map of current and historic panda distributions and revised some of the figures. The translation of this text and its subsequent publication was underwritten by the Seidel Fund of the Smithsonian Institution, without whom this would still be an unfulfilled desire. Two anonymous reviewers provided critical comments to the volume. Last, we thank Professor Pan Wenshi, not only for the original inspiration that created this work and the wonderful field team that has gone on to change conservation biology in China, but for his constant openness to new possibilities. He supported this effort from the beginning without a sure knowledge of whether the final product would reflect his ideals. We dedicate this effort to him and hope he finds it worthy of his aspirations.

Appendix

Because of potential confusion over names for bamboo species due both to frequent revision of nomenclature and the difficulty in identifying species in the field, in this appendix we provide the scientific names used in the text, the synonyms or revisions for these names, and their English translation (Table I.A.1). As an additional help to readers, we have translated into English the Chinese names that were given to each individual study subject (Table I.A.2).

Table I.A.1.

A LIST OF THE BAMBOO SPECIES REFERENCED IN THE TEXT AND THEIR COMMON NAMES (WHEN KNOWN), AS WELL AS REVISED NAMES AND SYNONYMS WHERE APPROPRIATE. SYNONYMS ARE INDICATED WITH AN ASTERISK.

Chinese name	Scientific nomenclature used by Pan et al.	Common name	Revised scientific nomenclature or synonym
刺竹子	*Bambusa arundinacea*		
箭竹	*Chimonobambusa monophylla*		
和刺竹子	*Chimonobambusa pachstachys*	Thorny bamboo	
八月竹	*Chimonobambusa szechuanensis*	Sichuan square bamboo	*Arundinaria szechuanensis**
油竹子	*Fargesia angustissima*	Oily bamboo	
缺苞箭竹	*Fargesia denudata*	Giant panda fodder bamboo	
华西箭竹	*Fargesia nitida*	Fountain bamboo	*Sinarundinaria nitida**
拐棍竹	*Fargesia robusta*	Walking stick bamboo	*Fargesia robusta*
糙花箭竹	*Fargesia scabrida*		
华桔竹	*Fargesia spathacea*	Umbrella bamboo	*Arundinaria sparsiflora**, *Thamnocalamus spathaceus**
青川箭竹	*Fargesia rufa*	Giant panda fodder bamboo	
箬竹	*Indocalamus longiauritus*	Long-ear cane	
石绿竹	*Phyllostachys arcana*	Arcana bamboo	
白夹竹	*Phyllostachys bissetii*	Bisset's bamboo	
水竹	*Phyllostachys heteroclada*	Fishscale bamboo	
白夹竹	*Phyllostachys nidularia*	Big node bamboo	
金竹	*Phyllostachys nigra var. henonis*	Striped culm black bamboo	
叶箪竹	*Qiongzhuea macrophylla*	Large-leaved Qiong bamboo	*Chimonobambusa macrophylla*
三月竹	*Qiongzhuea opienensis*		
筇竹	*Qiongzhuea tumidinoda*	Qiong bamboo	*Chimonobambusa tumidissinoda*
冷箭竹	*Sinarundinaria fangiana*	Fang's cane bamboo	*Bashania faberi/B. fangiana*
丰实箭竹	*Sinarundinaria ferax*	Prolific chinacane	*Fargesia ferax*
空柄玉山竹	*Yushania cave*		
峨嵋玉山竹	*Yushania chungii*	Large arrow bamboo	*Sinarundinaria chungii**
马边玉山竹	*Yushania mabianeusis*		
石棉玉山竹	*Yushania tineloata*		

Table I.A.2.

NAMES FOR GIANT PANDAS DESCRIBED IN THE VOLUMES IN CHINESE PINYIN, CHINESE, AND ENGLISH TRANSLATION.

Chinese pinyin	Sex	Chinese	Meaning of name
Jiaojiao	F	娇娇	Lovable
Huzi	M	虎子	Cub (also little tiger)
Xiwang	F	希望	Hope
Xiaosan	M	小三	Young number 3
Xiaosi	F	小四	Young number 4
Dabai	M	大白	Old white
Daxiong	M	大雄	Old male
Dahuo	M	大豁	Big broken nose
Xiaohuo	M	小豁	Small broken nose
Nüxia	F	女侠	Heroine
Jiaoshou	M	教授	Professor
Boshi	F	博士	PhD
Ruixue	F	瑞雪	Propitious snow
Baoma	F	豹妈	Mother leopard
Huayang	M	华阳	Sunny China
Yanghe	F	阳核	Central Yang (a local town)
Momo	F	嫫嫫	Ugly woman
Keke	F	可可	Cocoa
Xinxing	F	新星	Nova
Weiming	M	未名	Unnamed
129	M	129	129
Shuilan	F	水兰	Water orchid
Xiaof	F	小f	Young lass

References

Charlton, B., R. Swaisgood, Z. Zhang, and R. Snyder. 2012. "Giant Pandas Attend to Androgen-Related Variation in Male Bleats." *Behavioral Ecology and Sociobiology* 66:969–974.

Costello, C. M. 2010. "Estimates of Dispersal and Home-Range Fidelity in American Black Bears." *Journal of Mammalogy* 91(1): 116–121.

Dinerstein, E., C. Loucks, and Z. Lu. 2004. "Biological Framework of Evaluating Future Efforts in Giant Panda Conservation." In *Giant Pandas: Biology and Conservation*, edited by D. G. Lindburg and K. Baragona, pp. 228–233. Berkeley: University of California Press.

Fan, J., J. Li, Z. Quan, X. Wu, L. Hu, and Q. Yang. 2011. "Impact of Road Construction on Giant Panda's Habitat and Its Carrying Capacity in Qinling Mountains." *Acta Ecologica Sinica* 31(3): 145–149.

Fang, S., C. Ding, W. Feng, A. Zhang, G. Li, S. Li, J. Yu, and X. Li. 1996. 大熊猫 DNA 指纹分析材料的初步研究 [A Preliminary Study on the Material Resource of DNA in the DNA Fingerprinting Analysis of Giant Pandas]. *Acta Theriologica Sinica* 16(3): 166–170.

Feng, T. T., F. T. Manen, N. A. X. U. N. Zhao, M. Li, and F. Wei. 2009. "Habitat Assessment for Giant Pandas in the Qinling Mountain Region of China." *Journal of Wildlife Management* 73(6): 852–858.

Garshelis, D. L. 2004. "Variation in Ursid Life Histories: Is There an Outlier?" In *Giant Pandas: Biology and Conservation*, edited by D. G. Lindburg and K. Baragona, pp. 53–73. Berkeley: University of California Press.

Garshelis, D. L., W. Hao, W. Dajun, Z. Xiaojian, L. Sheng, and W. J. McShea. 2008. "Do Revised Giant Panda Population Estimates Aid in Their Conservation?" *Ursus* 19(2): 168–176.

Gong, M. H., and Y. L. Song. 2011. "Topographic Habitat Features Preferred by the Endangered Giant Panda *Ailuropoda melanoleuca*: Implications for Reserve Design and Management." *Oryx* 45(2): 252–257.

Gong, M., Z. Yang, W. Yang, and Y. Song. 2010. "Giant Panda Habitat Networks and Conservation: Is This Species Adequately Protected?" *Wildlife Research 37*(6): 531–538.

Harris, R. B. 2004. "Insights into Population Dynamics of Giant Pandas Gained from Studies in North America." *Chinese Journal of Zoology* 50:662–668.

———. 2008. *Wildlife Conservation in China: Preserving the Habitat of China's Wild West.* Armonk, NY: M. E. Sharpe.

Hu, J., and F. Wei. 1990. "Development and Progress of Research on the Giant Panda in the 1980s." In 大熊猫生物学研究与进展 [Research and progress in biology of the giant panda], pp. 9–18. Chengdu, China: Sichuan Publishing House of Science and Technology.

———. 2004. "Comparative Ecology of Giant Pandas in the Five Mountain Ranges of Their Distribution in China." In *Giant Pandas: Biology and Conservation*, edited by D. G. Lindburg and K. Baragona, pp. 137–148. Berkeley: University of California Press.

Hu, J., F. Wei, C. Yuan, and Y. Wu. 1990. 大熊猫生物学研究与进展 [Research and progress in biology of the giant Panda]. Chengdu, China: Sichuan Publishing House of Science and Technology.

Hu, Y., D. Qi, H. Wang, and F. Wei. 2010a. "Genetic Evidence of Recent Population Contraction in the Southernmost Population of Giant Pandas." *Genetica* 138(11): 1297–1306.

Hu, Y., X. Zhan, D. Qi, and F. Wei. 2010b. "Spatial Genetic Structure and Dispersal of Giant Pandas on a Mountain-Range Scale." *Conservation Genetics* 11(6): 2145–2155.

Ishibashi, Y., and T. Saitoh. 2004. "Phylogenetic Relationships among Fragmented Asian Black Bear (*Ursus thibetanus*) Populations in Western Japan." *Conservation Genetics* 5(3): 311–323.

Jin, K., C. Xue, X. Wu, J. Qian, Y. Zhu, Z. Yang, T. Yonezawa, M. J. C. Crabbe, Y. Cao, and M. Hasegawa. 2011. "Why Does the Giant Panda Eat Bamboo? A Comparative Analysis of Appetite-Reward-Related Genes among Mammals." *PloS ONE* 6(7): e22602.

Joshi, A. R., J. L. D. Smith, and D. L. Garshelis. 1999. "Sociobiology of the Myrmecophagous Sloth Bear in Nepal." *Canadian Journal of Zoology* 77(11): 1690–1704.

Kojola, I., P. I. Danilov, H. M. Laitala, V. Belkin, and A. Yakimov. 2003. "Brown Bear Population Structure in Core and Periphery: Analysis of Hunting Statistics from Russian Karelia and Finland." *Ursus* 14(1): 17–20.

Kovach, A. I., and R. A. Powell. 2003. "Effects of Body Size on Male Mating Tactics and Paternity in Black Bears, *Ursus americanus.*" *Canadian Journal of Zoology* 81(7): 1257–1268.

Li, Y., A. Viña, W. Yang, X. Chen, J. Zhang, Z. Ouyang, Z. Liang, and J. Liu. 2013. "Effects of Conservation Policies on Forest Cover Change in Giant Panda Habitat Regions, China." *Land Use Policy* 33:42–53.

Lindburg, D. G., and K. Baragona. 2004. "Consensus and Challenge: The Giant Panda's Day is Now." In *Giant Pandas: Biology and Conservation*, edited by D. G. Lindburg and K. Baragona, pp. 271–276. Berkeley: University of California Press.

Liu, J., Z. Ouyang, H. Zhang, M. Linderman, L. An, S. Bearer, and G. He. 2004. "A New Paradigm for Panda Research and Conservation." In *Giant Pandas: Biology and Conservation*, edited by D. G. Lindburg and K. Baragona, pp. 217–225. Berkeley: University of California Press.

Loucks, C. J., Z. Lü, E. Dinerstein, D. Wang, D. Fu, and H. Wang. 2003. "The Giant Pandas of the Qinling Mountains, China: A Case Study in Designing Conservation Landscapes for Elevational Migrants." *Conservation Biology* 17(2): 558–565.

Loucks, C. J., Z. Lü, E. Dinerstein, H. Wang, D. M. Olson, C. Zhu, and D. Wang. 2001. "Giant Pandas in a Changing Landscape." *Science* 294(5546): 1465.

Lü, Z., W. E. Johnson, M. Menotti-Raymond, N. Yuhki, J. S. Martenson, S. Mainka, S. Huang, Z. Zheng, G. Li, W. Pan, X. Mao, and S. J. O'Brien. 2001. "Patterns of Genetic Diversity in Remaining Giant Panda Populations." *Conservation Biology* 15(6): 1596–1607.

Lü, Z., and Y. Liu. 2004. "China's National Plan for Conservation of the Giant Panda." In *Giant Pandas: Biology and Conservation*, edited by D. G. Lindburg and K. Baragona, pp. 226–227. Berkeley: University of California Press.

Lü, Z., D. Wang, and D. L. Garshelis (IUCN SSC Bear Specialist Group). 2008. "*Ailuropoda melanoleuca.*" In IUCN Red List of Threatened Species. Version 2012.2. Accessed 22 July 2013. http://www.iucnredlist.org/details/712/0.

Ma, J., H. Chen, X. Xie, and F. Li. 1992. 中国熊类研究现状与展望 [Perspectives of bear research in China]. In 第二届东亚熊类会议论文集 [Proceedings of the 2nd Symposium on Northeastern Bears], pp. 1–9. Haerbin, China: Northeast Forestry University Press.

McLellan, B. N., and F. W. Hovey. 2001. "Natal Dispersal of Grizzly Bears." *Canadian Journal of Zoology* 79(5): 838–844.

Nie, Y., R. R. Swaisgood, Z. Zhang, Y. Hu, Y. Ma, and F. Wei. 2012a. "Giant Panda Scent-Marking Strategies in the Wild: Role of Season, Sex and Marking Surface." *Animal Behaviour* 84(1): 39–44.

Nie, Y., R. R. Swaisgood, Z. Zhang, X. Liu, and F. Wei. 2012b. "Reproductive Competition and Fecal Testosterone in Wild Male Giant Pandas (*Ailuropoda melanoleuca*)." *Behavioral Ecology and Sociobiology* 66:721–730.

Peacock, E., K. Titus, D. L. Garshelis, M. M. Peacock, and M. Kuc. 2011. "Mark-Recapture Using Tetracycline and Genetics Reveal Record-High Bear Density." *Journal of Wildlife Management* 75(6): 1513–1520.

Qi, D., S. Zhang, Z. Zhang, Y. Hu, X. Yang, H. Wang, and F. Wei. 2011. "Different Habitat Preferences of Male and Female Giant Pandas." *Journal of Zoology* 285(3): 205–214.

———. 2012. "Measures of Giant Panda Habitat Selection across Multiple Spatial Scales for Species Conservation." *Journal of Wildlife Management* 76(5): 1092–1100.

Reid, D. G., and J. E. Gong. 1999. "Giant Panda Conservation Action Plan." *Bears: Status Survey and Conservation Action Plan*, compiled by C. Servheen, S. Herrero, and B. Peyton, pp. 241–254. Gland, Switzerland and Cambridge: IUCN/SSC Bear and Polar Bear Specialist Groups.

Santangelo, P. 1998. "Ecologism Versus Moralism: Conceptions of Nature in Some Literary Texts of Ming-Qing Times." In *Sediments of Time: Environment and Society in Chinese Society and History*, edited by M. Elvin and T. J. Liu, pp. 617–655. Cambridge: Cambridge University Press.

Schaller, G. B. 1994. *The Last Panda*. Chicago: University of Chicago Press.

Schaller, G. B., J. Hu, W. Pan, and J. Zhu. 1985. *The Giant Pandas of Wolong*. Chicago: Chicago University Press.

Shapiro, J. 2001. *Mao's War against Nature: Politics and the Environment in Revolutionary China*. Cambridge: Cambridge University Press.

Shen, G., C. Feng, Z. Xie, Z. Ouyang, J. Li, and M. Pascal. 2008. "Proposed Conservation Landscape for Giant Pandas in the Minshan Mountains, China." *Conservation Biology* 22(5): 1144–1153.

Songer, M., M. Delion, A. Biggs, and Q. Huang. 2012. "Modeling Impacts of Climate Change on Giant Panda Habitat." *International Journal of Ecology* 2012:108752. doi:10.1155/2012/108752.

Sterckx, R. 2004. "Attitudes towards Wildlife and the Hunt in Pre-Buddhist China." In *Wildlife in Asia: Cultural Perspectives*, pp. 15–35. London: RoutedgeCurzon.

Steyaert, S. M. J. G., A. Endrestøl, K. Hackländer, J. E. Swenson, and A. Zedrosser. 2012. "The Mating System of the Brown Bear *Ursus arctos*." *Mammal Review* 42(1): 12–34.

Swaisgood, R. R., D. G. Lindburg, A. M. White, H. Zhang, and X. Zhou. 2004. "Chemical Communication in Giant Pandas: Experimentation and Application." In *Giant Pandas: Biology and Conservation*, edited by D. G. Lindburg and K. Baragona, pp. 106–120. Berkeley: University of California Press.

Swaisgood, R. R., F. Wei, W. J. Mcshea, D. E. Wildt, A. J. Kouba, and Z. Zhang. 2011. "Can Science Save the Giant Panda (*Ailuropoda melanoleuca*)? Unifying Science and Policy in an Adaptive Management Paradigm." *Integrative Zoology* 6(3): 290–296.

Swaisgood, R. R., F. Wei, D. E. Wildt, A. J. Kouba, and Z. Zhang. 2010. "Giant Panda Conservation Science: How Far We Have Come." *Biology Letters* 6(2): 143–145.

Tuanmu, M.-N., A. Viña, J. Winkler, Y. Li, W. Xu, Z. Ouyang, and J. Liu. 2012. "Climate-Change Impacts on Understory Bamboo Species and Giant Pandas in China's Qinling Mountains." *Nature Climate Change* 3:249–253. doi:10.1038/NCLIMATE1727.

Wan, Q., S. Fang, H. Wu, and T. Fujihara. 2003. "Genetic Differentiation and Subspecies Development of the Giant Panda as Revealed by DNA Fingerprinting." *Electrophoresis* 24(9): 1353–1359.

Wan, Q., H. Wu, and S. Fang. 2005. "A New Subspecies of Giant Panda (*Ailuropoda melanoleuca*) from Shaanxi, China." *Journal of Mammalogy* 86(2): 397–402.

Wang, D., S. Li, S. Sun, H. Wang, A. Chen, S. Li, J. Li, and Z. Lu. 2008. "Turning Earthquake Disaster into Long-Term Benefits for the Panda." *Conservation Biology* 22(5): 1356–1360.

Wang, T., X. Ye, A. K. Skidmore, and A. G. Toxopeus. 2010. "Characterizing the Spatial Distribution of Giant Pandas (*Ailuropoda melanoleuca*) in Fragmented Forest Landscapes." *Journal of Biogeography* 37(5): 865–878.

Wang, X., W. Xu, and Z. Ouyang. 2009. "Integrating Population Size Analysis into Habitat Suitability Assessment: Implications for Giant Panda Conservation in the Minshan Mountains, China." *Ecological Research* 24(5): 1101–1109.

Wei, F., Y. Hu, L. Zhu, M. Bruford, X. Zhan, and L. Zhang. 2012. "Black and White and Read All Over: The Past, Present and Future of Giant Panda Genetics." *Molecular Ecology* 21(23): 5660–5674.

Wei, F., Z. Zhang, and J. Hu. 2011. 野生大熊猫生态学研究进展与前瞻 [Research advances and perspectives on the ecology of wild giant pandas]. *Acta Theriologica Sinica* 31(4): 412–421.

Wildt, D. E., A. Zhang, H. Zhang, D. J. Janssen, and S. Ellis, eds. 2006. *Giant Pandas: Biology, Veterinary Medicine and Management*. Cambridge: Cambridge University Press.

Xu, M., Z. Wang, D. Liu, R. Wei, G. Zhang, H. Zhang, X. Zhou, and D. Li. 2012. "Cross-Modal Signaling in Giant Pandas." *Chinese Science Bulletin* 57(4): 344–348.

Xu, W., Z. Ouyang, A. Viña, H. Zheng, J. Liu, and Y. Xiao. 2006. "Designing a Conservation Plan for Protecting the Habitat for Giant Pandas in the Qionglai Mountain Range, China." *Diversity and Distributions* 12(5): 610–619.

Xu, W., X. Wang, Z. Ouyang, J. Zhang, Z. Li, Y. Xiao, and H. Zheng. 2009. "Conservation of Giant Panda Habitat in South Minshan, China, after the May 2008 Earthquake." *Frontiers in Ecology and the Environment* 7(7): 353–358.

Yong, Y., F. Wei, X. Ye, Z. Zhang, and Y. Li. 2004. 佛坪自然保护区野生大熊猫交配行为的观察 [Mating behaviors of wild giant pandas in Foping Natural Reserve]. *Acta Theriologica Sinica* 24(4): 346–349.

Yu, C. 2006. "The Effective Management of Reserves for the Conservation of the Giant Panda." In *Biology in Asia*, pp. 128–136. Kathmandu, Nepal: Society for Conservation Biology, Asia Section.

Yu, C., and X. Deng. 2004. "Management of Giant Panda Reserves in China." In *Giant Pandas: Biology and Conservation*, edited by D. G. Lindburg and K. Baragona, pp. 234–235. Berkeley: University of California Press.

Zhan, X., M. Li, Z. Zhang, B. Goossens, Y. Chen, H. Wang, M. W. Bruford, and F. Wei. 2006. "Molecular Censusing Doubles Giant Panda Population Estimate in a Key Nature Reserve." *Current Biology* 16(12): 451–452.

Zhan, X., Y. Tao, M. Li, Z. Zhang, B. Goossens, Y. Chen, H. Wang, M. Bruford, and F. Wei. 2009. "Accurate Population Size Estimates Are Vital Parameters for Conserving the Giant Panda." *Ursus* 20(1): 56–62.

Zhan, X. J., Z. J. Zhang, H. Wu, B. Goossens, M. Li, S. W. Jiang, M. W. Bruford, and F. W. Wei. 2007. "Molecular Analysis of Dispersal in Giant Pandas." *Molecular Ecology* 16(18): 3792–3800.

Zhang, B., M. Li, Z. Zhang, B. Goossens, L. Zhu, S. Zhang, J. Hu, M. W. Bruford, and F. Wei. 2007. "Genetic Viability and Population History of the Giant Panda, Putting an End to the 'Evolutionary Dead End'?" *Molecular Biology and Evolution* 24(8): 1801–1810.

Zhang, J., V. Hull, W. Xu, J. Liu, Z. Ouyang, J. Huang, X. Wang, and R. Li. 2011. "Impact of the 2008 Wenchuan Earthquake on Biodiversity and Giant Panda Habitat in Wolong Nature Reserve, China." *Ecological Research* 26(3): 523–531.

Zhang, Z., R. R. Swaisgood, S. Zhang, L. A. Nordstrom, H. Wang, X. Gu, J. Hu, and F. Wei. 2011. "Old-Growth Forest Is What Giant Pandas Really Need." *Biology Letters* 7(3): 403–406.

Zhao, H., J. R. Yang, H. Xu, and J. Zhang. 2010. "Pseudogenization of the Umami Taste Receptor Gene Tas1r1 in the Giant Panda Coincided with Its Dietary Switch to Bamboo." *Molecular Biology and Evolution* 27(12): 2669–2673.

Zhao, S., P. Zheng, S. Dong, X. Zhan, Q. Wu, X. Guo, Y. Hu, W. He, S. Zhang, W. Fan, L. Zhu, D. Li, X. Zhang, Q. Chen, H. Zhang, Z. Zhang, X. Jin, J. Zhang, H. Yang, J. Wang, J. Wang, and F. Wei. 2013. "Whole-Genome Sequencing of Giant Pandas Provides Insights into Demographic History and Local Adaptation." *Nature Genetics* 45(1): 67–71.

Zhu, C., and Z. Ouyang. 2004. "Restoring Giant Panda Habitat." In *Giant Pandas: Biology and Conservation*, edited by D. G. Lindburg and K. Baragona, pp. 187–188. Berkeley: University of California Press.

Zhu, L., Q. Wu, J. Dai, S. Zhang, and F. Wei. 2011a. "Evidence of Cellulose Metabolism by the Giant Panda Gut Microbiome." *Proceedings of the National Academy of Sciences of the United States of America* 108(43): 17714–17719.

Zhu, L., X. Zhan, T. Meng, S. Zhang, and F. Wei. 2010a. "Landscape Features Influence Gene Flow as Measured by Cost-Distance and Genetic Analyses: A Case Study for Giant Pandas in the Daxiangling and Xiaoxiangling Mountains." *BMC Genetics* 11(1): 72.

Zhu, L., X. Zhan, H. Wu, S. Zhang, T. Meng, M. W. Bruford, and F. Wei. 2010b. "Conservation Implications of Drastic Reductions in the Smallest and Most Isolated Populations of Giant Pandas." *Conservation Biology* 24(5): 1299–1306.

Zhu, L., S. Zhang, X. Gu, and F. Wei. 2011b. "Significant Genetic Boundaries and Spatial Dynamics of Giant Pandas Occupying Fragmented Habitat Across Southwest China." *Molecular Ecology* 20(6): 1122–1132.

Chapter 1

A Needed Conservation Perspective: The Question of Whether Our Understanding and Policies Are Consistent with Reality

Summary: Within the geographic area that giant pandas currently occupy, bamboo is sufficiently abundant to satisfy their nutritional requirements. Flowering and producing seeds on a large geographic scale is a normal part of the life cycle of bamboo. Bamboo flowering is not a threat to the persistence of giant pandas. Giant pandas, either in the wild or in captivity, are capable of normal reproduction. The short gestation length of female pandas, the small size of newborn cubs, and the slow pace of neonate development are all successful aspects of a reproductive strategy adopted from its ursid ancestors through the process of evolution. Because at present we lack sufficient understanding of implantation to attempt embryo transfer, we have no theoretical foundation for cloning pandas, so it is not a feasible option. The idea of cloning from another species is even more unrealistic. In the end, if we want to conserve giant pandas, the way forward does not lie in cryotechnology or extracting genes to replicate them in the laboratory, and it is certainly not in cloning. Rather, the path toward conservation of giant pandas is to maintain their genetic diversity in the wild as part of their biotic community and thereby to ensure their evolutionary potential.

Geological records make clear that all giant pandas (*Ailuropoda melanoleuca*) now in existence are the direct descendants of *Ailuropoda melanoleuca baconi* Woodward (Pan et al., 1988). From the middle to the late Pleistocene era, a duration of some 700,000 years, *A. m. baconi* experienced an unprecedented flourishing, spreading throughout the Pearl River region in southern China, as well as the Yangtze River drainage in central China and the Yellow River drainage in northern China (Pan et al., 1988). The fossil record also documents that giant pandas existed in the northern portions of present-day Vietnam and Thailand, as well as the eastern section of present-day Myanmar (Ginsburg et al., 1982; Thue, 1984). Throughout the 19th century and even into the 20th, pandas still lived in the mountains of the seven to eight counties of Hubei and Hunan Provinces bordering Sichuan, such as

Zhushan, Badong, Zigui, and Changyang Counties in Hubei, Lizhou and Dayonug Counties in Hunan, and Youyang in what is today Chongqing (Wen and He, 1981; He, 1989). It was not until the middle of the 20th century that giant pandas came to be restricted to the six mountain ranges in west central China where we find them today. From an evolutionary viewpoint, the rapid reduction in distribution and numbers of pandas to the few we have today has occurred in the blink of an eye.

The second half of the 20th century surely ranks as the most destructive in the species' entire evolutionary history in terms of loss of habitat and reduction in numbers. But viewed from the perspective of the efforts we have begun exerting to conserve the species and the progress we have made, we may yet count these years as the brightest for pandas. With China's rapid population growth (from 470 million in 1949 to 1.3 billion in 2000) leading to rapid agricultural and industrial development, deforestation and land clearing have spread throughout the geographic area that is home to the giant panda. At the Third National People's Congress in 1957, a decision was made to establish a nature reserve for pandas, heralding a new era for panda conservation. From 1963 to 1978, a total of 12 nature reserves were established specifically to conserve giant pandas, covering 20% of total panda distribution within China. In 1998, the State Council promulgated a logging ban in native forests within the upper and middle reaches of the Yangtze River in 1998. Thus, by 1999, the overwhelming majority of giant panda habitat had achieved protected status, which is a most auspicious development for the possibility of modern pandas persisting in the wild.

However, in recent years, wild pandas have been experiencing difficulties. At the same time that many researchers and agency staff have made great efforts to clarify the threats facing pandas and to develop effective management plans, the popular media has time and again attempted to link some "newly found crisis" to panda conservation and, through that, to arouse public sentiment. Exaggerated by the media in this way as a "crisis," many people have become quite pessimistic about the giant panda's future, and it has become fashionable to promote the adoption of some sort of "advanced technology" for pandas. For example, during autumn 1983, a large flowering and die-off of bamboo occurred in Sichuan's Wolong Nature Reserve. Almost overnight, newspapers and radio and television stations were full of reports that 'the bamboo flowering has put pandas at imminent risk of extinction' and of plans to help pandas avoid starvation by establishing 'captive breeding centers.' In 1985, the media quoted a researcher saying that 'the giant panda's reproductive behavior is so highly specialized that their fertility has actually become impaired, leading to reduced population size and bringing them to the edge of extinction.' By 1991, a belief had taken hold among some that the only way to increase panda numbers was through such techniques as bisectioning embryos, in vitro fertilization or test-tube babies, or embryo transfer. In the summer of 1997, when the cloned sheep Dolly was born in Britain, people were immediately quoted in newspapers and on television as suggesting that pandas be cloned as well. In 1998, the unrealistic notion that pandas could be produced by interspecific cloning was put forward in some circles. By 1999, some newspapers and television stations were declaring that 'the demise of the giant panda is imminent; in zoos, pandas only have another 50 years, and the wild, at most, 100 years.'

The truth is that at the present time, we really have no basis for making a scientific assessment of the future prospects for giant pandas. We do know that habitat fragmentation and the resulting small size of local populations is a reality in many panda areas, but we have not learned nearly enough about their reproductive behavior and physiology. Clearly, ensuring a future for giant pandas, aside from the biological requirements of wild habitats and evolutionary potential, will require that we sincerely care for and carefully manage them. We should not make decisions bearing on conservation actions based on speculations or assumptions. We are better off using reality as a basis for our understanding, attitudes, and views of the appropriateness of policy. This is a huge issue affecting the giant panda's future.

The Issue of Bamboo Flowering

In the autumn of 1983, it gradually became evident that the bamboo in and around Sichuan's Wolong Nature Reserve had begun a mass flowering. News reports in the media followed, with radio and television reports such as 'Because bamboo is the main food for giant pandas, the flowering and subsequent withering and death of bamboo will lead to food shortages, hunger, starvation, and death.' This type of news story was then followed by a myriad of media reports that featured stories of saving individual pandas from starvation.

In Wolong, however, in addition to the cane bamboo species (*Sinarundinaria fangiana*, now considered *Bashania faberi*) that had flowered, another very abundant species of bamboo, the umbrella bamboo (*Fargesia spathacea*, now considered *Thamnocalamus spathaceus*), also grows. Pandas eat both. Using radio-marked animals, researchers working in the area at the time documented not only that pandas responded to the flowering of cane bamboo by moving to areas in which the umbrella bamboo (*F. robusta*) was concentrated but also that they had found not a single panda that had starved to death. However, for many years afterward, the general public had no appreciation of these facts. Even now, many people firmly believe that the natural world is the panda's enemy and view pandas as living within the dark shadow of mass bamboo flowering events.

What is the real effect of the mass bamboo flowering and subsequent mortality on the giant panda? Is massive bamboo flowering some sort of abnormal phenomenon, or is it instead an inevitable or necessary stage in its life cycle? If mass flowering of bamboo occurs again, what should we do?

A Needed Conservation Perspective   29

Reasoning and Examples

Wen and Wang (1980) wrote a popular article concerning the effects of mass bamboo flowering on pandas. According to these authors, 4 years prior, during winter and spring, 138 individual pandas had been found dead in Wenxian County in Gansu and in Pingwu and Nanping Counties in Sichuan. An investigation disclosed that the large number of deaths had been caused primarily by starvation resulting from the flowering and subsequent mortality of the umbrella bamboo, *F. spathacea*. Flowering within Wenxian and Pingwu Counties occurred over an area exceeding 5,000 km². In this area, other species of bamboo are quite rare, so the effects of the die-off were catastrophic for the pandas (Wen and Wang, 1980).

In the next year, Yang et al. (1981) discussed the large panda die-off during 1976 in the Min Shan range of Gansu and Sichuan as well as the Motianling Mountains. Again, the mass mortality of the umbrella bamboo (*F. spathacea*) was implicated as the cause of the panda deaths. (Note that a number of species of the genus *Fargesia* flowered in the Min Shan range during the 1970s, not all of which were this species. Most flowering events were of other species within *Fargesia*.) Yang et al. pointed out that the region in which the mass flowering occurred is one of frequent earthquakes. Bamboo closer to the core of the earthquake area flowered earlier and over a larger area; bamboo toward the periphery of the earthquake area flowered later and within a more limited area.

Before it was possible to accurately assess the number and causes of panda deaths during the Min Shan range die-off of the mid 1970s, another mass flowering occurred in the early 1980s, this time of the cane bamboo (*B. faberi*) in the Qionglai mountain range. Because we had 'learned the lessons of the previous flowering event,' this next bamboo flowering was met with a campaign using the slogan 'Immediately initiate large-scale work to save the giant panda from extinction.' After this, the flowering of umbrella bamboo (*F. spathacea*) in the mid-1970s and the deaths of 138 pandas became a raison d'être for researchers and were elevated to the status of being "typical examples" for the media when they reported on bamboo and panda mortalities.

We need to point out that the area of the northern Min Shan range, in which the mass flowering had occurred in the 1970s, had also recently been affected by a very strong earthquake. Thus, the panda die-off of that time period was actually quite a complex phenomenon. Surveys done at the time were only able to obtain carcasses and bone fragments, so causes of death for most individual pandas could not be confirmed. Zhu et al. (1988) stated, "In June 1976, we documented 95 carcasses of pandas in the Min Shan of Sichuan and Gansu." However, they went on to say that "excepting individual cases of old age, infectious disease or hyperthermia caused by a sharp rise in temperature (caused by the earthquake), starvation from the flowing

30   Chapter 1

of bamboo was the cause of panda mortalities." They also estimated the affected area to be 5,250 km^2, within which they considered the most seriously affected area to be 1,900 km^2 and the lightly affected area to be 3,350 km^2 (Zhu et al., 1988). However, a relatively large proportion of the panda carcasses was in the 63% area they had characterized as lightly affected, and within this area, the bamboo had not yet completely died out. Thus, simply concluding that starvation due to food shortage was the cause of deaths is less than fully persuasive. We provide two examples that can help deepen our understanding of the issue.

Example 1

After the 1983 mass flowering of cane bamboo in Wolong, research indicated that the entire affected region was >2,600 m in elevation and accounted for some 75%–95% of the total cane bamboo area. At the time, some of the researchers termed this the "bamboo flowering disaster area." However, umbrella bamboo was still doing very well at lower elevations (<2,600 m) and was showing no sign of a mass flowering event.

From 1983 to 1985, Ken Johnson used radiotelemetry to obtain data on panda activity in the part of Wolong where the flowering had occurred (Johnson et al., 1988). He also compared his data with those collected prior to the bamboo flowering (Table 1.1).

With the flowering, cane bamboo in Wolong plummeted by 75% in total biomass, and its formerly contiguous habitat was transformed into isolated fragments. Yet it seems that pandas still had adequate food. It is true that the bamboo die-off did affect pandas' energetics; for example, they moved farther in search of food. But in the 2 years following the die-off, pandas

Table 1.1.

COMPARISON OF PANDA FEEDING BEHAVIOR BEFORE AND AFTER BLOOMING OF *SINARUNDINARIA* IN WOLONG. DERIVED FROM JOHNSON ET AL. (1988).		
Item	Before blooming of *Sinarundinaria*	After blooming of *Sinarundinaria*
Percent of bamboo leaf in panda's food	40.7 ± 36.8	43.0 ± 36.5
Percent wastage of *Sinarundinaria* shoot by panda	12.0	13.3
Time of activity per day (h)	13.5	13.5
Percent active		
Spring	62	62
Summer and autumn	53	49
Winter	59	57
Distance moved per day (m)		
Spring	495	567
Summer and autumn	393	489
Winter	481	543

A Needed Conservation Perspective   31

adjusted their movement patterns only slightly, and their patterns of seasonal dietary selection and behavior were essentially unchanged. Most critically, not a single panda starved to death in the Wolong Nature Reserve during this period.

George Schaller described this quite clearly in his book *The Last Panda*. Faced with a shortage of cane bamboo, pandas had three options: increase home range size, migrate to another location entirely, or forage for alternative species of bamboo; all they needed to do was to move upward or downward along the same mountain slope (Schaller, 1994).

Example 2

The former Changqing Forestry Bureau (now the Changqing National Nature Reserve) is located on the southern slopes of the Qinling Mountains of Shaanxi Province at middle to high elevations. In autumn 1983, investigations conducted by this bureau aimed at understanding the distribution and abundance of the two bamboo species eaten by pandas revealed that bamboo occupied some 91.8 km² (48%) of the 191.8 km² survey area at elevations of 900–2,900 m. Within the survey area, *B. faberi* was growing well in areas below 2,000 m and had not experienced a die-off. The other principal bamboo species in the region, *Fargesia* spp., growing at elevations of 2,000–2,350 m, had experienced a partial flowering, covering 21.6% of its total area. The investigation also revealed that 28.8% of the area in which the die-off had occurred was already showing signs of regeneration. Moreover, at elevations above 2,350 m, *Fargesia* growing on ridges within the Qinling continued to flourish and had not experienced flowering (Pan et al., 1988). The area of *Fargesia* flowering did not expand in either 1984 or 1985. In contrast, the years 1985 and 1986 saw increases in the area of recovery. As early as April 1987, we documented pandas eating newly regenerating *Fargesia* shoots at an elevation of 2,300 m within the die-off area.

From June 1987 to June 1998, radio tracking suggested that the panda population was stable regardless of whether or not *Fargesia* was flowering and dying. Pandas accommodated the die-off by adding a few extra hundred meters to their existing seasonal movements, which enabled them to pass through the die-off areas and find abundant food supplies. Not once during the course of our research in this area did we document evidence suggesting that pandas lacked food as a result of bamboo flowering.

The prevailing wisdom has been that the effect that cyclic flowering and die-off of bamboo has on pandas is primarily a function of the size of the area affected, as well as whether there are alternative species of bamboo in the vicinity (Schaller et al., 1985). In earlier times, perhaps even before the agricultural era when the geographic distribution of pandas was broad and many species of bamboo grew within panda habitat, flowering never constituted a threat to the persistence of pandas. Not only that, natural disturbances, of which bamboo flowering is an example, may even be helpful in maintaining long-term evolutionary processes in pandas. However, in the present situation in which pandas are restricted to six isolated mountain ranges and their habitats are increasingly fragmented, large-scale bamboo flowering events, particularly if they occur in areas with only a single bamboo species, may have local negative impacts. Theoretically, this view has a certain logic. But realistically, what do we know about panda distribution and bamboo distribution? We discuss the most important points in the following sections.

BAMBOO SPECIES

Many species of bamboo exist within current panda habitat. Using Qin (1985; for Sichuan) and our own team's early work in the Qinling (Pan et al., 1988), we have a complete picture of the status and distribution of bamboo within the panda's range nationwide:

- Min Shan range: 1,500–2,000 m: *Fargesia scabrida, F. rufa, F. angustissima*; 2,000–3,400 m: *Fargesia nitida, F. denudata,* and *Sinarundinaria fangiana*
- Qionglai mountain range: below 1,600 m: *F. angustissima, Phyllostachys nidularia, P. arcana,* and *Chimonobambusa pachystachys*; 1,600–2,000 m: *Fargesia robusta, Yushania chungii, Chimonobambusa pachystachys*; 2,000–2,600 m: *Fargesia robusta, Sinarundinaria fangiana, Yushania chungii, F. nitida,* and *S. ferax*; 2,600–3,600 m: *Sinarundinaria fangiana, Fargesia nitida,* and *Yushania chungii.*
- Xiangling mountain range: below 1,800 m: *Chimonobambusa szechuanensis* and *C. pachystachys*; 1,800–2,100 m: *Yushania tineloata*; 2,100–3,500 m: *Chimonobambusa szechuanensis, Sinarundinaria fangiana,* and *Yushania cava*
- Liangshan mountain range: below 1,800 m: *Chimonobambusa pachystachys, Indocalamus longiauritus, C. monophylla, Phyllostachys nidularia, P. heteroclada,* and *P. nigra* var. *henonis*; 1,800–2,100 m: *Chimonobambusa szechuanensis, Qiongzhuea tumidinoda, Q. opienensis, Q. macrophylla,* and *Indocalamus longiauritus*; 2,100–3,500 m: *Chimonobambusa szechuanensis, Qiongzhuea tumidinoda, Q. macrophylla, Indocalamus longiauritus, Sinarundinaria fargiana,* and *Yushania mabianeusis*
- The southern slopes of the Qinling Mountains: 800–2,000 m: *Bashania fargesii* and *P. nigra* var. *henonis*; 2,000–3,200 m: *Fargesia* and what local people call dragon's head bamboo

All existing panda populations are located in areas with numerous bamboo species, and in all cases, no fewer than two species are capable of providing ample forage for pandas. Excepting a situation in which multiple species flower and die off simultaneously, we believe pandas can relatively easily find alternative foods during a bamboo flowering. Thus, as a general rule, the mass flowering of a single bamboo species does not constitute a threat to the long-term persistence of a panda population. As pointed out by Schaller et al. (1985), most valleys in Wolong contain two or more species of bamboo. Thus, the large-scale flowering of cane bamboo of 1983 did not necessarily lead to mass starvation. In most mountain ranges, if one species flowers and dies, most pandas need only move vertically 200–300 m to find an alternative species of bamboo (Schaller et al., 1985).

Even if only a single bamboo species exists in a given area, a mass flowering and die-off should not be interpreted as meaning 100% mortality. As described by Johnson et al. (1988), pandas in Wolong evidently met their forage demands entirely from consuming large quantities of the dried stalks of dead bamboo following the large cane bamboo flowering. In autumn 1984, a field worker related to us that he had observed a group of five to seven pandas feeding together in a small patch of cane bamboo that had not flowered. In the spring 1984, we found large quantities of bamboo residue and evidence of bamboo within panda scats in an area where cane bamboo had earlier occurred over an extensive area. In fact, our research has suggested that yearly consumption of any given species of bamboo by pandas never exceeds 2.0% of that species' annual growth (Pan et al., 1988). Thus, regardless of whether we are considering Wolong or the Qinling, mass flowering and die-offs of bamboo have not constituted the primary threats to panda persistence.

The Issue of Panda Mortalities

During the Second National Conference for Conserving and Rescuing Giant Pandas (Lanzhou, December 1984), the following statistics were announced: Flowering events involving various species of bamboos had occurred in some 38.7% of existing panda habitat. Within this area, cane bamboo flowering and die-offs accounted for over 87%. Of 43 counties containing pandas, flowering events had occurred in 26, and disastrous consequences had occurred in 10 counties. A total of 30 pandas had been rescued from starvation since May 1983. Additionally, field workers were continuing to discover cases of starvation and had documented 33 carcasses of pandas that had died naturally (Meng, 1985).

In 1988, another article appeared that quoted statistics to the effect that in the previous 4 years, rescue efforts had been attempted for 88 starving pandas, 64 of which had survived. Meanwhile, over 70 carcasses and skeletons were discovered (Wang, 1988).

Now, if we calculate that during the period 1983–1988 there were 70 known mortalities and we add to this the 22 deaths of animals for which rescue efforts failed, we have a total of 92 mortalities during the 4 years, i.e., an average of 23 deaths/year. The official estimate of the panda population at that time was over 1,000 (we estimate 1,200–1,500). If we assume that pandas live to be about 20 years old, then in a population of 1,000 pandas we would expect about 50 deaths yearly for a stable population, considerably more than the number that were documented as perishing as a result of the bamboo die-offs in those years. Thus, the number of deaths occurring among free-ranging pandas during the period 1983–1988 is within the range of what we would expect to see in a stable population.

Within any panda population some individuals will always be at high risk of mortality due to disease, weakness, or injury. For example, during our studies in the Qinling Mountains in 1987–1995, we noticed a trend toward observing more illness among pandas at the end of winter and the beginning of spring. This time of year was also when bamboo plants had not yet grown new leaves, and their nutritional value for pandas was at its seasonal nadir. Also, pandas had been exposed to months of cold weather, and they were in generally poor condition. With such weakness, it was unsurprising that they became prone to illness and mortality. However, two species of bamboo native to the southerly slopes of the Qinling grew well during this time, suggesting a lack of correlation between bamboo flowering and panda mortality. As we can see, considering these various aspects of panda mortality together, simply assuming that panda mortalities result directly from bamboo flowering can only lead to confusion.

CYCLIC LARGE-SCALE FLOWERING OF BAMBOO IS NOT AN ANOMALY

Bamboo dies after flowering. This die-off is not, as suggested by newspaper reports, some type of genetic abnormality, but rather an essential part of the normal life cycle of the plant. Although we still have much to learn about the population genetics and environmental correlates of periodic mass flowering, the phenomenon must have been known to Chinese historians as early as 770 BC because the "Shan Hai Jing" recounts tales of bamboo flowering on a 60 year cycle. Other history books include documentation of bamboo species flowering at 120 year intervals. According to these records, cane bamboo in the Wolong area flowered twice, once in 1893 and then again in 1935, intervals of 42–48 years (Schaller et al., 1985). The 1983 flowering in Wolong was thus some 48 to 54 years subsequent to the previous event.

Within the overall panda distribution area with its numerous species of bamboo, one or another species flowers in any given year. Of course, the scope of these flowering events varies from an entire mountain slope to small, scattered patches. In 1975 and 1976, *Fargesia denudata* flowered over a large area in the Min Shan at elevations below 2,600 m; a partial flowering occurred again in 1982, this time at elevations above 2,900 m. Bamboo of the genus *Chimonobambusa* flowered in the Liangshan range in 1975, 1976, and again in 1980. In the Xiaoxiangling mountain range, a type of cane bamboo flowered in 1982. In the Qinling, the years 1980 and 1981 witnessed isolated flowerings of *B. faberi* on southerly slopes below 1,400 m. *Fargesia* spp.

A Needed Conservation Perspective   35

experienced a large-scale flowering in the Qinling at elevations of 1,850–2,350 m during 1983–1985. In 1993, bamboo flowered in Gansu's Baishuijiang Nature Reserve.

In short, pandas are entirely reliant on bamboo, and bamboo undergoes periodic mass flowering (and has since the Pleistocene era). Thus, our responsibility is not to attempt to adjust or interfere with what may appear to us a conflict. Nor should we (or, indeed, can we) capture all the giant pandas, put them in cages, and feed them artificially to see them through these bamboo flowering "crises."

The autumn of 1983 witnessed considerable exaggerated publicity regarding both the flowering and the efforts to save pandas; subsequently, this has gradually quieted down. Influenced by public sentiment, government bureaus at all levels are still engaged in promulgating emergency measures to save pandas, establishing small leadership teams to conduct rescue work, organizing patrol teams, energetically engaging in publicity, and building captive breeding facilities to take in sick and starving pandas. All of these have had positive effects for conservation to one degree or another.

Giant Panda Reproductive Biology

Living as they do in the rugged mountains, pandas are, perforce, not an abundant animal. Even for researchers who spend countless hours in the dense bamboo forests looking for them, observing them is very difficult. Notwithstanding our long fascination with the animals, we humans have never known much about the reproductive behavior of free-ranging giant pandas and even less about their reproductive physiology. To understand reproductive behavior, we have been limited to observations of captive pandas in zoos who are responding to artificial environments, with our fragmentary understanding supplemented from the occasional acquisition of carcasses of diseased pandas. Thus, we have emerged with a somewhat erroneous picture of whether pandas are capable of normal reproduction.

There is a general perception that modern pandas suffer from a sort of reproductive depression. Conventional wisdom suggests that male pandas have no interest in mating. Similarly, conventional wisdom would have it that the altricial nature of young cubs is evidence that the panda's reproductive ability has somehow declined; there is also the perception that the mortality rate of cubs is too high. Finally, there is a widespread notion that females have difficulty ovulating.

Reflecting these misperceptions, some researchers have held the erroneous view that both the reproductive and cub-raising abilities of pandas have developed to some sort of special stage and that this stage has progressed to the point where "these are the primary reasons that giant pandas stand on the brink of extinction" (Feng et al., 1985). We believe these misperceptions reflect a lack of understanding of the way in which evolution has molded the reproductive strategy of the giant panda.

To analyze these problems, we first need to avoid muddling the issue by talking about captive and wild, free-ranging pandas in the same breath. Additionally, when we do use information from captive pandas, we need to clearly distinguish among various reproductive situations that have been reported from individual zoos. Taking these steps will help us avoid further misunderstandings.

REPRODUCTION AMONG FREE-RANGING PANDAS

There are currently over 1,000 giant pandas spread among the six mountain ranges within west central China that still contain habitat for them. This number is far lower than during the 1950s, primarily because of excessive mortality and the sharp reduction of their habitat. Although many questions about panda reproductive biology remain unanswered, two research groups have been working since the 1980s on these issues and have obtained some preliminary results.

An international cooperative team worked from winter 1980 through summer 1983 in a 35 km^2 study area in Wolong observing the social life of pandas. Within their study area, this research group observed 2 older cubs, 4 independent subadults, 7 adult males, and 5 or 6 adult females, i.e., a total of 18 or 19 individuals. Generally, adult pandas were solitary, only occasionally coming into contact with each other. Except during the mating season, they rarely communicated with each other vocally, but rather maintained contact with each other using olfactory messages. The most frequently observed method of scent marking was leaving anal secretions, urine, or a claw mark on a tree trunk or protruding object, thereby producing a scent station that functioned alternatively to facilitate communication or mutual avoidance. In the Wolong region, female pandas came into estrus and mating aggregations occurred from mid-March to mid-May. The peak period of fertility among females lasted 1–3 days, during which time from two to five males would contest for dominance and mating privileges. However, it was also possible for the loser of such contests to mate with the female. During mating, males mounted females repeatedly, but we still lack adequate evidence that females are induced ovulators. During the Wolong research, two radio-collared females produced young. Additionally, in April 1982 the researchers had an opportunity to necropsy an adult female that had died. She was lactating at the time of her death (Schaller et al., 1985), and examination of her uterus revealed 2 placental scars.

Our research group was the second to work on free-ranging pandas, and we began our work in 1985 on the southern slopes of the Qinling Range, focusing on mating systems and reproductive strategies. We worked in an area of some 90 km² over 10 years, making observations of four adult females. We documented that most mating involves multiple males. Females gave birth every second year, and the combined time of gestation and lactation exceeded 1.5 years. Both birth rate and survival rates were fairly high. We estimated the annual population growth rate to be 3.5%. This is 10 times higher than the rate at which the human population was increasing in China. During the past 15 years, we have come to understand that the 650 km² of the southern Qinling Range that provided a home to pandas contained an estimated population of 150, and we assessed the population as being reasonably stable. In spring 1989, we put our first radio collar on a 4.5-year-old named Jiaojiao. By 1998, when she turned 14, she had produced five offspring. By then, her eldest son Huzi and her older daughter Xiwang had both produced offspring of their own (in 1996 and 1997, respectively). During 1989 to 1997 we documented no fewer than 16 cubs produced within the 40 km² area around our field camp that formed the core of our research area, of which 15 survived. In light of these examples, we found no basis for the idea that reproductive depression limits giant panda populations.

WHY ARE PANDA CUBS SO SMALL WHEN FIRST BORN?

Newly born panda cubs are indeed quite small relative to the size of an adult. This is one of the species' life history characteristics that has evolved through natural selection. Newborn cubs are only 1/900 the mass of their mothers. If we take mean litter size as 1.7 cubs, the weight of the entire litter is still tiny compared with the mother. At about 180 g, the total litter mass is barely 1/500 of the mother (Beijing Zoo, 1974; Liu, 1988). Presently, only eight species of bears exist, and understanding the close relationship between the evolution of pandas and that of the other bear species can help us better explain unique physiological characteristics and modes of behavior common to all. All bears give birth to small and altricial cubs, and total litter mass as a proportion of maternal mass is among the lowest of all eutherian mammals. In turn, pandas have proportionally the smallest litter mass of any of the bears. This reflects their evolutionary history.

Although some modern bears have spread to South America and South Asia, most ursids have evolved in the midlatitudes of the Northern Hemisphere. To cope with food shortages, bears never evolved the trait of seasonal migration, nor did they develop the ability to lower their body temperature in the way that true hibernators do. Some bears evolved a form of dormancy to help them resist the cold temperatures and food shortages of winter. This is a kind of deep sleep that nonetheless occurs at body temperatures close to the normal state. While in such a state, typically in a den, they neither eat nor drink. This is a severe challenge, particularly for

38   Chapter I

pregnant bears that face the twin pressures of satisfying the needs of their growing fetus while still maintaining themselves adequately to survive.

Of course, all mammals can supply energy by mobilizing fat stores when fasting. But free fatty acids produced by hydrolysis are incapable of passing through the placental barrier between mother and fetus. Thus, the mammalian fetus cannot benefit from the large volume of free fatty acids contained in the mother's blood and instead depends on oxidation of glucose for its basic sustenance. Only by transforming proteins into sugars can a fasting mother bear supply nutrition to her fetus and thus ensure its survival. This biochemical process, called gluconeogenesis, consumes a great deal of energy and can threaten the health of the mother if prolonged. If this process took too long, it would expend considerable energy and be dangerous for the mother's safety. Because of these physiological pressures, natural selection has favored animals that can adapt. If either mothers or cubs die during this time, those animals with maladapted characteristics would fail to pass them on.

Thus, bear species that have survived are those that have evolved a most ingenious solution: shortening gestation and producing altricial cubs. Once a cub is born, the mother can supply nutrition through lactation rather than via her placenta and in this way can keep the cub growing and developing. She is able to use hydrolysis to produce fatty acids from stored fat and in this way conserve protein and avoid a harmful buildup of ketones. This reproductive strategy allows her to maintain her condition during winter dormancy but results in small and altricial cubs (Beijing Zoo, 1974).

Conventional wisdom has it that ursids have a long history and had adopted a type of winter dormancy as early as the mid-Miocene, some 25 million years ago. It is hypothesized that dormancy evolved as an energy conservation strategy in response to seasonal shortage of food. Although the ursid radiation produced differing types of bears, including some that live in tropical regions where food is abundant and polar bears, which kill seals even in winter and thus forage year-round, pregnant females still tend to den up and fast during the months of lactation.

From the standpoint of systematic evolution, giant pandas are very similar to some other modern Ursidae. Although they face a very different dynamic in terms of their forage base than other bears living in temperate and cold climates, pandas are nonetheless able to make use of a stable food supply: bamboo. Where they differ from other true placental mammals is that rather than adopting a strategy of lengthy gestation and large cubs, they follow the ursid reproductive pattern. We can now say definitively that the patterns we see in giant pandas of short gestation length and low birth weight of altricial cubs are actually part of a successful reproductive strategy acquired from ursid ancestors through the process of evolution. These patterns are certainly not manifestations of "reproductive degeneration" or "particularly specialized reproductive behavior."

The Issue of Panda Cloning

Since 1997, when I. Wilmut from the University of Edinburgh in the United Kingdom made headlines with his famous sheep Dolly, cloning has become recognized as a potential advance in reproductive biology. Inspired by Dolly, a researcher in Beijing soon suggested cloning a giant panda (Chen, 1999), and before long, the mass media had aroused the interest of the public in this idea. Those who supported cloning pandas took the view that there would be no adverse consequence to any panda who acted as a donor, and in fact, that this reproductive supplementation could be seen as contributing to natural reproduction. Thus, they reasoned that cloning could function as an ideal approach to conserving the endangered panda. At the same time, proponents of cloning recognized that the science involved in having pandas act as clone recipients had yet to be worked out. They assumed that to clone a giant panda a surrogate mother would be required. Thus, in 1998, they promulgated the notion of transferring somatic panda cells into denucleated cytoplasm from embryos of rabbits, dogs, and black bears, after which such a restructured embryo would be transplanted back into the surrogate mother's body and a cloned panda would be produced via abdominal pregnancy.

This is a fascinating image, but it is neither practical nor necessary. From the standpoint of scientific significance, there is no doubt that whenever a new technology arises, new applications will emerge and that this technology will continue to develop. Science is a field in which free discussion is critical. The results and methods of scientific research are open to free inquiry and debate and should not be subject to meddling from other scientists. Thus, there is no reason to obstruct the work of a scientist who wishes to use cloning technology to save a species that everybody is so concerned about.

However, we feel that those who advocate cloning pandas have reversed the priorities appropriately placed on the objectives of panda conservation with those appropriately placed on advancing this technology. If we overemphasize the utility that cloning may have in conserving pandas and in so doing repeat the now-outdated and misleading claims of the species' reproductive degeneration, we fear adverse consequences.

What Is the Goal of Conserving Giant Pandas?

The end result of conserving pandas certainly cannot be mere frozen genetic material or carbon copies of existing animals. Faced with the realities of habitat destruction and populations under threat, we cannot comfort ourselves by simply raising the flag of unleashing high technology. An even worse course would be to allow technology to mislead the public consciousness, preventing them from seeing reality, convincing them that a cloned specimen is a sufficient achievement. History has a lesson for us that we ignore at our peril: The best way to rescue a species is to ensure its integrity, stability, and genetic diversity. Today, our bottom line must be protecting fully wild and free-ranging populations.

What can we do to realize these goals? First, we should not view the species as immutable specimens as though they were on display in a museum or zoo. Rather, we should regard each individual as a direct participant in the process of evolution. The main goal of conserving any species should be maintaining numbers and genetic diversity sufficient to ensure that necessary adaptation can occur and minimizing the probability that stochastic factors will lead to extinction. Only by keeping this in the forefront of our consciousness can we ensure that we will have maintained the species' evolutionary potential. If we cannot ensure their genetic resilience, we have no way to ensure that they will evade the threats posed by random factors.

Second, we must acknowledge that neither laboratories nor zoos are capable of maintaining the species' evolutionary potential. Each species has its own habitat needs and must maintain an optimum genetic balance. Who among us can possibly predict exactly what this optimum really is? The species' most appropriate genetic constitution can only be maintained by the process of natural selection. That is to say, only through allowing populations to increase in the wild can we increase genetic diversity. If we allow the insularization of species such as pandas, isolating populations on high ridges, then they will have minimal opportunity for genetic exchange. The best method to preserve genetic diversity and integrity is to knit together multiple populations over larger geographic areas in a metapopulation within which mating can occur. This is the only way to ensure that the natural processes of evolution can continue. It is also the reason for us to emphasize the importance of habitat protection and effective population size.

Within China, as a local refuge exists for pandas within the mountains of the Hanzhong Basin in southern Shaanxi Province, why then cannot we find a similar refuge somewhere within the much larger mountain chains around the Sichuan Basin and the eastern portion of the Tibetan Plateau? In summer 1998, the State Council banned logging in the natural forests within the upper reaches of the Yangtze River drainage. From the summer of 1999, just about all panda habitat has thus come under protection. This protection indeed represents a new hope for modern pandas. If we want to conserve the species, we must understand existing threats, and we must also understand where there is potential for hope. Let's not make our plans casually on the basis of assumptions. With that kind of thinking and those kinds of plans, we would certainly lose touch with reality.

References

Beijing Zoo. 1974. 大熊猫的繁殖及幼兽生殖发育的观察 [Observation on reproduction and cub development of giant panda]. *Acta Zoologica Sinica* 20:139–147.

Chen, D. Y. 1999. 圈养大熊猫的人工繁殖及放归 [Captive breeding of giant panda and its reintroduction to the wild]. In 大熊猫放归野外可行性国际研讨会 [International Workshop: the Probability of Releasing Captive-bred Giant Pandas to the Wild], 62–64. Beijing: China Forestry Publishing House.

Feng, W. H., T. Q. Hu, and F. Z. Bi. 1985. 大熊猫濒危原因剖析 [The reasons why giant panda is endangered]. *Animal World* 2:1–7.

Ginsburg, L., R. Ingavat, and S. Sen. 1982. "A Middle Pleistocene (Loangian) Cave Fauna in Northern Thailand." *Comptes Rendus de l'Academie des Sciences, Serie III, Sciences de la Vie* 294:295–297.

He, Y. H. 1989. 鄂、湘、川间大熊猫的变迁 [The changes of distribution of giant panda in Hubei, Hunan and Sichuan Provinces]. *Chinese Journal of Wildlife* 2:28–31.

Johnson, K. G., G. B. Schaller, and J. C. Hu. 1988. "Responses of Giant Pandas to a Bamboo Die-Off." *National Geographic Research* 4:161–177.

Liu, W. X. 1988. "Litter Size and Survival Rate in Captive Giant Pandas *Ailuropoda melanoleuca*." *International Zoo Yearbook* 27:304–307.

Meng, S. 1985. 第二次全国保护抢救大熊猫工作汇报会召开 [The Convening of the Second National Giant Panda Rescue Work Report Meeting]. *Chinese Journal of Wildlife* 2:26.

Pan, W. S., Z. S. Gao, Z. Lü, Z. K. Xia, M. D. Zhang, L. L. Ma, G. L. Meng, X. Y. She, X. Z. Liu, H. T. Cui, and F. X. Chen. 1988. 秦岭大熊猫的自然庇护所 [The giant panda's natural refuge in the Qinling Mountains]. Beijing: Peking University Press.

Qin, Z. S. 1985. 四川大熊猫的生态环境及主食竹种更新 [Giant panda's bamboo food resources in Sichuan, China, and the regeneration of the bamboo groves]. *Journal of Bamboo Research* 4:1–10.

Schaller, G. B. 1994. *The Last Panda*. Chicago: University of Chicago Press.

Schaller, G. B., J. C. Hu, W. S. Pan, and J. Zhu. 1985. *The Giant Panda of Wolong*. Chicago: Chicago University Press.

Thue, L. V. 1984. "On the Distribution of Pleistocene Giant Panda in Vietnam." *Inqua* 11:146.

Wang, H. M. 1988. 关于加强保护和管理大熊猫的几点对策 [Suggestions to better conserve and manage giant pandas]. *Chinese Journal of Wildlife* 5:17–18.

Wen, H. R., and Y. H. He. 1981. 近五千年来豫鄂湘川间的大熊猫 [The status of giant panda in Hubei, Hunan and Sichuan Provinces in the last five thousand years]. *Journal of Southwest China Normal University (Natural Science Edition)* 1:87–93.

Wen, Z., and M. H. Wang. 1980. 大熊猫与竹 [Giant panda and bamboo]. *China Nature* 1:12.

Yang, R. L., F. Y. Zhang, and W. Y. Luo. 1981. 1976 年大熊猫灾难性死亡原因的探讨 [A preliminary discussion on the cause of the disastrous 1976 giant panda mortality (*Ailuropoda melanoleuca*)]. *Acta Theriologica Sinica* 1:127–135.

Zhu, J., J. H. Qing, and Z. Long. 1988. "Mass Death of the Giant Panda in 1975–1976." In *Symposium of Asian-Pacific Mammalogy: Abstracts of Symposium*, edited by A. T. Smith. N.p.: .n.p.

Chapter 2

Distribution and Habitats of Giant Pandas in the Qinling

Summary: The current distribution of giant pandas in the Qinling Mountains is mainly within the montane and subalpine belt along the upper reaches of the Xushui, Youshui, and Jinshui Rivers at elevations of 1,300–3,000 m. This area forms a natural refuge for the species and is characterized by high mountain ridges running east to west, which act as a physical barrier, protecting the area from the cold north winds. The cool temperate climate of the area produces a dense forest-bamboo ecosystem. Agriculture in the area goes no higher than 1,350 m, and this altitudinal line also approximates the upper limit of permanent human occupation, thus leaving the habitat as the panda's last retreat. Our studies have revealed that they are restricted to four "mountain islands." There is potential exchange of individuals among the Xinglongling, Niuweihe, and Taibai Shan ranges, which together can be considered a single metapopulation. However, the Tianhua Range is separated from the other three by a highway that isolates pandas in this population from the others. Analyses of remote sensing data from satellites indicate that the primary vegetation types in winter are *Bashania*-dominated mixed conifer-broadleaf forests as well as *Bashania*-dominated oak forests. Summer habitats are characterized by Taibai larch, fir, and birch forests, all of which contain a strong component of the bamboo *Fargesia spathacea*. Habitat research conducted by staff of the Changqing Forestry Management District in Shaanxi makes clear that well-planned, modest levels of selectively harvested timber removal is unlikely to negatively affect pandas. In contrast, clear-cutting and the subsequent establishment of a fast-growing plantations are likely to have extremely destructive impacts on panda habitat.

The late Quaternary ice age, beginning 18,000 years ago, had a profound influence on giant pandas. From that time forward, pandas became extinct north of the Qinling, and the distribution to the south contracted suddenly. Following this ice age, some 12,000 years ago, the climate became warmer again, ushering in what is known as the postglacial period. During the Neolithic, some 5,000 years ago, the climate was at its warmest.

From Bala Caverns in Laibing, Guangxi Province, we know that fossils of such species as palm civets (*Paguma larata*), dholes (*Cuon alpinus*), and giant pandas were excavated alongside shards of pottery created by humans (Pan et al., 1988). This suggests that giant panda and humans shared the tropical and subtropical forest environment of the Zhujiang River Valley at that time. With a warming climate, the geographic distribution of giant pandas gradually moved northward, to Xichuan in present-day Henan Province. As another example, fossils of pandas were also found to be contemporaneous with scores of other species at the Xiawanggang site in Xichuan. In addition to the roughly 50% of species associated with southern climates and wide-ranging species, fossils unearthed there included such cold-adapted or alpine species as the raccoon dog (*Nyctereutes procyonoides*), Eurasian badger (*Meles meles*), and roe deer (*Capreolus* spp.). Ash from bamboo was also excavated from the same time period at this site, suggesting that the panda range in the Yangtze River drainage included not only subtropical forests but also warm and humid temperate forest, i.e., bamboo habitat (Jia and Zhang, 1977). Thus, we know that 5,000 years ago people and pandas shared these tropical, subtropical, and warm temperate forests.

However, in the subsequent millennia, both panda range and panda numbers contracted, producing the impression that the species was headed on a one-way trip toward extinction. A primary cause of this contraction was the increase in human populations in the river valleys and mountain foothills, with attendant development of cultivated agriculture that destroyed large areas of the forest and bamboo.

Giant Panda Habitats in the Qinling Range

The formation of the Qinling Range can be traced back as early as the Paleogene era, some 40 to 70 MYA, but mountains that we would recognize as precursors of today's Qinling appeared during the Quaternary period, about 2.4 MYA. The large elevational span of today's Qinling (2,000–3,000 m) is a product of crustal movements from 700,000 years ago. Tectonic forces, erosive forces, glaciation, and freeze-thaw dynamics have all contributed to the formation of the Qinling Range as we find it today.

FORMATION OF THE QINLING AND ITS PANDAS

The thrusting up of the Himalayas during the third stage of the mid-Pleistocene was among the most dramatic geologic events in Asia. The Qinling Range arose in association with this larger crustal movement. In the Qinling region, the mid-Pleistocene epoch (some 100,000–700,000 years ago) was characterized by a drying climate. This was also the time period in which the loess that characterizes so much of northern China began to

44   Chapter 2

accumulate (although in the Qinling area, primarily to the north of the range). This loess was largely carried along by winds in the atmosphere at elevations of less than 3,000 m. Consequently, the Qinling, which rose to approximately that elevation, constituted an effective barrier, separating areas in which loess was deposited from those where it was not.

The presence of the physical barrier created by the Qinling Range generated substantial differences in both the climate and fauna on either side of it. As the global climate gradually warmed about 11,000 years ago, the Qinling's climate did so as well. The tree line gradually rose to where we see it today, considerably higher than its position during glacial times. Forest today overlays what was once glacial outwash and boulder fields. Snow, ice, and permafrost are currently found only above 3,400 m, where we also find development of rubble fields and stone polygon soil. Below 3,400 m, landforms are dominated by erosive forces and are molded by water (Figure 2.1).

The physical presence of the Qinling peaks, as well as the vastly different climate regimes north and south of the range, formed a natural barrier that led to large, new areas to which the fauna adapted. Areas south of the Qinling came to be occupied by oriental fauna. The small form of the giant panda, *Ailuropoda microta*, lived as far north as the southern Qinling. However, a little earlier, the larger-bodied panda, *A. melanoleuca baconi*, had already reached the northern Qinling and, following the close of the mid-Pleistocene, immediately extended its distribution to northern China (Pei, 1985). However, under the influence of the Tertiary ice age, giant pandas became extirpated in northern China. As the Holocene began, pandas had become restricted to the southern slopes of the Qinling, which form the northernmost extent of their current distribution.

In winter 1986, an *A. microta* fossil was found at the mouth of the Jinshui River (a tributary of the Han River) in Yang County, a mere 30 km from the densest distribution of modern pandas. We visited this location

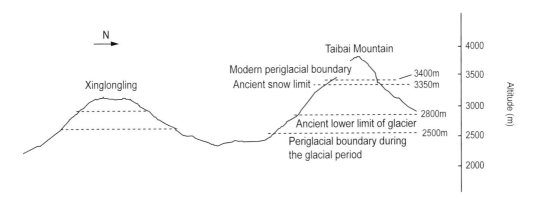

FIGURE 2.1. Present-day and ancient boundaries of glacially influenced landforms in the Taibai and Xinglongling mountain regions.

along the Jinshui River three times from 1986 to 1995 and noted that it was characterized by a belt of low hills at about 600 m. Mountain slopes there had become denuded, and previously forested land had been converted to agricultural fields planted with rice, corn, sesame, and other crops. We found fossil fragments of various species scattered among the fields and nearby ridges. Among these, we collected teeth and bones of wild boar (*Sus* spp.), deer (Cervidae), and other mammals. Geological examination indicated that the fossil originated from the early Pleistocene (Tang et al., 1987). Interestingly, Pleistocene fossils of the large form panda, *A. melanoleuca baconi* (Li, 1962), were found only about 45 km west of this, in Nijiadaba Valley in Yang County. Pandas currently live within the upper drainages of the Jinshui, Qiushui, and Weishui Rivers. Together, this evidence convinces us that giant pandas in the lower mountain slope region have a very long evolutionary history. The area bounded by the river valley and the foothills was formerly panda habitat, but currently, the last bastion of natural panda habitat is restricted to the area between the middle and high slopes.

Unique Physical Characteristics of Panda Habitat

Panda habitat in the Qinling is currently bordered to the north by the upper reaches of the Weishui and Hei Rivers north of the Xinglongling Range. To the south, it stretches to the northeastern edge of the Hanzhong Basin. The Qinling's core habitat begins at Landianziliang ridge in the Xinglongling Range and continues along the main Qinling spine toward the east, all the way to the Caiziping region in Ningshaan. To the west, panda habitat continues along the Xushui River as far as Liuba County. If it could be seen from far above, the Qinling Range would appear as a huge arc, with the panda distribution being situated just at the very apex of the arc, with the Xinglongling Range as its center.

The central section of the Qinling has four topographic characteristics. Massive ridges (>3,000 m) extend in an east-west direction in the central Qinling. This section constitutes the main spine of the entire Qinling, forming a natural barrier that prevents the cold north winds from incursions to the south. From here going in an easterly direction, the ridge gradually diminishes in elevation to about 2,000 m, with attendant weakening of the barrier's effectiveness. In winter, cold air can traverse the main Qinling ridge to its southern slopes, which acts to constrain panda habitat from extending farther eastward. In a westerly direction one finds the Jialing River valley extending north to south. Because this valley has its source in an area of the main Qinling spine that is only 2,000 m in elevation, winter winds from the north can easily surmount it, making it cold, dry, and generally inhospitable for pandas. Similarly, because the northern slopes of the Qinling are extremely steep and transition directly to the low elevation of the Wei

River plain, frigid winter temperatures prohibit luxuriant bamboo growth, once again making this poor panda habitat.

During the late Pleistocene, high-elevation areas within the Qinling were covered by glaciers. On the basis of the current conditions at the snowline of 3,350 m, we can infer that average temperatures during this time must have been about $5°C$ lower than currently. From palynological analyses, we know that the rift basin, on the western slopes of Mount Taibai at about 1,500 m, were dominated by dry forbs. Trees made up approximately 20% of the flora, of which cold-hardy pines and birches were the main representatives. This flora reflects the arid to semiarid grassland climate characteristic of the area at the time (Xu, 1984). It is possible that the southward movement of the panda distribution observed at this time resulted from the harsh environmental conditions produced by the cold, dry climate. For example, perhaps pandas living south of the Han River moved into Sichuan by way of the relatively low elevation Daba Shan Range. Alternatively, pandas may have established themselves in the warm climes of the Sichuan Basin via the Jialing River Valley. The fact that wooly rhinoceros fossils (Zhou, 1955) have been discovered in the Ziyang stratum from the late Pleistocene in Sichuan provides strong evidence that a cold-hardy fauna migrated into Sichuan along the Jialing River. However, it would have been difficult for pandas living north of the Han River to migrate to the south; from a geographic standpoint, the Han River would have constituted a formidable barrier to such movement. Pandas either already lived there or must have followed the river from locations to the east.

Our yearlong weather records from our Qinling research station (at 1,500 m) for 1989–1996 show a mean annual temperature of $8.4°C$, with $-2°C$ in January and $19°C$ in July. Mean annual temperatures in current panda habitat are $4°C–12°C$ on southerly slopes at 1,400 to 3,000 m in the Qinling and $5°C–11°C$ at 2,200 to 3,100 m in Wolong (in the Qionglai Mountains). Thus, a comparative analysis suggests ancient pandas could probably have found appropriate habitat at the close of the Pleistocene epoch extending from the Hanzhong Basin to the southern foothills of the Qinling. This would have allowed them to move seasonally from as low as 600 m to as high as 2,000 m in elevation. Thus, pandas likely have never gone extinct in this region.

The asymmetry of the Qinling means that the broad southern slopes obtain more solar radiation than the northern slopes. Furthermore, the east-west-flowing Han River allows seasonal southeasterly winds to drive straight into its upper reaches and glide upward through those broad valleys toward the alpine belt. Consequently, the southern foothills of Mount Taibai and the range extending from the Xinglongling through which the Weishui River flows south to the Hanjing River are all characterized by moderate climatic conditions. On midmountain slopes at about 2,000 m elevation, mean annual temperatures are in the range $6°C–9°C$, and annual precipitation is about 2,000 mm, some 200 mm greater than in the Hanzhong Basin and 50–100 mm more than in Foping County and Huayang Township on the periphery of panda habitat. This climatic regime allows the growing season to be approximately 140–160 days.

From the Han River Valley in a northerly direction, mountain heights increase gradually to about 2,000 m and climb to 3,000 m as one reaches the main ridge of the Xinglongling Range, extending through the upper reaches of the Weishui and Hei River drainages to Mount Taibai. The elevational difference between the Han Valley and the top of Mount Taibai is 3,200 m. This large elevational gradient creates a diversity of complex topographic, edaphic, hydrologic, climatic, and biogeographic conditions. One can identify six separate vertical zones corresponding to climatic and vegetation types (Table 2.1).

Bamboo is distributed from subtropical deciduous broadleaf forests in the low-elevation river valley all the way to the shade-tolerant conifer forests of the subalpine area. The area boasts no fewer than five bamboo species. Of the widespread and abundant bamboo species in panda habitat within the Qinling, we have a clear understanding of two from both taxonomic and biological perspectives: *Bashania fargesii* and *Fargesia* spp.

Table 2.1.

CLIMATIC ZONES AND VERTICAL VEGETATION ZONES BETWEEN THE HANJIANG RIVER VALLEY AND THE TOP OF MOUNT TAIBAI.

Elevation (m)	Climatic zone	Climatic characteristic	Elevational vegetation zone	Giant panda activity (%)	
				Winter and spring	Summer and autumn
3,750–3,350	Subalpine	Cold and semihumid (precipitation > evaporation)	Alpine shrub and meadow belt	0	0
3,350–2,400	Upper montane cold temperate	Cold and humid (precipitation > evaporation)	Midmontane mixed broadleaf-conifer zone (with *Fargesia*)	1.2	54.3
2,400–2,100	Lower montane cold temperate	Temperate and humid (precipitation > evaporation)	Midmontane cool mixed broadleaf-conifer zone (with *Fargesia*)	4.1	25.1
2,100–1,350	Montane temperate	Temperate and humid (precipitation > evaporation)	Midmontane cool mixed broadleaf-conifer zone (with *Bashania*)	93.5	20.5
1,350–800	Montane warm temperate	Temperate and humid (precipitation > evaporation)	Deciduous broadleaf zone (with *Bashania*)	1.1	0
800–500	Montane subtropical	Warm and humid (precipitation > evaporation)	Deciduous broadleaf and remnants of broadleaf evergreen zone (with *Bashania*)	0	0

48   Chapter 2

Bashania fargesii is a widely distributed species within midlatitude subtropical, northern subtropical, and warm temperate zones. Within the Xinglongling Range, it is found at elevations of 800–2,000 m, although production, both in terms of individual plants and bamboo groves, increases as one descends in elevation, with attendant changes in temperature and moisture regime. Its most favorable growth conditions are within the warm and moist mixed conifer-deciduous broadleaf forests at 1,200 to 1,800 m elevation. Shoots of *Bashania fargesii* grow only during a fairly short period in May (although somewhat later at higher elevations). During this time of year, some pandas follow the sprouting phenology of this species and specialize in eating it.

Fargesia is distributed in rather higher elevations, from 1,800 to 3,100 m. It finds suitable climatic conditions in the shade-tolerant conifer-fir belt in mountain (and even subalpine) zones. Following deforestation, whether human caused or natural, the typical seral sequence for these fir forests is that they are replaced by mixtures of *Fargesia*-fir and *Fargesia*-birch stands. *Fargesia* shoots and newly emerging stems constitute the primary forage for pandas, and these mixed stands are their primary habitats during summer (June through September).

A third possible species of bamboo, *Sinarundinaria* sp., is also broadly distributed in the Qinling. However, we limit our description and discussion of it here because so little is known of its morphology, taxonomy, and ecology.

On the basis of observations during the past 20 years, it appears that both *Bashania fargesii* and *Fargesia spathacea* have flowered and set seed within small areas annually. In contrast, there has been no documentation of mass flowering and mortality for either species or any indication of synchronous flowering and fruiting. The primary reason for this appears to lie in the complex topography and considerable variation in elevations in the Qinling, which form multiple microenvironments and microclimates. This habitat diversity allows any given species to live at differing elevations and in forests with multiple climatic conditions. Even among stands at the same elevation, habitat variation on a small scale arising from diversity in topography or forest structure may result in diversity in growth stages and thus a decoupling of life cycle stages on a spatial basis. Thus, both our observations and previous documentation in the Xinglongling Range show that even during times of wide-scale bamboo flowering, small areas of *Fargesia* vegetative propagation can be found.

During the 1952–1954 period, mixed *Fargesia* bamboo–fir forests in the upper reaches (2,400–2,800 m elevation) of the Dajian Valley in Taibai County of the northern Xinglongling Range were clear-cut. However, in the 30 years following, *Fargesia* has returned and flourished. In 1983, a small patch of about 20 ha of *Fargesia* flowered and set seed on a mountain that had been cleared of fir trees. We suspect that the synchronous nature of this flowering reflected the fact that it belonged to the same cohort that had been produced by vegetative propagation. During 1984, we noticed that about 10% of the bamboo began resprouting, but in the large area of bamboo growing on the southern slopes over the ridge, no wide-scale flowering occurred; what flowering we observed was restricted to isolated groves that exerted little influence on the overall condition of bamboo. Thus, pandas in the southern Qinling were provided with a stable food source by the generally asynchronous bamboo growth patterns. We suspect this is a characteristic of panda habitat in the Qinling.

THE UPPER LIMIT OF THE WARM TEMPERATE MONTANE BELT SERVES AS BARRIER TO HUMAN SETTLEMENT

Human history in the upper reaches of the Han River area can be traced back at least 700,000 years to the mid-Pleistocene. No fewer than six Paleolithic sites have been discovered within the Hanzhong Basin from the area starting in Mianxian County moving east along the Han River to the mouth of the Jinshui River in Yang County. Two of these contained panda fossils (Figure 2.2; Tang et al., 1987).

Repeated migrations across the Hanzhong Basin led to agricultural development and increases in population density in both plain and mountain areas, as the Hanzhong Basin became known as a land of plenty. As populations continued to increase, cultivated land on the basin's flatter areas became in short supply, and people increasingly turned toward mountain

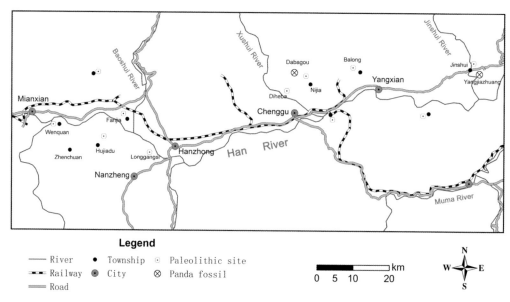

FIGURE 2.2. Locations of panda fossils and Paleolithic sites, Hanzhong Basin, China. We have indicated major landscape features referred to in the text.

areas to open up new farmland. Foothills that had originally been covered by dense forests were converted to agricultural fields. As needs increased for construction timber, smelting of iron, charcoal, and domestic firewood, the yearly rate of forest cutting on the Qinling's southerly slopes far exceeded what could be sustained, and forest cover rapidly declined. The lower-elevation slopes were the first to be cut, and they were transformed into barren ridges as the forest, consisting of a mosaic of mixed evergreen broadleaf and deciduous trees, disappeared, replaced by shrublands.

On middle slopes, the few originally existing patches of broadleaf evergreen forest had difficulty recovering after being harvested. Regeneration of broadleaf deciduous trees was also prevented. All of this deforestation increasingly led to a general drying of the environment. Bamboo, originally widespread, was also lost to varying degrees through the processes of vegetation succession. The lower the slopes on the mountains were, the more intense human activities were; naturally, these areas incurred the most serious losses of panda habitat. According to chronicles kept during the early part of the Qing Dynasty, southerly slopes in the Qinling were said to be "a land teeming with people," and reports described "people gathering from all corners, over-running the land, pulling out trees in order to push roads through . . . although the mountains are very high and steep, they too are transformed into croplands."

Currently, agriculture extends up to an elevation of 1,350 m: human habitation cannot extend up mountain slopes indefinitely. In the Xinglongling region, agricultural development can be traced back at least as far as the early Qin Dynasty. Cultivated fields that were abandoned during various historical eras can be found scattered in every corner of these mountains. From the lowest-elevation river valley to the highest mountain ridge, we find evidence of historical human presence: the remnants and relics of ancient temples, tombs, roads, bridges, and broken walls. These remnants tend to be particularly concentrated in certain areas, such as where old roads intersected, in broader valleys, and at the mouths of rivers. We even find evidence of more recently existing road houses, workshops, and pawnshops, all suggesting that these places were once thriving agricultural and commercial centers with large human populations. Of course, these areas were eventually abandoned, particularly during the last hundred or so years.

These indications of past human activity suggest that the Xinglongling area had become a center of human activity. Although there are relics of temples and inns as high as 2,600 m elevation on mountain passes and a Buddhist abbot lived in one of the temples until the 1930s, permanent inhabitance by farmers did not extend above 1,500 m (e.g., the Sanguan Temple in Foping County). The higher elevations escaped the influence of past human activity and this can be seen clearly from the large expanse of healthy *Bashania fargesii* growing on abandoned agricultural fields in the surrounding foothills.

At present, very few people live in the vast expanse of the southern Qinling. Foping County has a mean population density of 9 people/km^2, Huayang in Yang County has a somewhat higher density of 20 people/km^2, but Pindu Township has a density of only 2.7 people/km^2. Only 60 people live within an area of hundreds of square kilometers in the village of Sanguanmiao in Yueba Township, Foping County. In contrast, human populations are concentrated in central towns and among the river valleys.

Historical records indicate that a post road was constructed through the steep and rugged mountains of Qinling as early as the Wei (AD 220–265) and Jin (AD 265–420) Dynasties. This road connected the Hanzhong Basin with the central Shaanxi plain. Human populations along the road ebbed and flowed with the road's popularity but ultimately declined because of geography: It was simply too close to the high and forbidding terrain of the main Qinling peaks, with their rough topography, sheer canyons, and difficult access from outside. To get to Zhouzhi, north of the Qinling, from Huayang or Houzhenzi in Yang County required trudging over innumerable watersheds, fording countless streams, and entering many deep canyons. This was the most dangerous of the four ancient post roads. The Tang poet Can Shen described it thusly in his poem "Traveling on the Luogu Road":

> Thousands of cliffs follow one after the other,
> Why only one path to cross them?
> The cart wheels slip on the ice,
> While dense bamboo stabs our gowns.
> We pass the night in deep forest,
> The road planks soar into the emptiness,
> Gusts of snow shrivel my horse's mane,
> Piercing winds tear at my skin.

Beyond simply being inaccessible, arable land on flat slopes and dykes in the Xinglongling Range is very rare and scattered; most remaining land that might be cultivated is on extremely steep slopes, which bring their own set of difficulties. Despite the promulgation of past policies aimed at attracting people to cultivate these remote areas (even resulting in the aphorism "Young and old, thousands of groups, coalescing into an endless stream"), ultimately, relatively few were willing or able to "sacrifice valley living in exchange for living among the cliffs, spending their days in reclamation work." Because the inherent productivity is so low, these montane areas were capable of providing only limited crops.

The area near the source of the Dong River was formerly one of the most important transportation routes from Foping and Yang County to Zhouzhi. The remains of a string of ancient hamlets can be seen along the now-abandoned trail leading from Liangfengya to the Dong River, which stretches to the Foping Nature Reserve's Sanguanmiao Guard Station (in the concentrated area of panda-dense distribution). Among the evidence still visible are signs of a shop where wooden steamers for cooking vegetables were produced as well as an old stock post (at about 1,800 m). As the name implies, this was evidently a place where itinerant merchants stopped to rest, and there was also a workshop (probably located here to make use of the abundant birch and poplar as raw materials for producing the steamers). The ruins of row upon row of houses extend for kilometers. Some of the elderly people in the area still have memories of thriving commerce in this area until the late Qing Dynasty. Evidently, bandits and famines took their toll, and the area gradually declined. The area is currently uninhabited, and large pines and firs 60 to 70 years old grow amidst the ruins of the houses, along with *Fargesia* and *Bashania fargesii* expanding to fill large areas. Thus, this area has recovered its status as good habitat for pandas.

There are seven villages under the jurisdiction of Huayang Township in Yang County (at 1,130 m). In 1935, these villages contained 900 families (3,500 people), but over the next half century the number increased to 950 families (4,440 people). This rate of population increase for Huayang, which is situated just at the upper limit of the mountain foothills, is actually quite moderate, and population increases have been even slower in the montane zones. This rate stands in clear distinction from the rapid population increases observed among human communities in the lower-elevation valleys and in the Hanzhong Basin.

QINLING MONTANE VEGETATION IS CAPABLE OF RAPID RECOVERY

Putting aside for the moment the fact that the lower-elevation foothills have supported agriculture with human densities similar to what we see today, higher up in the montane belt, one finds primary forest and vegetation growing only in the portion near the main Xinglongling crest. Without exception, secondary vegetation covers the vast remaining portions, from ridges (at 2,500 m) down the slopes to valleys at 1,400 m. This secondary vegetation is characterized by mixed conifer-broadleaf forest with an understory of *Bashania fargesii*; this understory is expanding outward at rate of about 2 m annually. These secondary forests are where the pandas live.

We see bamboo recovery evident in areas such as Sanguanmiao, Zhenglongchang, Baiyangping, Shanshuping, and others. Sanguanmiao is located in the center of the Foping Nature Reserve, and in 1985 it had a population of 57 residents embedded within panda habitat. However, just 5 km downslope from this village at Sanxing Bridge an old stone monument erected in honor of those donating to the building of the bridge can be found, documenting that at that time, the population exceeded 2,000 people, suggesting much more intensive human activity at the time. Now, however, the area has recovered to become excellent wildlife habitat.

Summarizing, the refuge that pandas currently enjoy in the southern Qinling has resulted from unique geographic, bioclimatic, and vegetation characteristics, as well as pandas' disinclination to live near human activity. Together, these have provided modern pandas a unique habitat, allowing them a place to avoid the intrusions of humans in the lower-elevation valleys. It is just these interrelated connections and constraints that have allowed giant pandas to make it through the vicissitudes of life in the Hanzhong Basin over these thousands of years and to persist to the present.

Discovery of Giant Pandas in the Qinling and Their Distribution

PANDA FOSSIL RECORD

The fossil record of giant pandas in the Qinling can be traced back to the mid-Pleistocene, some 700,000 years ago. In late 1986, fossils of the small-bodied *A. microta* were unearthed among sediments deposited on the third Han River terrace at the mouth of the Jinshui River in Yang County (located 100 m above the river, at 600 m elevation). This is the earliest record of any giant panda living in the southern Qinling and demonstrates the long history of their occupancy in the area. Other mammal fossils unearthed at the same time included wolves (*Canis lupus*), bears (*Selenarctos thibetanus*), the extinct elephant-ancestor stegodon (*Stegodon* spp.), the extinct Chinese rhinoceros (*Rhinoceros sinensis*), the strangely clawed (and now extinct) perissodactyl known as a chalicothere (*Chalicotheriidae* spp.), Rusa deer (*Rusa* spp.), gazelles (*Gazella* spp.), and water buffalo (*Bubalus* spp.). Most of these species were associated with southern subtropical forests, providing further evidence that panda habitat in the southern Qinling was warm and moist at the time. In 1988, when we studied this excavation site in more detail, we discovered two old stone tools, suggesting that *A. microta* lived alongside early humans.

HISTORICAL RECORDS AND MODERN DISCOVERY

The climate on the central Shaanxi plain was warmer and moister 2,500–3,000 years ago than it is now. Historical records show that bamboo grew luxuriantly on the banks of the Weishui River. We know that some 2,000 years ago, pandas lived in present-day Henan, Hubei, Guizhou, and Yunnan (Wen and He, 1981). Panda skulls and teeth were among the funerary objects found in the tomb of Empress Dowager Bo, the mother of the Western Han Dynasty's fourth emperor Wen (Wang, 1977). It is said that the Tang emperor Taizong (AD 649–672) granted a panda fur to each of 14 officials at a grand banquet (Wang, 1977). Later, Emperor Wu Zetian provided a gift of two living pandas to Japan to mark the two countries' friendly relationship.

Taken together, these instances all imply not only that pandas were regarded as valuable and precious even from ancient times but that obtaining pandas in the Changan (present day Xi'an) region was not too difficult during the Han and Tang Dynasties. As recently as the middle of the 19th century, there still existed a belt of panda habitat in the mountains of Zhushan ("Bamboo Mountain") County in the middle Han River drainage, and pandas were known from southern Shaanxi, northern Hubei, western Hunan, and eastern Sichuan.

Local people in the southern Qinling have long known of the existence of the so-called patterned bear (i.e., the giant panda), a term that also occurs in local historical records. On the basis of descriptions from local hunters obtained during their expedition of Mount Taibai in 1932, Sowerby (1937) published evidence that pandas lived in the Qinling in 1937. During May–October 1959, the Shaanxi Agriculture, Forestry and Animal Husbandry team conducted a survey of the middle portion of the southern Qinling, documenting the fauna they encountered. In their 1960 report, they included clear descriptions of hunter reports of patterned bears roaming the mountains of northeastern Yang County. During 1958–1959, student groups from the biology departments of Northeast University and Beijing Normal University both obtained panda skins during their forays into Chaijiagou in Ningshaan County and Yueba in Foping County. This group provided proof that pandas still existed in Taibai County during the 1964 survey of Taibai County. Also in 1964, Zheng Guangmei and colleagues collected a panda specimen from Yueba in Foping (Zheng, 1964). These materials served to prove the existence of pandas, as well as the counties that still contained them, within the southern Qinling. In 1973, Zhang Jishu obtained eight panda pelts and two skulls from Longtanzi, Sanguanmiao, Taiguoping, Dachenghao, and Xiaolanping and also from Yueba in Foping County. These specimens further demonstrated that pandas in the Qinling were not merely scattered individuals but, in fact, that the southern Qinling contained a relatively concentrated population. The next year, Shi Dongchou and colleagues obtained additional specimens during their ecological survey of pandas. Also in 1974, Shaanxi Province organized an enormous biological resource survey team in which some 475 staff participated, conducting surveys for pandas, golden monkeys, and takin in Foping, Taibai, Yang, Zhouzhi, and Ningshaan Counties. The Foping Nature Reserve was formally established in 1980 (Shaanxi Biological Resource Investigation Group, 1976).

In 1981, Zhang Jinliang and colleagues reported that the area around Mount Taibai contained pandas living in the birch, fir, and *Fargesia spathacea* forest zones from 2,216 to 3,000 m but that they were few in number. Also in 1981, Yong Yange and colleagues conducted a preliminary survey of panda ecology in the Foping Nature Reserve. They accumulated a great deal of information on the distribution, abundance, habitat, and breeding of pandas within their study area of approximately 100 km^2 by walking trails and direct observations (Yong, 1981).

Spatial Distribution Pattern of Qinling Pandas

QINLING'S MOUNTAIN ISLANDS

The insularization of forests in the southern Qinling into four "mountain islands" has occurred only during the last 20 years, a period of time that equates to no more than four panda generations. Logging also stopped in forest areas of the Qinling following the 1998 logging ban, and vegetation has begun to recover, which may allow these mountain islands to become reconnected and gene flow among pandas to recover as well. Thus, we regard pandas living within the four mountain islands as members of four local populations (Plate 3).

Xinglongling

This area is located south of the main Qinling ridge and is centered in Xinglongling. Beginning in the upper reaches of the Jinshui River, it stretches southward toward a line connecting Huayang, Jiuchiba, and Yueba at 1,300 m and northwest to the Xushui River valley. The total area is 960 km^2. Here we find the core of the panda distribution of the entire Qinling, encompassing the national-level Changqing, Foping, and Laoxiangcheng Nature Reserves as well as portions of forests administered by Shaanxi's Taibai Forestry Management District. Among all panda habitat patches in the Qinling, this is the largest in area and also has the densest panda population. Consequently, it also has the largest number of pandas.

Niuweihe River

This area is primarily located along the upper reaches of two tributaries of the Xushui River, Guanyin Gorge, and the Niuweihe River. Its western region, which includes the Xushi and Taibai River watershed (which itself is an upper tributary of the Baoshui River), is a *Fargesia* bamboo forest. The southern boundary is the northern portion of Motianling. Its eastern border is delimited by a highway. This area is administered by Shaanxi's Taibai County and totals 320 km^2.

Mount Taibai

Although the northern border of this area has a mean elevation of 3,300 m, forest and bamboo vegetation suitable for panda occupancy is found only below the 3,100 m contour level (Shaanxi Mount Taibai National Nature Reserve Management District, 1993). Its southern border extends to the Xushui River valley. Here, however, because agriculture has increasingly taken over the east-west-trending valleys along the upper reaches of Xushui

56 Chapter 2

River, the extent of panda habitat has moved upward to elevations higher than the 1,350 m found on the southern slopes of the Xinglongling area. To the east, it extends to Wanquan valley south of Baxiantai (which is the main peak of Mount Taibai, all of this area lying within the upper reaches of the Heishui River branch of the Wei River). In the west, it includes the Hongshui River area, and in the south it ends at the highway. Totaling 200 km², this area includes part of Mount Taibai Nature Reserve, as well as portions of Shaanxi's Taibai Forestry Management District.

Tianhua Mountains

Centered in the Tianhua Mountains and including the eastern portion Shaanxi's Longcaping Forestry Management District and the western portion of the Ningxi Forestry Management District, the Tianhuashan local population is located east of the Zhoucheng Highway in an area of about 190 km². We know very little about this area. Although the 1989 World Wildlife Fund (WWF) survey documented a panda presence here, our only direct information comes from a single, old panda scat that we saw on a ridge connecting the three counties of Foping, Ningshaan, and Zhouzhi during a short survey in May 1995 (Forestry Ministry of the People's Republic of China and WWF, 1989).

Giant Panda Habitat in the Xinglongling Mountain Island

In 1987, we used remote sensing methods to classify panda habitats into various types. These remote sensing images not only made clear the diversity of vegetation but also reflected environmental heterogeneity. We integrated landscapes formations with vegetation communities and, using visual interpretation, recognized a total of 14 categories. Additionally, on the basis of bamboo resources within the habitat, landscape conditions, and levels of human disturbance, we classified four habitat conditions (Table 2.2).

We repeated our analysis of panda habitats and vegetation types using remote sensing in 1997, and with an additional 10 years of heavy logging, vegetation within panda habitat had undergone considerable change. Under the influence of continued human activity, entire areas had been transformed into small, heterogeneous, habitat patches. At this point, vegetation characteristics bore little relationship to topography or elevation; thus, we considered only habitat types in our analyses. We focused our analyses on the Landianzi Ridge located in the southern Qinling, a patch 30 × 30 km. It was bordered on the south by a line connecting Huayang village with Jiuchiba and on the north by a line connecting the Xushui River with the old county town. We used three Thematic Mapper (TM) scenes from August 15, 1994, produced by the Landsat 5 satellite. We produced a composite image from three wavelength bands, TM3 (blue), TM4 (red), and TM5 (green), using IDRISI software. We then produced a supervised classification based

Table 2.2.

VEGETATION TYPES FOR PANDA HABITATS IN THE XINGLONGLING MOUNTAIN ISLAND IN 1987, GROUPED INTO FOUR HABITAT CONDITIONS BASED ON HUMAN IMPACTS TO THE ELEVATION BAND.

Habitat type	Habitat condition
1. Midmontane and mountain top *Fargesia* forest, shrub meadow (>2,900 m) 2. Midmontane *Fargesia* forest, Taibai larch forest (>2,600 m) 3. Midmontane *Fargesia* forest, fir forest (>2,300 or 2400 m) 4. Midmontane *Fargesia* forest, birch forest (>2,200 or 2,300 m)	Vast areas from 2,200 to 2,900 m on upper midmountain slopes never logged. Gentle slopes are suitable for pandas. Marked elevational zones of vegetation communities. Two species of *Fargesia* bamboo grow well under the forest canopy. Excellent and stable summer panda habitat.
5. Midmontane *Fargesia* forest, *Bashania fargesii*, Huashan pine, Chinese hemlock, mixed broadleaf forest (1,800–2,300 m) 6. Wood bamboo, mixed broadleaf forest (2,000–2,300 m)	At elevations of 1,800–2,300 m the heads of valleys are currently subjected to large-scale logging or have previously been logged. Thus, these habitats have suffered considerable disturbance, both before and after 1987.
7. Low, midmontane *Bashania fargesii*, oak-pine mixed forest (1,350–2,100 m)	At elevations of 1,350–1,900 m mostly abandoned agricultural fields except in steep gullies. Following earlier logging, few mature trees remain, but shrubs have proliferated. Large patches of healthy and expanding *Bashania fargesii* bamboo grove, providing good winter habitat for pandas. However, cutover areas and numerous types of human activities limit the available habitat for pandas to some extent.
8. Low, midmontane *Bashania fargesii*, oak forest (800–1,350 m) 9. Low, midmontane oak, Chinese pine (800–1,350 m) 10. Low, midmontane shrub (800–1,350 m) 11. Bamboo cutting sites, mixed shrubland 12. Clear-cut area 13. Agricultural fields 14. Low, midmontane *Bashania fargesii*	The lower slopes have long been dominated by human activity. Although a variety of shrubs and bamboo groves exist on agricultural fields, previously clear-cut areas, and recently (5-year-old) logged areas, these areas do not provide suitable habitat for pandas.

on ground-truthing selected training areas. In the final analysis, we used a combination of image characteristics and ground surveys to produce a map of vegetation types and distribution in panda habitat within the Xinglongling area (Plate 2; Table 2.3).

Obviously, towns, agricultural fields, and grasslands on low slopes are not panda habitat; also, alpine areas above 3,100 m have no bamboo and thus are not suitable for pandas either. In fact, our observations and radio-locations substantiate this observation: in habitat types 7, 8, 9, and 10 (from Table 2.2), we never observed pandas. In fact, no bamboo is growing in any of these types, so we consider them unsuitable for pandas.

History of Logging in the Changqing Forestry Management District

The provincially administered Changqing Forestry Management District was established in 1970. Its primary resource base was the nationally owned

Table 2.3.

VEGETATION TYPES OF THE PANDA HABITAT IN THE XINGLONGLING AREA IN 1994.

Vegetation type	Description
Subalpine mountain shrub (including *Fargesia*)	On mountain tops above 2,900 m
Taibai larch forest (including *Fargesia*)	In the upper montane, typically on cool slopes, above 2,600 m
Fir forest (including *Fargesia*)	Below larch forests, relatively limited areas at 2,500–2,700 m
Birch forest (including *Fargesia*)	Mainly consisting of red birch and buffalo hide birch at 2,200–2,500 m
Mixed coniferous-broadleaf forest	Mixed pine and birch forests in montane zones, mixed pine and oak forests at lower elevations, understories include *Bashania* and *Fargesia*
Oak forest (including *Bashania fargesii*)	At low elevations
Low-elevation sparse shrub	Mainly in oak and newly regenerating shrub communities on warm slopes
Valley mixed forest	Along valley bottoms, mainly broadleaf trees
Subalpine meadow	Including natural meadows and secondary meadows created following forest cutting
Artificial conifer forest	At high elevations, mainly Japanese and northern Chinese larch, firs, and dragon spruce; at lower elevations, mainly Chinese pine
Plantation	Planted following logging
Low-elevation grassland	Along valley bottoms
Agricultural fields	Two types: dry land and wet fields

forests in the Xinglongling, and initial lumber removals occurred in the montane and alpine forests of Huayang and Maoping. To accomplish this, the District's first activities included rebuilding and lengthening roads from lower elevations through the montane belt to the alpine, allowing the placement of timber camps deep in the mountains (Editorial Office of Changqing Forestry Management District, 1986). Timber harvesting on a large scale commenced in 1972, continuing until the government-ordered shutdown in July 1994. In 1996, the operation was transformed into the national-level Shaanxi Changqing National Nature Reserve. We identify two separate periods of logging activity on the basis of the type of logging undertaken as well as the effects on giant panda habitat.

PERIOD 1, 1972–1989

During this period, timber harvests were conducted under plans that strictly controlled their volume and type, as directed by the various regulations and policies of the National Forestry Industry. Consequently, the logging's effect on wildlife and its habitat was quite minor.

Timber harvesting was conducted under reasonable and specified plans that were based on the volume of standing timber and annual growth rates.

To this end, a survey of the entire area on fixed sample plots was undertaken every 4 years to ensure that the offtake rate did not exceed the growth rate. During 1981–1986, the management district produced a total of 150,000 m³ of timber while the standing volume of forests was 230,000 m³. Surveys showed that the standing timber volume did not decrease during this time period (Editorial Office of Changqing Forestry Management District, 1986).

During this time period, cutting was restricted to trees that were mature or overmature, had some evidence of disease or rot, or, because of competition with other trees, had little potential for future growth. Cutting occurred only on slopes between 15° and 35°. All other types of trees, particularly younger ones, were retained. Surveys showed that the removal rate was about 36% of the standing stock, that canopy cover was 42%, and that 79.7% of the forest potentially subject to timber harvest area was treated in this way. Because of the light touch of this type of logging in which old, large, and diseased trees were removed, sunlight, water, and nutrients were mobilized in the service of younger trees, allowing for rapid recovery of a mature forest. On slopes greater than 35°, newer timber harvest methods were adopted, the primary objectives of which were to maintain normal forest growth characteristics and to encourage ecological functionality. The targets of extraction here were old, dead, diseased, and pest-infected trees. Extraction rate was about 28% of the standing volume, and the area was about 12.5% of the total area potentially subject to harvest. The original forest environment was not fundamentally altered by logging of this type. To be sure, small patches of clear-cutting of mature and overmature trees occurred on some of the gentler slopes, but their extent was strictly limited, for example, to patches not exceeding 3 ha; younger trees were retained within these patches, and the area clear-cut totaled 7.8% of the entire area.

PERIOD 2, 1990–1994

During this period, timber harvesting was essentially devoid of management or planning, viewed either from the perspective of the volume of cutting or the approach to timber harvesting. This period was characterized by its sole focus on economic efficiency, with essentially no consideration whatever given to ecological function.

Administrative rearrangements in Shaanxi during the early 1980s led to the Longcaoping Forestry Agency being refashioned as an independent entity. Foping Nature Reserve was established in 1979, and as part of this, Yueba and other areas were carved off from the Changqing Forestry Management District. These two changes substantially reduced the resources controlled by Changqing Forestry. At the same time, mature timber within the Huayang District had been subject to high consumption rates for some 20 years. Thus, the Changqing Management District's resource base was much lower than it had been at its inception. Rather than the number of workers contracting to reflect this reality, it had, in fact, increased. Including retirees, the total work force at the time of the logging ban in 1994 was 2,000. This exerted great pressure on the resource. Spurred on by market forces, timber production was driven higher and higher. To meet these higher production targets, large numbers of temporary workers were hired from the local population. Timber harvests conducted by these recently hired, unprofessional workers were not strictly designed or planned. Trees were cut solely on the basis of their ability to fetch a profit, regardless of their age. Extraction methods were primitive, and in situations where it was too difficult to haul trees down the mountain to market, many were simply abandoned, left to rot on the mountainside. We often observed cases in which workers had simply left trees of over 1 m diameter at breast height on the hillside. At its peak, the number of unprofessional loggers exceeded 2,000, more than the total number of official staff of the Changqing Forestry Management District.

The earlier objectives of sustainable use and ecological function encapsulated by the motto "extract and cultivate, log selectively" were simply abandoned. In their place, the cultivating second-growth forestry strategy was adopted as a response to market demands. In practice, this meant clear-cutting in more areas, followed by a strategy of fast growing, high-yield plantations in order to shorten the production cycle.

Forest management under the cultivating second-growth forestry strategy emphasized rapid establishment of monocultures designed for short rotations on areas that had previously been clear-cut in order to shorten rotation lengths. Among tree species planted were exotics, including Japanese larch. In order to ensure growth of the selected tree species, all other vegetation, importantly including bamboo as well as grasses, would be cleared from the plantation yearly for a period of 5 years prior to planting. As a result, the planted species came to dominate the area entirely, with essentially no other trees present. Even today, one finds no bamboo growing in the understory of Japanese larch plantations.

According to their plans, the Changqing Forestry Management District intended to produce a total of 30,000 m³ of timber during 1988. However, with the punching of the road through to Wanfang Gorge in July 2003, some 20,000 m³ of forest was cut within just the right-hand branch of the gorge in less than a month. With the right-hand portion of Wanfang Gorge completely cut over, barely a tree remaining, workers then began building a switchback road up the gorge's left-hand branch. Only the central government's sudden logging ban saved this very last piece of primary forest from suffering the same fate as the others had.

A subsidiary effect was that uncontrolled logging was generally associated with a lack of postharvest treatment, leaving a tremendous amount of slash on the landscape. This lack of treatment not only adversely affected the normal successional process of the vegetation but also made animal movement difficult. Additionally, although exotic trees had the advantage of quickly attaining maturity, their ability to dominate stands reduced forest biodiversity. We note in particular the example of bamboo being unable to grow under the canopy of Japanese larch.

Additionally, to allow transport of logs down the mountainside, roads were built higher and higher into the mountains. One aspect is that road building generally involved using explosives, and this process alone had a significant impact on panda habitat in addition to frightening away many species of birds and mammals. In interviews, one elderly mountain resident told us that he believed the reason that dholes (*Cuon alpinus*) were no longer found in the area was because of the explosions associated with road building. Although it is difficult to confirm such stories, they do illustrate that the effects of road building should not be overlooked. However, an even more important effect of road building was that in facilitating access, human activity increased on these mountain slopes, including both an intensification of agriculture and commerce, exacerbating other negative influences on the habitat.

RECOVERY OF PANDA HABITAT AFTER TIMBER HARVEST

In March 1999, we conducted a survey in a valley in the southern Qinling at 2,300 m, which strengthened our optimism about the ability of panda habitat to recover: Following clear-cutting during 1991–1992, this area was reforested with a local species of fir. At the time of our survey, we found the valley covered with a belt of birch trees about 4 m in height. Under these trees we found thickly growing *Fargesia* bamboo groves. If recovery of this nature is possible at this elevation, we can well imagine that at lower elevations, where the climate is warmer, recovery would be even easier.

References

Editorial Office of Changqing Forestry Management District. 1986. 长青林业局志 [Local history of Changqing Forestry Management District]. Internal report.

Forestry Ministry of the People's Republic of China and WWF. 1989. 中国大熊猫及其栖息地综合考察报告 [Comprehensive research report on giant pandas and their habitats in China]. Internal report.

Jia, L. P., and Z. B. Zhang. 1977. 河南淅川县下王岗遗址中的动物群 [Zoological aggregations from the Xiawanggang remains, Xichuan, Henan]. *Cultural Relics* 6:41–49.

Li, Y. H. 1962. 汉水上游哺乳类化石的新线索 [New clues regarding mammal fossils in the upper reaches of the Hanjiang]. *Vertebrata PalAsiatica* 6:29–34.

Pan, W. S., Z. S. Gao, Z. Lü, Z. K. Xia, M. D. Zhang, L. L. Ma, G. L. Meng, X. Y. She, X. Z. Liu, H. T. Cui, and F. X. Chen. 1988. 秦岭大熊猫的自然庇护所 [The giant panda's natural refuge in the Qinling Mountains]. Beijing: Peking University Press.

Pei, W. Z. 1985. 第三个 "巨猿" 下颚骨的发现 [The discovery of the third lower jaw of "giant ape"]. *Journal of Paleo-Vertebrates* 2:183–200.

Shaanxi Biological Resource Investigation Group. 1976. 陕西秦岭地区大熊猫初步调查 [Preliminary survey of giant pandas in the Qinling region, Shaanxi Province]. *Biological Investigations (Zoological Portion)* 3:91–104.

Shaanxi Mount Taibai National Nature Reserve Management District. 1993. 太白山自然保护区大熊猫调查报告 [Research report on giant pandas in Mount Taibai Nature Reserve]. Internal report.

Sowerby, A. de C. 1937. "The Giant Panda's Diet." *China Journal* 26:209–210.

Tang, Y. J., G. F. Zong, and Y. L. Lei. 1987. 汉水上游旧石器的新发现 [On the new materials of paleoliths from the Hanshui Valley]. *Acta Anthropologica Sinica* 6:55–60.

Wang, X. L. 1977. 两千年前西安存在过大熊猫吗？ [Were there pandas in Xi'an two thousand years ago?] *Fossil* 1:16.

Wen, H. R., and Y. H. He. 1981. 近五千年来豫鄂湘川间的大熊猫 [The status of giant panda in Hubei, Hunan and Sichuan Provinces in the last five thousand years]. *Journal of Southwest China Normal University (Natural Science Edition)* 1:87–93.

Xu, Q. Q. 1984. 华北更新世人和哺乳动物的进化与气候变迁的关系 [Relationships between the evolution of Pleistocene humans and other mammals and the climate change]. *Prehistoric Research* 2:93–98.

Yong, Y. G. 1981. 佛坪大熊猫的初步观察 [Preliminary observations of giant pandas in Foping]. *Chinese Journal of Wildlife* 4:10–16.

Zheng, G. M. 1964. 秦岭南麓发现大熊猫 [Giant pandas in the southern Qinling Mountains]. *Chinese Journal of Zoology* 6:3.

Zhou, M. Z. 1955. 关于大熊猫的化石 [On giant panda fossils]. *Bulletin of Biology* 10:41.

Chapter 3

Relationships among Roads, Forest Management, and Pandas in the Qinling

Summary: Most forests we observe currently functioning as panda habitat are in fact secondary forests, with primary forests remaining only in small patches above 2,600 m elevation. Abandoned fields can provide good habitat for giant pandas because bamboos grow densely and dominate the understory because of relatively thick soils and gentle slopes. It is evident that although bamboo recedes under agricultural encroachment, the intervals between successive uses allow for panda occupancy. Where traffic volume, the extent of the road system, and human activities that result from access are limited, roads have a relatively minor effect on panda habitat.

The earliest uses of forests on the Qinling's southern slopes by local residents were in support of agriculture (extracting wood ash for fertilizer) and to provide fuel for home use. Throughout this long historical period, although the accumulated consumption of forest resources was considerable, it was not comparable to the use of forest resources in the modern sense of exploiting forests for providing commercial timber or specialty forest products. Later, as paper production and furniture finishing developed in the county, commercial harvesting of timber and bamboo as raw materials gradually increased. As of 1986, human activity on the southern slopes of the Qinling had advanced to a certain elevation but stopped there, providing giant pandas and other wildlife species a bit of space. This has allowed for a kind of truce and some degree of coexistence among people, forests, and wildlife.

Roads and Panda Distribution

The effects of motorized access into giant panda habitat are multifaceted. It would be simplistic to assert that every road produced habitat fragmentation. For example, the road originally developed by the Changqing Forestry Bureau in Yang County did not fragment panda habitat. On the other hand, the isolation of panda habitat in the Niuweihe River area from that on Mount Taibai as well as that between the Xinglongling and Tianhua Mountain areas can be blamed on roads. Whether or not any roads function to fragment a population depends both on the extent of the road system and on traffic volume.

Despite the well-known aphorism "Getting to Sichuan is more difficult than getting to heaven," roads passing over the Qinling were never completely obstructed or abandoned. Various roads existed during differing historical periods, but throughout all periods, there were five main east–west routes (Xin, 1996): the Chencang Road, the Baoxie Road, the Tangluo Road, the Heishuipu River Road, and the Ziwu Road (Figure 3.1). Of these, the Tangluo Road passed directly through what is today the center of the panda distribution and was both the shortest of the ancient routes and also the most dangerous because of its topography.

The effects produced by modern transportation differ from those of ancient roads in several respects. First, the intensity of impact is greater for modern roads, to the point where forests can be completely lost. Second, modern roads have a longer-lasting impact because they are accompanied by economic activity, and thus, their effects are less well contained. All roads constructed under the auspices of state-controlled forestry operations, although built for the purpose of timber production, have had the effect of improving standards of living for local agriculturalists and also have created the potential for forest destruction. The effects are not easily separable.

In the Qinling area, highway construction began in 1934 with the planning and construction of the Bao-Han Highway, the first to cross over the Qinling.

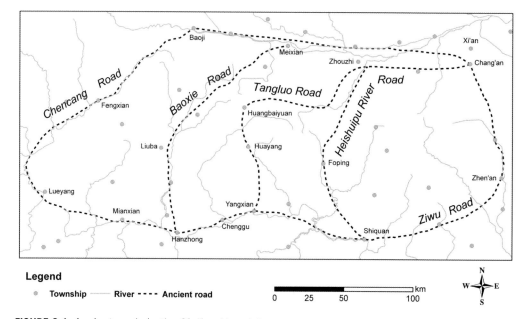

Legend

• Township ⎯ River - - - - Ancient road

0 25 50 100 km

FIGURE 3.1. Ancient roads in the Qinling Mountains.

66   Chapter 3

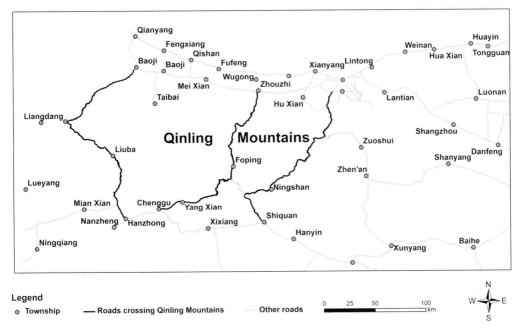

Legend
⊙ Township   — Roads crossing Qinling Mountains   — Other roads   0 25 50 100 km

FIGURE 3.2. Distribution of roads in the Qinling Mountains in 1985. We have indicted the three major roads with a thicker line. The three roads from west to east are Bao-Han Highway, Zhou-Cheng Highway, and Xi-Wan Highway. In the 1936–1949 period there was a single major road across the area (Bao-Han Highway).

It was formally opened in 1936. Afterward, road construction activities in the Qinling slowed, and the existing Bao-Han Highway was damaged and had to be repaired numerous times because of floods and war (Editorial Committee of History of Roads in Shaanxi Province, 1988; Figure 3.2). After the establishment of the People's Republic in 1949, additional roads over the Qinling were built to satisfy the demands of economic development, including the Zhou-Cheng Highway and the Xi'an-Wanyuan (Xi-Wan) Highway. By 1985, the main arterial roads of the Qinling were already well developed (Shaanxi Public Road Bureau, 1993; Figure 3.2). From this foundation, a multitude of smaller branch roads were built by various local governments and forestry bureaus to meet economic and social needs, ultimately forming a complex network of roads throughout the Qinling. The Zhou-Cheng Highway was opened in 1971, although some sections were still unpaved and of very poor quality. In 1980 a flood completely washed away 27 km of road surface, and repairs were not completed until 1986 (Shaanxi Public Road Bureau, 1993). This road has been a major factor in fragmenting giant panda habitat. Construction of the Xi-Wan Highway also had a substantial effect on pandas: where it passes through the Huoditang area, pandas existed 30 years ago, but now they are gone.

Why, then, is there a gap in the road network in the Xinglongling area? As the network of roads took shape, some apparently chance factors came into play. To the north, the Taibai Forestry Bureau had already built a road to Xinglongling pass at 2,600 m, and to the south, only 1 km away, a road was built by the Changqing Forestry Bureau. However, in part to control illegal harvesting within their administrative zone and in part to control transportation access, these forestry agencies never linked their roads to one another. That this shortcut over the Xinglongling between Taibai County and Hanzhong might have seen considerable use is made clear by the fact that people have been known to transport motorcycles, tractors, and even small farm vehicles over it (despite the lack of a road) by breaking their equipment down and later reassembling it. Had these roads been physically connected, the impacts on panda habitat would have been that much greater.

We summarize the effects of road access on panda habitat as follows: where traffic volume, the extent of the road system, and human activities that result from access are limited, roads have a relatively minor effect. This conclusion is supported by the alpine mountain road through panda habitat in the Changqing Forestry Management District. Unfortunately, these conditions are met for only a few roads; thus, most roads must be judged as degrading panda habitat and fragmenting panda populations.

Forest Regeneration and Panda Habitat Use

During April–May 1987, we conducted a survey of panda activity on 27 small patches (sample plots) of forest that had not yet been cut (but were planned for harvest during 1988), ranging in elevation from 1,850 to 2,350 m (Table 3.1), in order to compare them with forest patches that had already been harvested. Simultaneously, we examined 33 sample plots at elevations of 1,660 to 2,500 m in small forest patches that had been subjected to selective harvest during 1983–1985, recording conditions of forest growth and panda presence (Tables 3.2 and 3.3).

In 1987, pandas were present in 69.4% of forest regions harvested during 1981–1982, (Table 3.2), in 70.5% of areas logged during 1983–1985 (Table 3.3), and in 69.5% of areas not yet harvested. Thus, it appears that logging had no effect on panda habitat selection, suggesting that the Changqing Forestry Bureau's harvesting and reforestation activities on the southern slopes of the Xinglongling from 1981 to 1985 did not threaten panda persistence

Winter and spring habitat for pandas on southerly Xinglongling slopes is *Bashania fargesii* forest at 1,350–2,000 m. Currently, the successional stage in these areas is a mixed conifer-broadleaf forest, with a complex understory of *B. fargesii*. Here we can classify human disturbance into three categories: The first consists of forest patches that have not yet been logged but where the subcanopy herb and shrub layer has been cleared within the previous three years, the second consists of logged forests that have been replanted,

Table 3.1.

PRESENCE OF GIANT PANDA ACTIVITY WITHIN SAMPLE PLOTS IN UNCUT FOREST PATCHES (SPRING 1987). OF 363 HA OF FOREST SURVEYED, PANDA ACTIVITY WAS RECORDED IN 252 HA AND NOT DETECTED IN 111 HA.

Sample plot	Area (ha)	Elevation (m)	Canopy density	Panda activity
1	10.4	1,900	0.8	Yes
2	11.7	1,950	0.7	No
3	20.0	2,000	0.7	No
4	9.5	2,100	0.65	No
5	12.9	2,150	0.7	No
6	15.0	2,200	0.65	Yes
7	14.2	2,150	0.7	Yes
8	20.8	2,250	0.7	Yes
9	9.4	2,200	0.7	Yes
10	17.0	2,300	0.7	Yes
11	24.5	2,350	0.8	Yes
12	12.2	2,200	0.7	Yes
13	15.5	2,200	0.75	Yes
14	10.3	2,200	0.8	No
15	12.3	2,150	0.8	No
16	11.6	2,150	0.7	No
17	17.1	2,150	0.7	No
18	17.2	2,000	0.65	No
19	14.9	2,000	0.8	Yes
20	13.3	1,900	0.8	Yes
21	10.2	1,850	0.8	Yes
22	16.5	1,950	0.6	Yes
23	9.0	1,900	0.8	Yes
24	5.0	1,850	0.7	Yes
25	6.0	1,850	0.7	Yes
26	13.5	1,850	0.7	Yes
27	13.0	1,750	0.7	No

and the third consists of logged forests that are recovering naturally. To study the effects of these three management prescriptions on pandas, we examined panda activity in three adjoining valleys in the Baiyangping region where all three were present. We walked up both sides of each valley, starting at the mouth and working upward in elevation, and established a total of thirty 2 × 2 m plots, spaced every 25 m. In each, we recorded the number of scats, bedding sites, and feeding sites (Table 3.4).

The geographic positions, elevations, and forest types examined were all very similar (Table 3.4). Differences in canopy cover, bamboo age structure, height, and stem diameter were caused solely by differences in forest

Table 3.2.

PRESENCE OF GIANT PANDA ACTIVITY IN SAMPLE PLOTS WITHIN FORESTS CUT DURING 1981 AND 1982 (AS DOCUMENTED IN SPRING 1987). OF 324 HA OF FOREST SURVEYED, PANDAS WERE DETECTED IN 225 AND NOT DETECTED IN 99 HA. ALL PATCHES WERE HARVESTED IN 1982, EXCEPT THOSE INDICATED WITH AN ASTERISK (*), WHICH WERE HARVESTED IN 1981.

Sample plot	Area (ha)	Elevation (m)	Percent harvested	Canopy density		Panda activity
				Before harvest	After harvest	
1*	7.3	2,200	54	0.8	0.4	No
2*	11.6	2,250	50	0.65	0.4	No
3*	4.1	2,200	55	0.7	0.4	No
4*	7.1	2,250	50	0.75	0.4	No
5*	7.9	2,300	40	0.85	0.5	Yes
6*	6.3	2,450	40	0.8	0.5	Yes
7*	7.0	2,500	40	0.7	0.5	Yes
8*	11.0	2,450	55	0.7	0.4	Yes
9*	5.9	2,500	55	0.7	0.4	Yes
10*	6.1	2,450	45	0.8	0.4	Yes
11*	6.6	2,400	45	0.7	0.4	Yes
12	5.8	1,800	60	0.7	0.4	Yes
13	9.8	1,850	58	0.75	0.4	Yes
14	15.2	1,900	60	0.75	0.4	Yes
15	12.9	2,050	58	0.7	0.3	No
16	17.0	1,850	59	0.7	0.4	Yes
17	18.4	2,100	50	0.7	0.4	No
18	16.7	1,900	52	0.75	0.4	Yes
19	15.8	2,000	60	0.8	0.4	No
20	12.2	1,950	48	0.7	0.3	Yes
21	13.9	1,850	56	0.75	0.4	Yes
22	13.5	1,660	57	0.7	0.3	Yes
23	6.0	1,750	70	0.6	0.18	No
24	5.5	1,900	68	0.7	0.22	Yes
25	8.0	1,950	50	0.7	0.35	Yes
26	4.5	1,900	66	0.8	0.27	Yes
27	3.0	1,800	70	0.6	0.18	No
28	8.0	1,850	59	0.7	0.29	Yes
29	17.5	1,800	69	0.7	0.22	Yes
30	19.0	1,750	68	0.6	0.19	Yes
31	7.5	1,700	45	0.65	0.36	Yes
32	8.5	1,800	70	0.6	0.18	No
33	4.5	1,850	70	0.6	0.18	No

Table 3.3.

PRESENCE OF GIANT PANDA ACTIVITY IN SAMPLE PLOTS WITHIN FOREST PATCHES CUT DURING 1983–1985 (AS DOCUMENTED IN SPRING 1987). ALL FORESTS WERE HARVESTED DURING 1983–1985. OF 766 HA OF FOREST SURVEYED, 540 HA CONTAINED EVIDENCE OF PANDAS, AND PANDAS WERE UNDETECTED IN 226 HA.

				Canopy density		
Sample plot	Area (ha)	Elevation (m)	Percent harvested	Before harvest	After harvest	Panda activity
1	14.4	1,900	45	0.8	0.4	No
2	8.9	1,950	40	0.6	0.4	No
3	15.8	1,950	45	0.7	0.4	No
4	9.9	2,000	50	0.8	0.4	No
5	12.3	2,050	40	0.7	0.49	No
6	8.0	1,900	50	0.8	0.4	No
7	8.4	1,800	50	0.8	0.4	No
8	10.5	2,200	40	0.7	0.49	No
9	19.7	2,300	55	0.8	0.4	No
10	6.4	2,400	40	0.7	0.48	No
11	11.7	2,500	40	0.7	0.53	Yes
12	12.6	2,400	35	0.7	0.55	Yes
13	19.3	2,300	40	0.8	0.52	Yes
14	24.4	2,300	40	0.7	0.48	Yes
15	10.9	2,000	60	0.7	0.4	Yes
16	17.7	2,350	60	0.8	0.4	Yes
17	11.0	2,400	60	0.7	0.3	Yes
18	6.3	2,200	60	0.7	0.4	Yes
19	12.9	2,300	60	0.8	0.4	Yes
20	14.3	2,250	55	0.7	0.4	Yes
21	14.3	2,200	60	0.8	0.4	Yes
22	8.4	1,950	60	0.7	0.4	Yes
23	9.9	2,000	57	0.6	0.3	No
24	6.9	2,100	60	0.7	0.3	No
25	26.2	2,300	60	0.7	0.4	Yes
26	5.5	2,150	55	0.7	0.4	Yes
27	23.6	2,050	59	0.8	0.4	Yes
28	41.6	2,100	60	0.8	0.4	Yes
29	18.5	2,200	60	0.7	0.3	Yes
30	26.6	2,300	60	0.7	0.4	Yes
31	8.3	2,200	46	0.8	0.5	Yes
32	15.8	1,950	56	0.7	0.4	No
33	11.6	1,950	54	0.8	0.4	No
34	18.3	1,850	53	0.8	0.4	Yes
35	21.9	1,700	59	0.7	0.4	Yes
36	18.4	1,800	56	0.8	0.4	Yes
37	6.2	1,700	59	0.8	0.4	Yes
38	5.1	1,650	46	0.6	0.33	Yes
39	20.6	1,850	59	0.7	0.4	Yes
40	28.9	1,950	55	0.8	0.38	Yes
41	23.3	1,850	50	0.8	0.3	No
42	15.0	1,700	54	0.7	0.4	Yes
43	4.6	1,800	60	0.7	0.4	Yes
44	10.7	1,850	60	0.7	0.4	Yes
45	10.6	1,850	60	0.8	0.4	Yes
46	18.0	1,900	60	0.7	0.4	Yes
47	13.5	1,950	60	0.7	0.4	No
48	29.4	2,050	60	0.7	0.35	No
49	24.8	2,000	60	0.8	0.4	Yes
50	12.2	1,950	60	0.6	0.4	Yes
51	10.3	1,900	60	0.8	0.4	Yes

Relationships among Roads, Forest Management, and Pandas   71

Table 3.4.

THE PRESENCE OF PANDA SIGN IN HARVESTED-REPLANTED FORESTS, HARVESTED–NATURALLY RECOVERING FOREST, AND UNCUT FORESTS SUBJECTED TO UNDERSTORY TREATMENTS. WE RECORDED THREE TYPES OF SIGN: PILES OF SCAT, BEDDING SITES, AND FORAGING (SHOOTS OR LEAVES). ALL THREE TYPES OF MIXED FOREST WERE IN EAST–WEST VALLEYS, WITH ELEVATIONS OF 1,645–2,668 M. UNHARVESTED AND HARVEST-REPLANTED FORESTS HAD THEIR *BASHANIA* BAMBOO AND SHRUB UNDERSTORY CLEARED 3 YEARS PREVIOUSLY, WHEREAS THE HARVESTED–NATURALLY RECOVERING FOREST (HARVESTED 7 YEARS PRIOR) HAD NO UNDERSTORY MANIPULATION. ALL FORESTS HAD PATCHES OF RECOVERING BAMBOO, BUT THEY DID DIFFER IN CANOPY COVER (UNHARVESTED, 75%; REPLANTED, 39%; NO REPLANTING, 45%)

					Harvested				
	Unharvested (4.2 km^2)			Replanted (1.3 km^2)			No replanting (1.2 km^2)		
Plot (4 m^2)	Scats	Bedding	Foraging sign	Scats	Bedding	Foraging sign	Scats	Bedding	Foraging sign
1	0	0	0	0	0	0	0	0	0
2	0	1	20	3	1	24	0	0	0
3	0	0	0	0	0	0	0	0	0
4	0	0	0	0	0	0	0	0	19
5	1	0	0	0	0	16	1	0	57
6	0	0	0	1	0	15	1	0	42
7	2	1	21	0	0	8	0	0	46
8	2	1	24	0	0	0	0	0	45
9	0	0	0	0	0	0	0	0	0
10	0	0	0	1	0	40	0	0	9
11	0	0	14	1	0	46	0	0	0
12	1	0	13	2	1	54	0	1	21
13	0	0	20	2	0	30	0	0	0
14	0	0	0	0	0	8	0	0	0
15	1	0	31	1	0	11	0	0	2
16	0	0	15	0	1	18	0	0	0
17	0	0	10	0	0	13	2	1	28
18	0	0	0	2	1	46	1	0	15
19	0	0	0	1	1	49	0	0	0
20	1	0	0	0	0	25	0	0	6
21	0	0	12	0	0	23	1	1	36
22	0	0	26	0	0	0	1	0	14
23	0	0	0	0	0	27	1	0	5
24	0	1	17	0	1	32	0	0	0
25	2	1	45	1	1	43	3	1	61
26	0	1	18	0	0	0	1	0	25
27	1	0	18	1	1	52	0	0	0
28	2	0	23	0	1	32	0	0	15
29	1	1	19	1	0	23	0	0	0
30	2	1	41	1	0	25	0	0	8
Total	16	8	387	18	9	660	12	4	454

treatment. Areas regenerating naturally after harvest were characterized by a bamboo understory of variable age structure that was taller, thicker, and older than in the other two valleys. In contrast, bamboo age in the replanted areas was similar to that in the cleared areas, in both cases about 3 years old or younger. Canopy cover in the uncut areas was twice that in harvested cut areas. On the basis of our preliminary observations of these three valleys, it appears that the naturally regenerating bamboo has grown the highest. However, pandas evidently had their own view of the bamboo situation. From winter 1986 to April 1987, panda signs (scats, bedding sites, bamboo stem fragments, and leaves of eaten bamboo) were most common on replanted sites, followed by unharvested but cleared forests. Panda sign was least common in naturally regenerating forests.

Another aspect to consider is that the rank order of panda selection among these different bamboo forests changes seasonally. In May, when pandas forage for new sprouts of *B. fargesii*, they gravitate toward naturally regenerating areas because bamboo shoots there are thicker than those available in the other two areas. Thus, in spring, pandas rarely visit the other two types of regenerating forest.

In late May 1987, we radio-collared a male panda called Shanshan. Over the next month, in June 1987, his travels took him into areas that had been logged from 1981 to 1984 and again in 1986. Radio tracking indicated that he moved through valleys that were in the process of being logged. During July and August he entered a *Fargesia* bamboo forest near the main Xinglongling ridge that had been cut in 1954–1956 and had since progressed to a mixture of larch and *Fargesia* bamboo. Shanshan's monthly home ranges during 1987 provide a direct illustration of his use of previously logged areas (Pan et al., 1988; Figure 3.3).

Our 190 km² study area provides a good case study of these panda-forest management dynamics. Pandas in our study area enjoyed a decade of stability until the 1970s, at which time industrial forest production increasingly began impacting forests and bamboo. As the 1980s ended, some 75% of the study area's forests had been affected in some way by industrial activity. During the period 1989–1993, the completely untouched section was a portion of the Taibai larch stands above 2,800 m in elevation; all other habitat has suffered from human disturbance to some degree. Although our telemetry data illustrate that pandas do use these areas, there is little doubt that habitat quality has been greatly diminished.

The destructive logging practices that began in the early 1990s were not restricted to the southern Xinglongling slopes around the Xinglongling mountain island. Similar practices occurred in the Longcaoping Forestry Management District to the east, the Taibai Forestry Management District to the north, and the Hanxi Forestry Management District to the west. All these forestry bureaus sought greater economic "benefits" from forest

Relationships among Roads, Forest Management, and Pandas   73

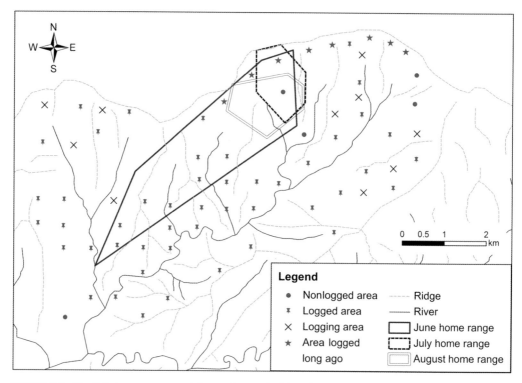

FIGURE 3.3. Monthly home ranges of the male panda Shanshan during summer 1987, Xinglongling Mountains, Shaanxi.

production and gave little consideration to potential forest regeneration. These logging practices of the time, destructive and unprecedented, presaged the beginning of Qinling's forested ecosystem collapse. Such could easily include the loss of pandas and biodiversity more generally. The stark reality is that pandas face a much worse situation than 20 years ago when we started sloganeering about "protecting the giant panda."

The greatest problems facing the Qinling pandas are all human caused. Logging activities of the Changqing Bureau ended only in October 1993 when the state council ordered a halt to all cutting on natural forests in the middle and upper Yangtze River drainages, forcing some 2,400 workers to find alternative livelihoods. Thus, as the final summer of the century began, panda habitat that might ensure the long-term persistence of this population in the Qinling Mountains had once again entered into a period of renewal and recovery.

References

Editorial Committee of History of Roads in Shaanxi Province. 1988. 陕西公路史, 第一册: 近代公路 [The history of roads in Shaanxi province. Vol. 1, Roads in modern times]. Beijing: China Communications Press.

Pan, W. S., Z. S. Gao, Z. Lü, Z. K. Xia, M. D. Zhang, L. L. Ma, G. L. Meng, X. Y. She, X. Z. Liu, H. T. Cui, and F. X. Chen. 1988. 秦岭大熊猫的自然庇护所 [The giant panda's natural refuge in the Qinling Mountains]. Beijing: Peking University Press.

Shaanxi Public Road Bureau. 1993. 陕西省公路局史 [The history of Shaanxi Public Road Bureau]. Beijing: China Communications Press.

Xin, D. Y. 1996. 古代交通与地理文献研究 [A review of studies on transportation and geography in ancient China]. Beijing: Zhonghua Book Company.

Chapter 4

Home Ranges of Giant Pandas

Summary: In this chapter we describe our experiences with radio-collaring and tracking free-ranging giant pandas, as well as the methods we used to collect and analyze telemetry data. We followed the movements of 22 radio-marked pandas over a period of 8 years. Mean home range size was 10.6 km^2, varying from 3.3 to 28.9 km^2 among individuals. These were generally larger ranges than found in Wolong (3.9–6.2 km^2). Male pandas in the Qinling generally had larger home ranges (7.6–28.9 km^2) than females (3.3–20.2 km^2). Home ranges overlapped one another. Patterns of home range use differed by sex, with adult males tending to move widely within their ranges, whereas females tended to concentrate their use in certain portions, called core areas. Female core areas were separated from each other, and juveniles and subadults tended to establish home ranges adjacent to those of their mothers. Adult males used their home ranges in ways that maximized their opportunities for breeding; adult female strategies focused on successfully raising cubs. When cubs were relatively young, females acted in a territorial manner, driving away other pandas, especially adult and subadult females that could pose a threat to them.

Introduction

Pandas in the Qinling reside year-round in mixed bamboo-forested areas in the midmontane and subalpine belts. Because bamboo is abundant and available throughout the year, pandas have little need to roam widely to fulfill their nutritional needs. However, specific patterns of movement and home range use are affected by gender, developmental stage, social dynamics, habitat quality, seasonal forage resources, and human disturbance.

Fitting Animals with Radio Collars

Pandas reside in dense bamboo groves, so observing them directly is very difficult. To study their daily activity and movement, as well as to monitor them long term, we turned our efforts to capturing them, fitting them with radio collars, and using radio telemetry to track them. Because of the dense vegetation, we generally needed to approach animals to within 10 m to use a dart gun for immobilization. Thus, we elected not to use powerful dart guns that might injure the animal. We used a duplex dart gun from the United States that came equipped with 10 adjustable power positions that projected darts from 3 to 72 m, allowing us to adjust the power in accordance with our distance from the animal. We immobilized pandas with ketamine hydrochloride, an anesthetic often used in zoos and among the safest of immobilizing drugs. Injected with the appropriate dose, animals generally regain consciousness within 20–30 minutes (Wang, 1978; Zheng, 1980).

We used radio collars and receivers (Telonics, Inc.) in the 150- to 152-MHz range. Collars were also designed to drop off with wear sometime after battery depletion. In actual use, we encountered situations in which some collars were shed prior to battery depletion. Although some potential data were thereby lost, this characteristic minimized the chance that collars would remain on pandas for long durations after they were no longer functioning.

Upon encountering a panda, we would determine the appropriate dose of ketamine, load it into the syringe dart, and insert the charge into the dart gun. After a successful hit, we remained motionless, neither standing up nor approaching, to avoid startling the animal until it was clear that the anesthetic had taken effect and the animal was asleep. We then determined the animal's gender, examined its teeth and overall body condition, fitted it with a radio collar, switched the collar on by removing the magnet, and moved away without delay. From a safe distance, we then monitored the panda's recovery from the drug.

Radio-Tracking Methods

After fitting a panda with a radio collar, one of our daily tasks became locating it. Our most common radio-tracking method was triangulation. We obtained radio signals from three different points, determined the bearing to the animal from each, and created a triangle by drawing straight lines from each point in the direction of the signal. We took the center of the resulting triangle as the actual location of the panda. However, the complexity of the terrain often resulted in signal bounce, making it impossible for field staff to determine the panda's location from only three points. In these cases, we increased the number of points from which we listened to the signal and obtained bearings. This allowed us to assess which bearings were misleading and to exclude these when estimating the actual position of the panda.

We established and mapped a number of fixed points from which we radio tracked along the two roads through the forest within the study area. We marked each of these "position points" in the field with durable and easily visible markers. Our daily protocol was to travel along each road until we reached the point, turn on the receiver, listen for each panda's signal, and then estimate and record its bearing.

We also used direct observations to obtain panda locations. As we became more familiar with the typical activity patterns of specific collared pandas, we were able to approach them closely and watch them, and we recorded their locations on maps.

Because of topographic complexity, radio signals were often reflected or refracted. Thus, we sometimes obtained radio signals from some position points but not from others. In some cases, the bearings we obtained resulted in a very large error triangle. During the early stages of the study, we discarded such data. Later, as we accumulated experience, we discovered that these situations generally resulted from pandas being at specific locations, and we gradually gained the ability to determine where those locations were.

During early stages of our research, we manually transferred all raw data to 1:50,000 scale maps. As our database increased in size and our electronic capability improved, we began digitizing all data and storing it electronically. First, we overlaid a rectangular coordinate system on the study area map, on which we documented the date of each location, latitude and longitude, and the name of the recorder. We also entered the main ridges and water sources on the base map. Later, we developed a program in Turbo Pascal to automate the process of extracting and mapping digital location data for individual pandas. Acquisition of commercial geographic information system (GIS) software made this all much faster and easier, but calculation of home range size and spatial use was still conducted with software that we developed.

In any field research, the precision of the data depends on the attitude of the field workers, methods used, and equipment available, among other factors. Despite our best efforts, a multitude of uncontrollable factors conspired to make field work imprecise and led us in a never-ending search for better methods. We scrutinized the assumptions underlying all methods, confirmed their appropriateness, and supplemented them where needed before applying them in the field.

Methods of Estimating Home Range Sizes

MINIMUM CONVEX POLYGON METHOD

The minimum convex polygon method (MCP) was our most frequently used approach to delineating home range areas. The MCP ranges are produced by connecting the outermost points such that all outside angles are convex. The MCP approach has been used widely in other wildlife research, and has been reported in all studies of panda home range thus far (Schaller et al., 1985; Pan et al., 1988). It is also simple to produce.

However, we noted some disadvantages of calculating home ranges using the MCP method. First, MCP home range sizes tended to increase with

Home Ranges of Giant Pandas   79

sample size. This was not surprising because the addition of even a single location outside the existing outer periphery increases the MCP area. Thus, home range area was a function of the number of panda locations, hardly an objective measure. Second, MCP home ranges are defined entirely by the outermost locations, and these may represent nothing more than occasional movements; all other locations within the polygon are ignored. That is, location points are not considered equally.

We considered two approaches to surmounting these difficulties. The first was simply to increase the sample size of locations, to the extent possible achieving uniformity in the timing and spatial patterns of data acquisition. The second was to search for an alternative home range model.

ELLIPSE METHOD

We also made use of the Jennrich-Turner bivariate normal ellipse model, which has gained in popularity among wildlife studies (White and Garrott, 1990). This model regards all the animal locations as being distributed on a two-dimensional plane and distributed normally on both planes, yielding an elliptical shape. We could also formulate the ellipse to encompass less than all of the locations (e.g., the smallest area encompassing 90% of the locations).

We found that with increasing sample size, MCP home range areas increased, whereas elliptical home range areas did not change in a consistent manner. The range of variation in home range sizes with sample size was more restricted using elliptical home ranges than MCP home ranges. Given this result, if home ranges of different pandas (or the same panda in different seasons) are to be compared, the relocation frequency (i.e., days between relocations) should be roughly similar; otherwise, comparisons will not be reliable. However, the frequency of relocations of different pandas varied markedly, making MCP home ranges difficult to compare across individuals. However, relying solely on elliptical home ranges to estimate home range size was also problematic. First, it was not obvious what proportion of radiolocations should be included. Second, because previous research work on pandas used only MCP home ranges, elliptical home ranges would not be useful for comparing the two.

A very important point is that the elliptical home range method does not completely eliminate the problem of sampling error arising from limited sample size. Although one can always create an ellipse and from it calculate an area, its reliability increases as the number of locations increases (White and Garrott, 1990). [Editor note: As it turned out, the ellipse technique soon went out of favor and was replaced by a series of other techniques that also consider the distribution of internal home range points. Biologists are often faced with the dilemma of using improved analytical techniques but obtaining results that can be compared to past as well as future studies.]

In our study of panda home ranges, we used both MCP and ellipse methods to estimate home ranges for each panda and season. We considered MCP home ranges to be an intuitive estimate of home range size, whereas we considered the 90% elliptical home ranges to best represent home range area corrected for unequal sample sizes (relocation frequency). We used only MCP estimates when estimating area of home range overlap.

Home Range Area

From January 1989 to September 1997, we radio tracked 22 giant pandas (Table 4.1, Figure 4.1) and obtained a total of 5,973 radiolocations (Table 4.2). Radio tracking was one of our primary objectives during the early stage of our study. Beginning in about 1993, although the number of collared pandas had increased, we obtained fewer relocations for each animal because behavioral observations became our main priority.

We used relocation data to calculate seasonal and year-round home ranges of 20 pandas, using both the MCP and ellipse methods (Table 4.3). We used the ellipse method to compare seasonal home ranges sizes among individuals (Table 4.4) and by sex (Table 4.5). Males had significantly larger home ranges than females, year-round and by season ($P \leq 0.01$).

Year-round home ranges were larger than the sum of seasonal home range areas for each individual because pandas migrated seasonally. The area of transition between summer and autumn-winter ranges, which is used only

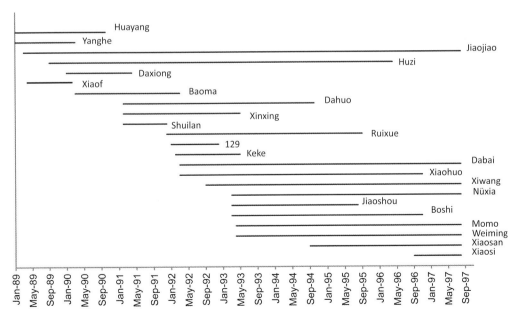

FIGURE 4.1. Timeline showing months of radio tracking for each of the 22 pandas followed in the Qinling study area, January 1989 to May 1997.

Table 4.1.

GIANT PANDAS TRACKED IN THE QINLING STUDY AREA, MARCH 1989 TO AUGUST 1997, SHOWING PANDA NUMBER, NAME, SEX, DATE TRACKING BEGAN, AGE (IN YEARS) WHEN FIRST MONITORED, DATE TRACKING ENDED, CAUSE OF DATA CESSATION, AND NOTES.

Panda code	Panda name	Sex	Date tracking began	Age when tracking began	Date tracking ended	Cause of data cessation	Notes
f01	Jiaojiao	F	March 1989	4	August 1997	Battery depleted	
m02	Huzi	M	September 1989	0	April 1996	Dropped collar	Jiaojiao's cub
f03	Xiwang	F	September 1992	0	August 1997	Battery depleted	Jiaojiao's cub
m04	Xiaosan	M	September 1994	0	August 1997	Battery depleted	Jiaojiao's cub
f05	Xiaosi	F	September 1996	0	August 1997	No collar[a]	
m06	Dabai	M	March 1992	8	August 1997	Battery depleted	
m07	Daxiong	M	January 1990	15	April 1991	Mortality	
m08	Dahuo	M	February 1991	14	October 1994	Mortality	
m09	Xiaohuo	M	March 1992	7	November 1996	Battery depleted	
f10	Nüxia	F	March 1993	4	August 1997	Battery depleted	
m11	Jiaoshou	M	March 1993	5	August 1995	Battery depleted	
f12	Boshi	F	March 1993	2	November 1996	Battery depleted	
f13	Ruixue	F	December 91	6	September 1995	Battery depleted	
f14	Baoma	F	March 1990	5	March 1992	Battery depleted	
m15	Huayang	M	January 1989	21	October 1990	Mortality	
f16	Yanghe	F	February 1989	19	March 1990	Dropped collar	
f17	Momo	F	April 1993	4	August 1997	Battery depleted	
f18	Keke	F	February 1992	5	May 1993	Left study area	
m19	Xinxing	F	February 1991	6	May 1993	Dropped collar	
m20	Weiming	M	April 1993	7	August 1997	Battery depleted	
m21	129	M	January 1992	17	December 1992	Dropped collar	
f22	Shuilan	F	February 1991	3	December 1991	Mortality	
f23	Xiaof	F	January 1990	4	March 1990	Dropped collar	

[a]This animal was not collared but was followed as long as it was associated with its mother, who was collared.

82   Chapter 4

Table 4.2.

NUMBER OF GIANT PANDAS RADIO TRACKED AND TOTAL NUMBER OF RADIO RELOCATIONS EACH YEAR, 1989–1997, QINLING STUDY AREA.

	Year									
	1989	1990	1991	1992	1993	1994	1995	1996	1997	Total
Animals tracked	4	7	8	11	14	13	12	10	8	22
Radio relocations	820	969	946	867	492	921	423	490	45	5,973

temporarily by pandas, is included within our year-round home range estimates but not our seasonal home range estimates.

Factors Influencing Home Range Area

Home range sizes varied considerably among individuals, from as small as 3.3 km² to as large as 28.9 km², a factor of almost 9; had we used sizes of all yearly home ranges (Table 4.3) instead of averages within individuals (Table 4.4), variation would be even greater. After removing outlier data that represented unusual movements and double-checking the accuracy of each data point as well as our calculation methods, it is clear that these differences are real. We are unable to develop a full explanation for these interindividual differences, but we examine some relevant factors below.

GENDER

In general, home ranges of males exceeded those of females seasonally and year-round (Table 4.6). However, considered as individuals, the situation was more complex. Omitting individuals who made exceptional movements, male year-round home ranges varied from 7.6 to 28.9 km², and female year-round home ranges varied from 3.3 to 20.2 km². Some female home ranges were larger than male home ranges (Figure 4.2). Among 56 year-round range sizes, about 80% were <15 km², and about half were in the 5–15 km² range. Most male home range sizes (n = 28) were 5–15 km², and few were 0–5 km². For females, the 0–5 km² category was the most common (about 40%).

OTHER FACTORS

To further investigate factors influencing home range size, we selected six representative individuals with relatively large sample sizes for comparison (Figure 4.3). Year-round home range sizes of some individuals (e.g., Jiaojiao, Huzi, and Xiwang) varied considerably among years, whereas year-round home range sizes of others (e.g., Dabai and Ruixue) were relatively stable. Trends in home range size by year differed among individuals. Home range size tended to increase with age as individuals developed from subadults to adults. This is shown by Huzi (m02) and Xiwang (f03). The

Table 4.3.

ESTIMATES OF SEASONAL AND YEAR-ROUND HOME RANGE (HR) SIZES OF QINLING PANDAS, 1993–1997, USING BOTH MCP AND ELLIPTICAL HOME RANGE ESTIMATION METHODS (SEE TEXT).

Panda name (Code)	Age	Year	Summer Relocation frequency	Summer MCP HR area (km^2)	Summer Elliptical HR area (km^2)	Autumn and winter Relocation frequency	Autumn and winter MCP HR area (km^2)	Autumn and winter Elliptical HR area (km^2)	Year-round Relocation frequency	Year-round MCP HR area (km^2)	Year-round Elliptical HR area (km^2)
Jiaojiao (f01)	4	1989				1.40	5.13	5.92			
	5	1989–1990	1.34	0.76	1.25	1.53	6.07	6.21	1.51	9.19	8.24
	6	1990–1991	1.96	0.87	1.31	1.68	9.80	11.88	1.71	13.30	14.00
	7	1991–1992	2.00	0.71	1.25	2.42	7.75	7.73	2.38	10.10	9.49
	8	1992–1993	4.67	0.74	2.64	2.74	3.40	3.16	2.84	5.81	4.47
	9	1993–1994	5.14	1.00	3.17	7.68	10.20	11.01	7.33	12.10	11.74
	10	1994–1995	5.14	0.87	2.66	2.08	6.35	3.81	2.22	11.20	5.33
	11	1995–1996	15.00	0.01	0.06	11.69	7.23	12.50	12.17	8.02	13.34
	12	1996–1997				4.74	0.50	0.38			
Huzi (m02)	2	1991–1992	2.43	0.34	0.74	2.45	9.01	8.85	2.46	12.00	13.17
	3	1992–1993	3.75	1.17	2.00	4.15	8.46	6.87	4.08	10.50	8.69
	4	1993–1994	6.60	0.59	1.38	6.77	3.03	4.80	6.75	4.41	6.23
	5	1994–1995	5.50	3.50	8.44	4.97	6.74	7.38	5.05	13.20	13.39
	6	1995–1996							6.74	10.60	17.39
Xiwang (f03)	1	1993–1994	5.14	0.92	2.94	7.41	3.20	4.18	7.12	5.22	6.96
	2	1994–1995	3.18	1.41	3.14	3.92	2.69	3.93	3.85	5.84	5.51
	3	1995–1996							9.03	7.50	12.06
Dabai (m06)	8	1992				3.39	2.41	3.60			
	9	1992–1993	4.07	2.04	3.63	5.33	4.46	5.57	5.06	6.32	7.83
	10	1993–1994	9.10	1.12	2.84	7.42	3.84	4.70	7.79	6.48	7.96
	11	1994–1995	8.10	2.21	5.27	7.19	2.72	4.42	7.38	4.97	6.93
	12	1995–1996							12.86	3.59	7.72
	13	1996–1997							20.13	3.95	9.98

Name	No.	Period									
Daxiong (m07)	15	1990				1.67	11.62	8.05			
	16	1990–1991	1.69	5.22	6.05	1.94	8.60	12.07	1.89	21.20	28.93
Dahuo (m08)	14	1991				2.53	6.29	7.53			
	15	1991–1992	2.41	4.32	5.33	3.16	9.03	10.95	2.94	18.50	25.86
	16	1992–1993	3.38	1.72	3.11	5.13	6.40	5.07	4.60	16.80	22.23
	17	1993–1994	6.50	2.64	4.77	11.50	1.09	2.19	9.34	6.53	11.08
Xiaohuo (m09)	7	1992				2.43	6.52	9.02			
	8	1992–1993	6.36	1.13	2.77	6.55	10.58	11.21	6.53	12.50	13.71
	9	1993–1994	12.20	1.88	8.52	8.63	8.17	12.04	9.10	15.80	19.39
	10	1994–1995							7.87	10.80	13.42
	11	1995–1996							20.21	4.12	13.10
Nüxa (f10)	4	1993				5.39	0.21	0.32			
	5	1993–1994	13.50	0.55	2.12	6.71	2.02	2.62	7.60	4.89	5.15
	6	1994–1995	9.38	0.60	1.97	5.25	1.95	1.65	5.86	3.76	3.32
	7	1996–1997							5.85	1.54	1.15
Jiaoshou (m11)	5	1993				3.21	2.04	3.47			
	6	1993–1994	7.85	2.89	5.86	6.65	3.83	5.72	6.96	8.58	10.33
	7	1994–1995	5.71	6.95	11.76	6.49	2.24	4.53	6.26	11.30	14.04
Boshi (f12)	2	1993							7.00	8.21	14.93
	3	1993–1994							61.33	5.76	20.64
	4	1994–1995							5.24	2.05	3.01
	5	1996							9.65	1.48	2.25
Ru xue (f13)	6	1992				4.03	2.69	2.34			
	7	1992–1993	4.00	1.76	3.23	5.98	1.43	2.01	5.42	3.92	4.17
	8	1993–1994	9.11	1.03	3.57	6.02	2.29	2.23	6.55	4.10	3.30
	9	1994–1995	11.43	0.71	2.11	4.82	2.67	2.25	5.64	3.09	2.63
Baoma (f14)	5	1990				1.51	1.00	1.13			
	6	1990–1991	1.59	4.24	4.08	5.06	2.39	3.01	2.99	7.68	9.39
	7	1991–1992	2.56	2.97	4.07	2.95	3.06	2.81	2.79	7.96	9.47

(continued)

Table 4.3. (continued)

Panda name (Code)	Age	Year	Summer			Autumn and winter			Year-round		
			Relocation frequency	MCP HR area (km²)	Elliptical HR area (km²)	Relocation frequency	MCP HR area (km²)	Elliptical HR area (km²)	Relocation frequency	MCP HR area (km²)	Elliptical HR area (km²)
Huayang (m15)	21	1989				1.62	11.11	11.48			
	23	1990	2.29	3.11	4.69	5.22	0.54	1.58	2.98	6.67	10.55
Yanghe (f16)	19	1989				1.29	6.49	6.81			
	20	1989–1990	2.02	3.11	3.07	1.62	7.89	4.50	1.75	19.90	16.85
Momo (f17)	4	1993				5.50	1.07	2.77			
	5	1993–1994	7.56	1.59	2.67	6.68	1.80	1.99	6.96	4.65	5.69
	6	1994–1995	4.60	1.22	2.48	2.80	1.55	1.51	3.06	3.41	2.74
	7	1995–1996							12.19	0.79	1.28
Keke (f18)	5	1992							4.69	17.10	22.52
	6	1993							5.08	0.63	1.39
Xinxing (m19)	6	1991				1.81	13.29	24.73			
	7	1991–1992	2.39	7.78	9.85	2.22	15.27	10.42	2.28	29.10	28.57
	8	1992–1993	14.33	2.29	8.12	11.09	3.56	5.77	11.86	12.10	15.17
Weiming (m20)	7	1993				4.58	2.26	5.50			
	8	1993–1994	8.69	3.97	7.33	7.58	4.96	6.66	7.91	12.40	13.89
	9	1994–1995	7.53	3.60	5.68	8.26	2.70	4.89	8.02	7.04	9.45
	10	1996							11.41	1.09	1.51
129 (m21)	17	1992	6.63	3.05	4.90				4.67	7.71	10.20
Shuilan (f22)	3	1991	2.32	13.19	15.31				3.51	50.40	60.35
Xiaof (f23)	4	1990							3.42	1.27	2.08

Table 4.4.

AVERAGE HOME RANGE (KM^2) ESTIMATION OF EACH PANDA RADIO TRACKED DURING THE STUDY. AREA IS ESTIMATED USING THE ELLIPSE METHOD (SEE TEXT FOR DETAILS). FOR ANIMALS FOLLOWED FOR A SINGLE SEASON OR SPORADICALLY OVER A SINGLE YEAR WE GIVE A SINGLE VALUE WITH NO ESTIMATE OF VARIABILITY. SD = STANDARD DEVIATION.

Panda name	Winter and spring		Summer		Whole year	
	Mean	SD	Mean	SD	Mean	SD
Jiaojiao	7.31	3.32	2.05	0.79	9.47	3.49
Huzi	6.98	1.45	2.75	2.88	11.78	3.91
Xiwang	4.06	0.13	3.04	0.10	7.75	2.55
Xiaosan					8.22	
Dabai	4.90	0.49	3.91	1.01	7.61	0.40
Daxiong	12.07		6.05		28.93	
Dahuo	6.08	3.66	3.22	1.52	19.73	6.29
Xiaohuo	9.75	2.68	5.53	2.35	14.99	2.55
Nüxia	2.14	0.48	2.05	0.08	4.24	0.92
Jiaoshou	5.37	0.83	6.28	0.42	12.41	2.53
Boshi					4.54	3.04
Ruixue	2.05	0.27	2.33	1.24	3.36	0.63
Baoma	6.48	5.04	4.08	0.00	9.44	0.04
Huayang	13.60		3.49	0.30	24.68	
Yanghe	4.50		3.07		18.69	
Momo	4.65	4.09	2.72	0.06	3.26	1.81
Keke			6.71		20.17	
Xinxing	7.62	2.02	8.99	0.86	21.88	6.70
Weiming	5.77	0.88	6.51	0.82	9.02	4.17
129			3.74		10.25	
Shuilan					60.39	

Table 4.5.

AVERAGE HOME RANGE (KM^2) OF MALE AND FEMALE PANDAS DURING EACH SEASON AND THE WHOLE YEAR BASED ON THE ELLIPSE METHOD (SEE TEXT FOR DETAILS) FOR HOME RANGES PRESENTED IN TABLE 4.3. THESE DATA EXCLUDE SUBADULT FEMALE SHUILAN, WHOSE MOVEMENT WAS CONSIDERED A DISPERSAL. SD = STANDARD DEVIATION.

Season	Female		Male		All	
	Mean	SD	Mean	SD	Mean	SD
Summer	2.51	1.01	5.28	2.77	4.27	3.05
Winter	4.18	3.3	7.68	4.49	6.02	4.34
Year-round	7.69	5.81	13.66	6.65	10.62	6.92

Home Ranges of Giant Pandas　87

Table 4.6.

JIAOJIAO'S MONTHLY HOME RANGE SIZE (KM^2), 1989–1997. THE ESTIMATE IS BASED ON THE ELLIPSE METHOD (SEE TEXT FOR DETAILS). A DASH (—) INDICATES DATA WERE NOT AVAILABLE.

Year	January	February	March	April	May	June	July	August	September	October	November	December
1989	0.10		0.28	2.18	2.69	2.80	0.77	0.19	0.05	0.08	0.71	0.20
1990	0.10	0.20	0.38	2.03	2.78	1.43	0.56	1.21	1.29	0.75	0.68	1.87
1991	0.23	1.56	5.38	0.73	3.35	1.20	0.62	0.29	0.28	0.63	0.44	0.86
1992	0.40	0.43	2.09	1.74	4.56	0.95	1.09	0.03	0.00	0.05	0.15	0.09
1993	0.04	—	0.36	0.59	1.11	0.20	0.18	—	—	—	—	—
1994	0.06	0.20	1.81	3.17	4.82	1.16	1.53	0.00	0.00	0.01	0.36	0.07
1995	0.20	0.00	0.39	2.83	0.11	—	—	—	—	—	—	—
1996	—	—	2.42	1.53	—	—	—	0.02	0.02	0.02	0.06	—
1997	—	—	0.05	—	—	—	—	0.10	^{a}			
Mean	0.17	0.48	1.46	1.85	2.78	1.29	0.79	0.26	0.27	0.26	0.40	0.62

^{a}Monitoring ended in August 1997.

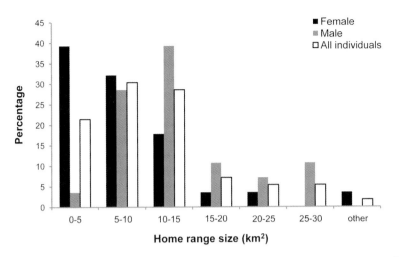

FIGURE 4.2. Histograms illustrating home range sizes of females, males, and both sexes in the Qinling study area, 1989–1997.

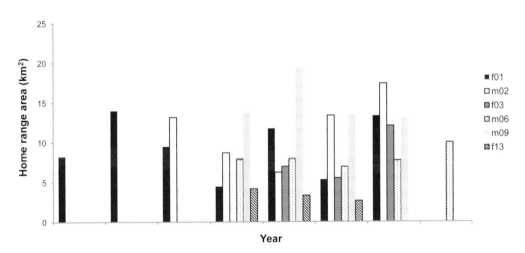

FIGURE 4.3. Sizes of six selected individual panda home ranges by year, 1989–1996, Qinling study area.

relatively large home range of Huzi in 1991 when he was 2 years old was likely influenced by the fact that he had not yet completely disassociated from his mother at that time. Home range sizes of the older individuals, Huayang (m15) and Yanghe (f16), were not clearly smaller than other individuals, suggesting no diminution of home range size with age.

The fluctuations in Jiaojiao's home range size were clearly related to her reproductive status: the three troughs in 1989, 1992, and 1994 all coincided with years in which she bore cubs. However, this pattern was not seen in Ruixue. That said, we believe there is a tendency for home ranges of adult females to shrink in years in which they give birth.

It is generally held that home range size is positively related to the body size of the animal (Schaller et al., 1985). The data we collected were insufficient to fully address this question. However, Xiaohuo (m09) occupied a much larger home range than Dabai (m06), even though Dabai was much larger. It appears that the home range area is not strictly related to body size on an individual level.

Home range sizes must be as large as necessary to provide needed resources, but not so large as to expend excess energy. For pandas, needed resources include forage and reproductive opportunities. Because bamboo is characterized by high biomass, rapid regeneration, and relative stability, forage resources, in general, should not be considered an issue for pandas. With forage resources relatively stable, males will tend to increase their home range size in order to gain access to more females, but they must balance the potential gains of more mates against the increased energy expenditure. Of all our study subjects, Dabai, who was the largest and heaviest, occupied a dominant position when contesting for mates. However, at 7.6 km², his home range size was not the largest and, in fact, was smaller than the overall mean of 10.6 km². However, his home range overlapped the areas of the most intensive use (core areas) of four estrous females (Figure 4.4).

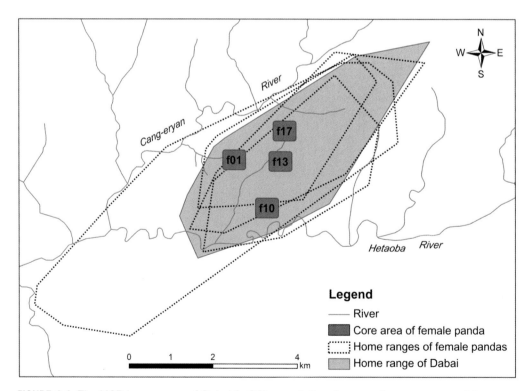

FIGURE 4.4. The MCP home range of Dabai (m06), an adult male, as well as core areas of four adult females, Qinling study area, 1989–1997.

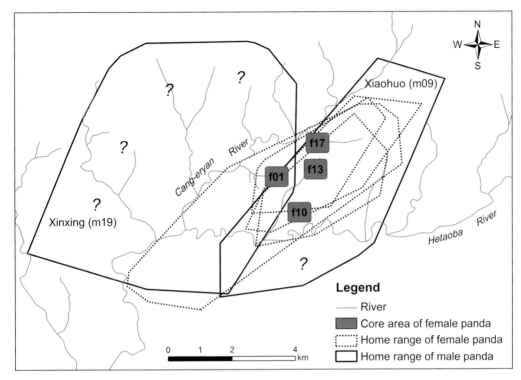

FIGURE 4.5. Home range of two males, Xiaohuo (m09) and Xinxing (m19), showing overlap with core areas of four females, Qinling study area, 1989–1997.

In contrast, other males who failed to occupy such a dominant position in the hierarchy may have had little choice but to search for mating opportunities, which naturally led to an increase in home range size. For example, the male Xinxing (m19) occupied a home range of 21.9 km², and the male Xiaohuo (m09) had a home range size of 15.0 km², not only larger than Dabai's home range but also larger than the year-round mean for males (Figure 4.5).

Thus, it appears that a dominant position was not necessarily manifested in larger home range size, but rather in efficient overlap with estrous females. We noted that Dabai optimized his movements by following females only during their peak of estrus, whereas younger, inexperienced males tended to spend resources following females who had just entered estrus.

Patterns of Home Range Use

Regardless of whether we employed the MCP or elliptical method, we obtained only an estimate of the geographical extent of activity within some defined time period. Within this boundary we were interested in what areas were used most intensively, what areas were used as travel routes, and what patterns of home range use were affected by gender, age, or physiological

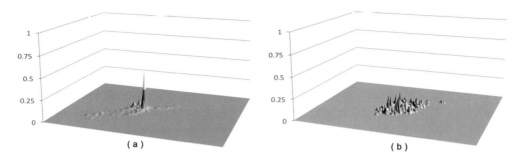

(a)   (b)

FIGURE 4.6. Home range use patterns as depicted using the grid cell method (see text) for (a) Jiaojiao and (b) Dabai.

condition. To address these questions, we needed to use other methods to examine use within the home range.

We adopted a nonparametric approach based on individual location data. We first attempted to match each location point with a 50 × 50 m pixel. Recognizing that this was too fine grained for the error in our locations, we enlarged the grain of our grid to 250 × 250 m (with the original grid positioned in the center of each of these larger cells) but differentially weighted each of the smaller cells within the larger ones. We assigned a weight of 9 to the smaller-sized cell in which the radiolocation seemingly occurred, a value of 4 to each of the 8 grid cells surrounding this central one, and a weight of 1 to each of the 16 cells surrounding these. For each cell within each home range, we then obtained a habitat use value from the sum of weighted location points. However, this value was influenced by the number of location points, which varied among home ranges. We therefore calculated a relative use value by dividing each cell's value by the total accumulated value for each home range and from that produced three-dimensional depictions of home range use.

Using these three-dimensional representations of home range, we noticed two patterns among pandas: centralized use and dispersed use (Figure 4.6). In the centralized use pattern, the panda spent the overwhelming majority of its time in only one or two locations, rarely visiting other places. In the dispersed pattern, the panda distributed its movements much more evenly across its home range.

Factors Influencing Patterns of Home Range Use

GENDER

Males generally tended to adopt a dispersed pattern of use, whereas females tended to adopt a more concentrated use pattern. The female Jiaojiao (Figure 4.7) provided a good model of the concentrated use pattern, whereas

92   Chapter 4

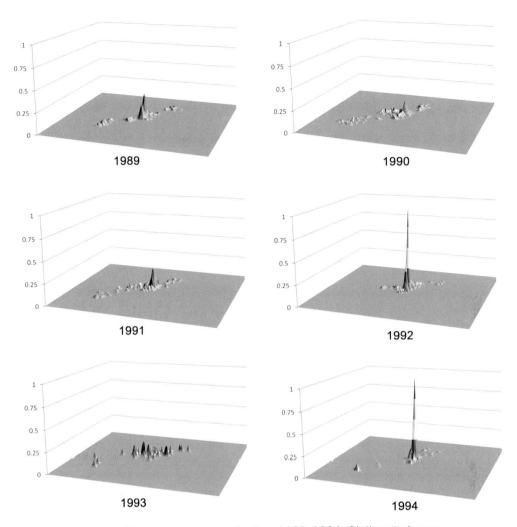

FIGURE 4.7. Patterns of home range use by Jiaojiao, 1989–1994, Qinling study area.

Huzi (Figure 4.8) and Dahuo (Figure 4.9) represented the more dispersed patterns typically seen among males. These patterns seem logical. After reaching adulthood, females spend almost half the year involved with reproduction, caring for a cub born that year or an older cub. During this time, it would be maladaptive for her to range widely; rather, she is better served by focusing on habitats that provide optimal and stable conditions. Obviously, this is not the case for males, for whom the most important resource is females. After taking care of their most elemental subsistence needs for food, their tendency is to attempt to patrol, defend, and determine the status of females within their home range, with the objective of enhancing their probability of successful reproduction.

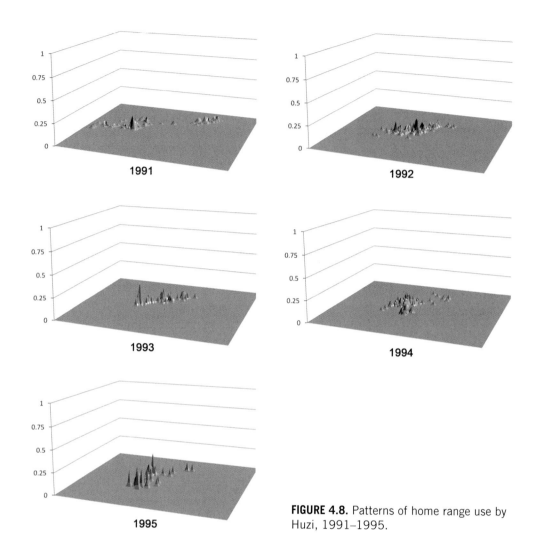

FIGURE 4.8. Patterns of home range use by Huzi, 1991–1995.

AGE

The illustrations of the changes in Huzi's range use pattern over time (Figure 4.8) make clear that he transitioned from centralized use to dispersed use as he grew from a cub to a subadult. In contrast, as the female Xiwang moved from the subadult to the adult age class, she gradually changed her use pattern from dispersed to centralized. Finally, Dahuo, as he transitioned from a prime-aged male to an elderly one changed his use pattern from dispersed back to centralized (Figure 4.9). A cub's range use was obviously limited by its mother's influence as well as its own mobility, leading to a centralized pattern; conversely, subadults establishing their own home range tended to range widely, and old individuals may have adopted a centralized use pattern as a result of their own limitations.

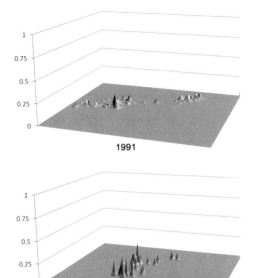
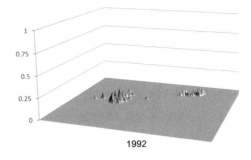

1991

1992

1993

FIGURE 4.9. Patterns of home range use by Dahuo, 1991–1993.

CHANGES IN HOME RANGE USE WITH TIME

Another way to examine changes of home range use is to look at the portion of the home range used each month. Taking Jiaojiao as a case study, we present monthly MCP home ranges within a 12-month calendar year (Figure 4.10), reflecting her response to bamboo. We present estimates of Jiaojiao's home range size each year for 1989–1997 (Table 4.6 and Figure 4.11a) and contrast her usage with changing patterns of home range use for the male Dabai (Figure 4.11b).

On average, Jiaojiao's home range was largest during April–June and smallest during August–October. This pattern was directly related to parturition, cub rearing, food resources, and migratory movements. During April–June, Jiaojiao was required to search widely for that year's *Bashania* bamboo shoots, resulting in a rather large area of activity. Parturition was in mid- to late August, and for the 3 months after that she was restricted to the area in and around the den site, resulting in a very small home range size.

In contrast, Dabai's home range size in March was larger than during the bamboo shoot season of April–June. Adult males expend energy seeking mates in March, the estrous season for females, so they expand their home range size. Additionally, Dabai's movements peaked in July, when he was trudging from his low-elevation winter range to higher-elevation summer range, which naturally resulted in a larger home range size during that month. This characteristic was not evident in Jiaojiao's pattern of home range use because her summer and winter ranges were quite close to one

Home Ranges of Giant Pandas   95

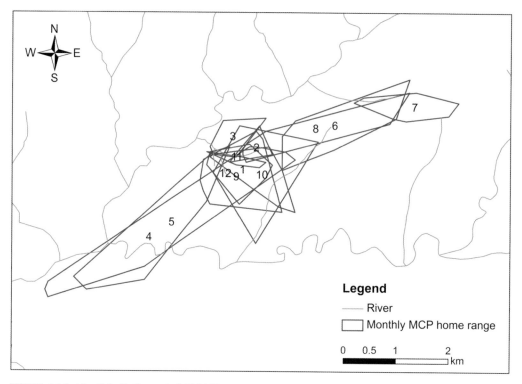

FIGURE 4.10. Monthly (1 through 12) MCP home ranges for the adult female Jiaojiao.

another. These comparisons of gender-specific monthly home range sizes reinforce the point that females' movement strategies center on cub rearing, whereas males prioritize increasing their chance of successful mating.

Home Ranges and Territoriality

Home range is defined as the area used in the course of normal activities, including foraging, mating, and raising young. This concept emphasizes normal activity as opposed to migratory movements (White and Garrott, 1990). Pandas spend almost all their time in what could reasonably be considered normal activities; seasonal migrations occur over very short time periods, and pandas continue foraging and caring for cubs during these movements. Thus, we consider all areas used by pandas as within their home ranges.

Panda home ranges appear to differ among mountain ranges: Wolong pandas had home range sizes of only 3.9–6.2 km^2 (Schaller et al., 1985), all smaller than the mean of 10.6 km^2 that we documented in the Qinling. [Editor note: Ellipse and MCP ranges are not directly comparable. With an increasing number of relocations, ellipse ranges tend to get smaller (because, as a probabilistic model, the estimated area becomes more certain), whereas MCP ranges get larger (because larger samples increase the chance of obtaining a location outside the main area of use). In this study, ellipse

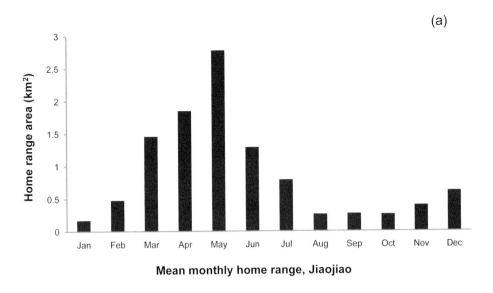

(a)

Mean monthly home range, Jiaojiao

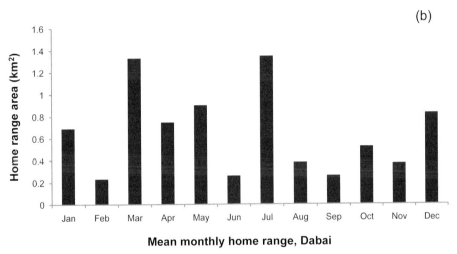

(b)

Mean monthly home range, Dabai

FIGURE 4.11. Mean monthly size (in km^2) of (a) Jiaojiao's and (b) Dabai's home ranges.

ranges were larger than MCP ranges, annually and seasonally. The comparison with Wolong should have used MCP ranges, as used by Schaller et al. (1985), to ensure that any differences detected were not simply an artifact of methodology. Average Qinling annual range size was 10.6 km^2 using the ellipse but only 8.7 km^2 based on MCP; the MCP average in Qinling was still 80% larger than at Wolong (60% of Qinling MCP ranges were larger than the largest MCP range in Wolong). Home ranges among the Wolong sample did not vary much, even though both sexes were included, but the sample was small (two males and three females). The Qinling sample was large but highly variable, so the pooled average may be a bit misleading, as it included pandas of different sexes, ages, year of observation, and number of radiolocations (although one young dispersing female with an exceptionally large range [f22, Table 4.3] was omitted). A t test assuming unequal variances for the two samples was highly significant ($P = 0.002$), indicating that Qinling pandas likely did have larger ranges.]

Among our 22 radio-marked pandas, we documented unusual, long-distance movements in only two individuals. We considered these to be dispersal movements and did not include the areas used during these travels within the home ranges we calculated.

Territory is defined as an area defended in some way against intrusions by other animals. Methods of defense to deter intruders may include aggressive behavior but may also be in the form of warning calls or chemical signals (Alcock, 1993; Krebs and Davies, 1993). For pandas, some heavily used portions of home ranges seemed to elicit a stronger defensive response. Females exhibited defensive behavior frequently when cubs were young and unable to fully defend themselves. Because females frequently moved among denning sites when cubs were young, territorial areas within their home ranges were not fixed. As cubs developed and gained strength and were able to follow their mother over a larger areal extent and particularly as they gained the ability to climb trees for protection, females transitioned their defensive behavior from one based on aggression to one based on evasion. We were unable to quantify the area actively defended by females, but in December 1994, we observed Jiaojiao fighting with Ruixue about 150 m from her den site at the time. We estimate that she defended an area less than 1 km^2.

Home Range Overlap

Consider individuals A and B, their home range areas within some specified time period as M_a and M_b, and the area of overlap M_c. We define the home range overlap ratio between them as $M_c^2/(M_a M_b)$. If the home ranges of two individuals are similarly sized and the proportional overlap is relatively high, then this ratio will be high. However, this ratio is not a good index of overlap for ranges that are very different in size, including cases when one home range is mostly or entirely encompassed within the other.

Because we obtained the most location data on Jiaojiao, we provide examples of the use of this home range overlap ratio for her (Figure 4.12). It is unsurprising that Huzi, who was Jiaojiao's oldest offspring that we knew of, had a high degree of home range overlap with Jiaojiao. Her next highest overlap was with Daxiong (m07) and Dabai (m06; Figure 4.13). It is worth noting that these two were the males we considered to occupy the dominant social positions during the time period based on our frequent observations. Jiaojiao was more compatible with males than she was with other females. Among females, Jiaojiao had the highest overlap ratio with her daughter, Xiwang (f03), but even this value was lower than her overlap

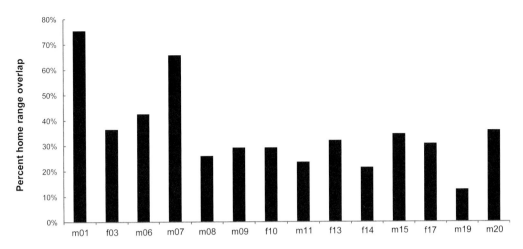

FIGURE 4.12. Home range overlap ratios between the female Jiaojiao and other pandas.

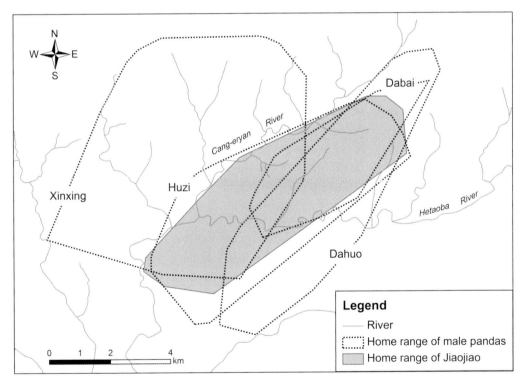

FIGURE 4.13. Spatial juxtaposition between the home range of Jiaojiao and male pandas Xinxing, Dabai, Dahuo, and Huzi.

with the dominant males. To test these conclusions, we selected Jiaojiao's home range during 1992–1993 and 1994–1995 and calculated home range ratios with other individuals. Results were similar, showing less overlap with other females (mean 21%, range 13%–28%) than with males (mean 37%, range 10%–58%).

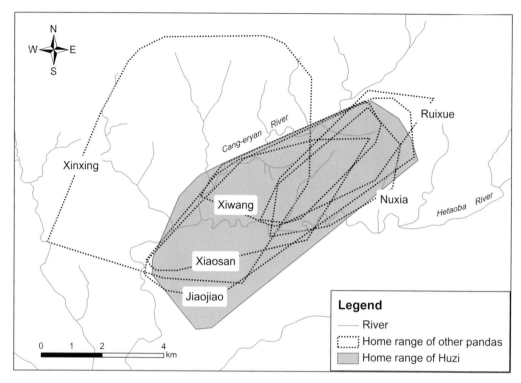

FIGURE 4.14. Spatial juxtaposition of male panda Huzi with other pandas.

We use the methods above to analyze the movements and overlap of Huzi (m02) in relation to other pandas. Huzi was most closely related to Jiaojiao (his mother), Xiaosan (his younger brother), and Xiwang (his younger sister; Figure 4.14). His home range overlap with other males exceeded that with females. This likely reflected the larger home range sizes of males. In addition to the obviously high degree of overlap between Huzi and his mother Jiaojiao and sister Xiwang (which we discuss elsewhere; Figure 4.14), in 1996, Huzi's home range overlapped with that of the female Nüxia more than with any other unrelated female. In March, Nüxia mated with Huzi and produced a cub in August.

Spacing of Individuals

SEPARATION AMONG FEMALE CORE RANGES

Most females concentrated their activities within the core areas of their home ranges. Although home ranges overlapped broadly, core areas were separated from one another (Figure 4.4). All pandas mark frequently during their daily travels. With the exception of the breeding season, during which marks function to advertise the individual's status within the hierarchy as well as their physiological condition, these marks aid in spatial separation.

100   Chapter 4

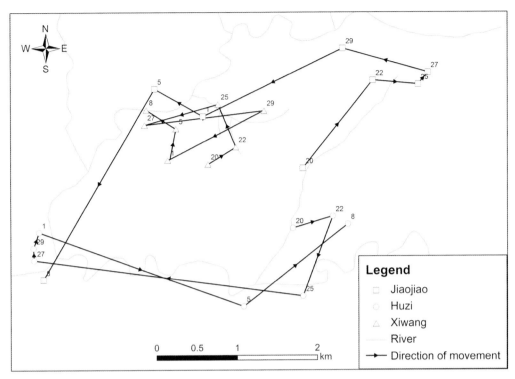

FIGURE 4.15. Movements of adult female Jiaojiao (squares), her male offspring Huzi (circles), and her female offspring Xiwang (triangles), March 20 to April 8, 1996. Numbers adjacent to each symbol represent day of the month.

By encountering these marks on trees, individuals can distinguish one another, determine each others' whereabouts, and thus maintain separation from each other. On the morning of April 8, 1995, while near the Cangeryan River at 1,450 m, we observed Jiaojiao studiously sniffing a mark left on the trunk of a pine tree as well as a scat, both left by Huzi the evening before. Within overlapping home ranges (such as these, involving a mother and offspring), separation was often achieved by staggering the times at which they used the same place. We illustrate the movements of Jiaojiao in relation to those of her male offspring Huzi and female offspring Xiwang during spring 1996 (Figure 4.15).

AGGRESSIVE BEHAVIOR

On occasion, when passive avoidance was unsuccessful and pandas did encounter one another, they were sometimes aggressive. We observed the following aggressive interactions:

- In spring 1990, in Liaojia Gorge, we were observing Jiaojiao with her son Huzi. The female Xiaof (f23) inadvertently entered her territory, which elicited an attack from Jiaojiao. Xiaof initially responded by

Home Ranges of Giant Pandas 101

climbing a tree, but Jiaojiao drove her from it, chasing and striking her on the face and forehead, drawing blood, and ultimately causing Xiaof to flee.

- In Shuidong Gorge on December 14, 1994, we observed Jiaojiao, who was traveling with her second son, Xiaosan, fighting with the female Ruixue, who finally left the area.
- Also in Liaojia Gorge, we observed Huzi chasing an unidentified panda on December 11, 1993.
- On February 6, 1995, we observed Huzi attacking an unidentified panda and ultimately forcing it to retreat from the mouth of Huangping Gorge.
- We observed Huzi fighting with an unidentified panda on August 13, 1996, in upper Shuidong Gorge.

Because these instances occurred outside the mating season, we interpret them to be related to territorial defense. In addition to their territorial function, these types of aggressive interactions were likely related to maintaining the social hierarchy.

Aggregation during Estrus

Although pandas normally tend to avoid each other, an important exception took place each spring during the mating season, when multiple males gathered together at fixed locations to contest for mating privileges with an estrous female. These places had some similar characteristics to leks, as described for other polygamous mammals and birds (Gibson and Bradbury, 1985), but we will refer to these as mating aggregations until more research can be conducted. Ovulation among adult female pandas did not occur completely synchronously, and thus, they traveled to these aggregations at differing times. Consequently, despite the overall 1:1 sex ratio within the Qinling panda population, there were almost always more males than females at these mating encounters, inevitably leading to fierce fights among them, which provided females an opportunity to select the optimal mate.

On April 10, 1994, we observed Jiaojiao at the mating aggregation. We noticed that there were are a number of males in the vicinity, but her son Huzi was not among them. We observed that Jiaojiao mated with Dahuo, who was at least 18 years old at the time (and who died the following spring), evidently preferring him over an unmarked male suitor who appeared to be younger and stronger.

A common mating aggregation site in our study area was located on a ridge between Shuidong Gorge and Chaijiawan at 1,700–1,900 m, encompassing 30 to 40 panda home ranges (Figure 4.16). We discovered other similar sites scattered along the southern slopes of the Qinling that functioned

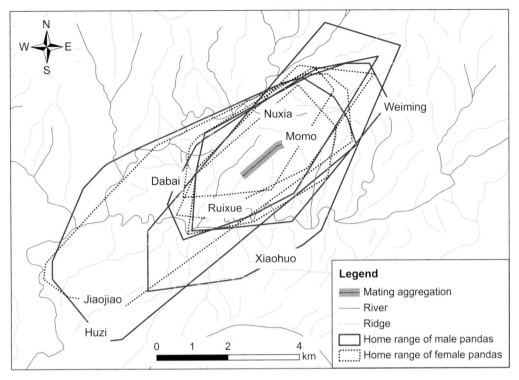

FIGURE 4.16. Home ranges of a number of pandas in the study area (males in solid polygons, females in dashed polygons), showing the location of the mating aggregations described in the text.

to provide opportunities for normally solitary pandas to aggregate during the breeding season and thus increase their reproductive success.

Social Relationships among Qinling Pandas

It is conventional wisdom that pandas are solitary, but our research has revealed that relationships among individual pandas are much closer and stronger than had earlier been imagined. From 1989 to 1998, Jiaojiao produced five cubs and thus formed five different mother-cub associations. We followed the first four of these family groups continuously over a period of 8 years, giving us a preliminary understanding of panda social structure and family relations. We use the case study provided by Jiaojiao and her offspring to describe the most elemental constituent of panda society: the process of formation, maintenance, and ultimate dissolution of the mother-cub family unit.

Jiaojiao's first cub, Huzi, was born in August 1989. Her subsequent estrus occurred when Huzi was 31 months of age, and her bond with Huzi began to dissolve during her next pregnancy. However, Huzi continued to live within his mother's home range until she gave birth to her second cub. Even after this time, telemetry data as well as direct observations indicated that the home ranges of Huzi and Jiaojiao continued to have substantial overlap.

Jiaojiao's second cub, daughter Xiwang, was born in August 1992. This bond began to split when Xiwang was 18 months old, at which time Jiaojiao entered a new reproductive cycle. However, Xiwang continued to spend a considerable amount of time within Jiaojiao's home range.

In August 1994, Jiaojiao gave birth to her third cub, a male we named Xiaosan. We noted that on October 14, 1994, Xiwang moved directly into the core area occupied by Jiaojiao and Xiaosan, at one point being no farther than 10 m from the then 2-month-old cub. This proximity, however, elicited no obvious defensive behavior on the part of Jiaojiao; in fact, she and Xiwang called to each other gently and continued to forage and rest close to one another for almost 5 hours. On July 16, 1996, we observed Xiwang and her younger brother Xiaosan resting in the same tree (one above the other), while Jiaojiao was 200 m distant preparing for her fourth parturition.

In August 1996, Jiaojiao gave birth for the fourth time, this time to a daughter. On September 9, 1996, we observed an interesting phenomenon: Xiwang was not only traveling with her younger brother Xiaosan (then 2 years old) but also positioned herself between Xiaosan and us, so as to prevent us from getting close to him. Even more surprising to us was that not more than 10 m away was another panda, quite impressive in size, foraging, and seemingly ignoring all of us. We guessed that this was Huzi, although we could not be sure because Huzi's collar had fallen off. However, this occurred within the area that Huzi frequently used, and it had also been our experience that no animals other than Huzi, Xiwang, and Xiaosan would tolerate our close presence. Jiaojiao and all four of her offspring spent time in close proximity to one another, with home ranges that overlapped and core areas that were quite close to one another.

In August 1997, when Xiwang was 5 years old, we observed that she moved to the west of her mother's home range, crossed a river, and on a ridge at an elevation of 1,800 m began to search for an appropriate denning location to give birth. This area was directly adjacent to, but not strictly part of, Jiaojiao's home range (the river seeming to form her home range boundary). Our radio telemetry data indicated that Xiwang had become crepuscular, and her percentage time active had fallen to 20%, suggesting she was pregnant.

Huzi evidently had achieved sexual maturity in March 1994 at the age of only 3; his testicles had enlarged greatly, and he had begun frequenting the mating aggregation and contesting for mating privileges. By March 1996, Huzi had developed into an adult male weighing 150 kg. At the mating aggregation, he followed Nüxia closely and drove other males away from her by roaring and clapping his jaws together. In August of 1996, Nüxia gave birth on the eastern slope of Shuidong Gorge, at a den site only 200 m from

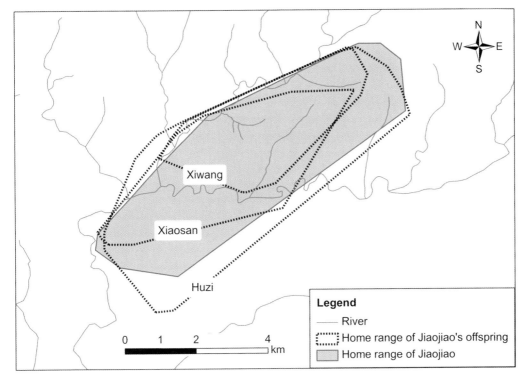

FIGURE 4.17. Home range polygons of Jiaojiao and three of her four offspring.

where Jiaojiao had given birth to Huzi on the Shuidong Gorge's western slope. From the time of his birth to this point, at which he was a dominant 7-year-old male, Huzi never really left his natal area. Further, his offspring would later be born within this natal area as well. Thus, in assessing the home range overlap between Huzi and other pandas, we obviously need to consider the familial relationships involved. Huzi's home range overlapped with his mother Jiaojiao (75%), his younger brother Xiaosan (48%), and his younger sister Xiwang (35%), whereas his overlap with unrelated individuals varied from 24% to 33%.

With this description of the process by which family groups form and dissolve, we tend to the viewpoint that, among pandas, male offspring are philopatric, remaining near their natal area. They inherit a portion of their mother's home range, establishing their own home ranges very close to their mother. In contrast, the young female pandas disperse when they are ready to do so, establishing their own home ranges in a new location.

References

Alcock, J. 1993. *Animal Behavior: An Evolutionary Approach*. 5th ed. Sunderland, MA: Sinauer.

Gibson, R. M., and J. W. Bradbury. 1985. "Sexual Selection in Lekking Sage Grouse: Phenotypic Correlates of Male Mating Success." *Behavioral Ecology and Sociobiology* 18:117–123.

Krebs, J. R., and N. B. Davies. 1993. *An Introduction to Behavioural Ecology.* 3rd ed. London: Blackwell Scientific.

Pan, W. S., Z. S. Gao, Z. Lü, Z. K. Xia, M. D. Zhang, L. L. Ma, G. L. Meng, X. Y. She, X. Z. Liu, H. T. Cui, and F. X. Chen. 1988. 秦岭大熊猫的自然庇护所 [The giant panda's natural refuge in the Qinling Mountains]. Beijing: Peking University Press.

Schaller, G. B., J. C. Hu, W. S. Pan, and J. Zhu. 1985. *The Giant Pandas of Wolong.* Chicago: Chicago University Press.

Wang, B. Q. 1978. 氯胺酮—安定麻醉大熊猫20 例 [Ketamine—20 examples of anesthetizing giant pandas]. 中国动物园年刊 [*China Zoological Yearbook*], 1:182–186.

White, G. C., and R. A. Garrott. 1990. *Analysis of Wildlife Radio-Tracking Data.* San Diego: Academic Press.

Zheng, J. Z. 1980. 氯胺酮麻醉大熊猫试验20例 [20 experiments with using ketamine on giant pandas]. *Chinese Journal of Zoology* 1:30–32.

Chapter 5

Panda Movements and Activity

Summary: Pandas living on the southern slope of the Qinling exhibited seasonal migrations, ascending to higher elevations in the late spring or early summer and descending in late summer or early autumn. Summer ranges were 2,000–2,900 m in elevation in Fargesia bamboo forest. We observed a great deal of variation in the amount of time individual pandas resided in their summer range (28 to 116 days) unrelated to gender. The rest of the year pandas lived primarily in Bashania bamboo forests at elevations of 1,350–2,000 m. These migratory movements tracked the palatability of bamboo as well as temperature and snowfall. Young female pandas also made extensive travels that appeared to be dispersal movements; subadult males did not disperse and remained close to home. Daily distances traveled by pandas averaged just over 400 m (more for males, less for females), which was similar to pandas in Wolong. Unlike in Wolong, however, where pandas were active in early morning and late afternoon, pandas in the Qinling displayed a unimodal pattern of diurnal activity, which peaked during midafternoon, 1300 to 1700 hours. Overall, Qinling pandas were active 49% of the time, but this varied seasonally. Pandas were most active in spring, when they were foraging on fresh Bashania bamboo shoots, and least active during summer.

Introduction

The movement and activities of animals reflect their needs to find sufficient forage and mates. A variety of movement and activity patterns result from differences related to age, sex, and reproductive status of individuals.

In 1985, we began recognizing that the elevational distribution of pandas on the southern slopes of the Qinling varied seasonally: in winter, we could easily find evidence of pandas in bamboo groves at low elevations, whereas in summer we had to travel up to bamboo at much higher elevations to find them. When we started radio-collaring, we found that they indeed made distinct and regular seasonal movements, which fit the definition of migrations. In this chapter we investigate these migrations, look for patterns and causes, and also examine daily movements and activity rhythms of pandas in the Qinling.

Migrations between Seasonal Ranges

From January 1989 to August 1991, we radio tracked 12 pandas to study their movements and plotted their locations on 1:50,000 scale topographic maps. The pattern of seasonal elevational change was consistent among these years (Table 5.1, Figure 5.1). The same pattern continued through the duration of our study (until 1997) and for nearly all of our 22 radio-monitored pandas. These pandas showed distinct use of summer (high-elevation) and winter (low-elevation) habitats, defined by the 2,000-m contour (Figure 5.2). Upward summertime movements began as early as May 21 and as late as July 1, with a mean of June 12. Descending movements began as early as July 8 and as late as October 1, with a mean of August 24 (Table 5.2, Figure 5.3).

During the 9.5 months from late August until mid-June of the next year, our radio-marked pandas lived primarily in *Bashania* forests at elevations of 1,350–2,000 m. During the 2.5 months from mid-June to the latter part of August, they lived above 2,000 m in *Fargesia* forests. Thus, we consider *Fargesia* forests the summer range and *Bashania* forest the fall, winter, and spring range for pandas. Pandas typically required 1–3 days to make the journey upward from winter to summer range and 1–2 days to make the reverse trek (although some migrations lasted up to 20 days).

We noted the following exceptions to these general patterns, which indicate that some pandas remained at higher elevations year-round:

- April 17, 1985: While surveying Guangtou Mountain, we discovered a large accumulation of fresh panda scats lying on snow at an elevation of 2,850 m.
- April 17, 1986: We observed a fight between pandas on the south side of the main Xinglongling ridge at an elevation of 2,500 m.
- February 26, 1989: We encountered the footprint of a panda in 50-cm-deep snow in a *Fargesia* bamboo grove at Wowodian at an elevation of 2,290 m. Later that day, at 2,300 m, we found additional tracks and feces with *Fargesia* bamboo stalks of a panda that evidently was descending from the mountain slope to lower elevation.
- March 15, 1990: We found evidence of a panda that we judged to be about a month old in a snow-covered *Fargesia* bamboo grove in Daping at an elevation of 2,400 m. We received information that panda calls had been heard at about 2,400 m on both March 9 and March 29.
- April 4, 1991: We observed three pandas near Huorenping at 2,900 m in an area that was still covered with thick snow.

Table 5.1.

ELEVATIONS (M) OF RADIO-COLLARED PANDAS IN THE QINLING STUDY AREA, JANUARY 1989 TO AUGUST 1991, SHOWING MAXIMUM, MINIMUM, AND MEAN FOR EACH MONTH, AS WELL AS THE NUMBER OF PANDAS RADIO TRACKED DURING THAT MONTH (N). ENTRIES IN BOLD INDICATE *BASHANIA-FARGESIA* MIXED BAMBOO FORESTS (GENERALLY AT 2,000–2,200 M), ITALIC ENTRIES INDICATE *FARGESIA* BAMBOO GROVES (GENERALLY >2,800 M), AND ENTRIES IN PLAIN TEXT INDICATE *BASHANIA*-TIMBER FORESTS, GENERALLY <2,000 M.

Month	1989				1990				1991			
		Elevation				Elevation				Elevation		
	n	Max	Min	Mean	n	Max	Min	Mean	n	Max	Min	Mean
January	1	1,700	1,500	1,600	6	1,900	1,400	1,625	4	1,900	1,500	1,640
February	2	1,800	1,500	1,620	5	1,950	1,350	1,645	7	2,000	1,400	1,657
March	3	1,800	1,500	1,700	6	1,950	1,300	1,725	6	1,900	1,300	1,751
April	3	1,900	1,500	1,680	4	**2,200**	1,300	1,695	6	2,000	1,300	1,687
May	3	1,900	1,500	1,670	4	2,000	1,300	1,656	5	2,300	1,400	1,750
June	3	2,800	1,500	**2,125**	4	2,700	1,400	**2,175**	6	2,800	1,500	2,267
July	3	*2,750*	*2,350*	*2,525*	4	*2,900*	*2,300*	*2,600*	6	2,900	1,900	2,454
August	3	*2,750*	1,950	*2,275*	4	*2,900*	1,800	*2,263*	6	*2,950*	1,850	*2,350*
September	4	2,600	1,800	**2,130**	4	2,600	1,550	1,950				
October	3	2,000	1,400	1,800	4	2,300	1,550	1,831				
November	3	**2,200**	1,400	1,800	3	1,950	1,500	1,775				
December	5	1,900	1,400	1,750	3	1,950	1,500	1,767				

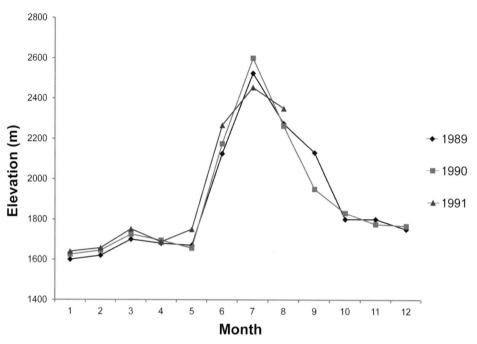

FIGURE 5.1. Mean elevations of radio-tracked pandas, January 1989 through August 1991, Qinling study area, Shaanxi.

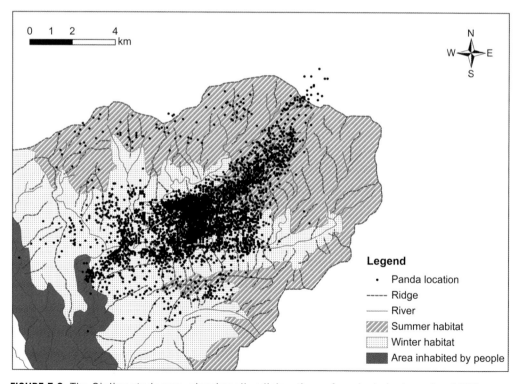

FIGURE 5.2. The Qinling study area, showing all radiolocations of marked giant pandas, 1989–1997, as well as our delineated summer and winter habitats and the area inhabited by people.

Table 5.2.

YEARLY DATES OF UPWARD AND DOWNWARD MIGRATION OF RADIO-COLLARED PANDAS IN THE QINLING STUDY AREA, 1989–1994, INCLUDING NUMBER OF DAYS SPENT IN SUMMER AND WINTER RANGES.

Panda	Year	Date of upward migration	Date of downward migration	Days in summer range
Jiaojiao	1989	June 22	August 1	39
	1990	June 15	August 4	49
	1991	June 22	July 27	34
	1992	June 21	July 20	28
	1993	June 26	August 2	36
	1994	June 14	July 21	36
	1995	June 16	August 16	60
	1996		August 11	
Huzi	1991	June 22	July 27	34
	1992	June 16	August 16	60
	1993	May 26	August 1	66
	1994	June 16	August 11	55
Xiwang	1993	June 26	August 2	36
	1994	June 25	August 01	35
Dabai	1992	July 1	September 1	61
	1993	June 1	September 1	91
	1994	June 11	September 1	81
Daxiong	1990	June 20	August 21	61
Dahuo	1991	June 21	September 19	89
	1992	June 21	September 11	81
	1993	June 16	September 16	91
Xiaohuo	1992	June 16	August 26	70
	1993	July 1	September 1	61
Nüxia	1993	June 10	September 1	81
	1994	June 11	August 26	75
Jiaoshou	1993	May 21	September 1	102
	1994	May 26	September 1	97
Ruixue	1992	June 11	September 1	80
	1993	June 10	July 8	28
	1994	June 6	August 26	80
Baoma	1990	June 6	October 1	116
	1991	June 7	October 1	115
Huayang	1989	June 19	September 1	73
	1990	June 21	September 1	71
Yanghe	1989	June 12	September 24	103
Momo	1993	June 1	October 1	121[a]
	1994	June 6	August 15	69
Xinxing	1991	June 1	September 20	110
	1992	June 5	September 1	86
Weiming	1993	June 1	September 23	113
	1994	June 1	September 23	113
129	1992	June 1		126[b]
Shuilan	1991	June 7		88

[a] Momo gave birth in the summer range in 1993.

[b] Panda 129 undertook a long-distance migration in 1992.

Panda Movements and Activity   111

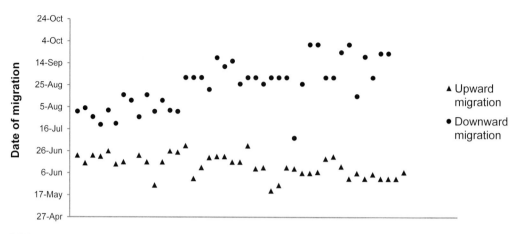

FIGURE 5.3. Dates of upward and downward migration of individual pandas in the Qinling Study area.

- March 11, 1992: We observed a fresh panda track in the snow in a *Fargesia* bamboo grove at 2,850 m.
- 1992–1993: The radio-marked female Keke was only occasionally found in the *Bashania* forests at 1,750 m during these years. Most of her time was spent on Guangtou Mountain at about 2,800 m. Her time spent at lower elevations was restricted to spring; she lived at higher elevations during all other seasons.
- April 23–26, 1994: We observed unmarked pandas contesting for mating privileges on the south side of the main Xinglongling ridge at 2,800 m.
- October 17, 1995: We observed a group of uncollared pandas fighting in a birch grove at 2,300 m, a time when most of our collared pandas had already moved to lower elevations.

The dates of vertical migrations varied among individual pandas (Table 5.2). However, within individuals, the dates of upward movement in spring tended to be rather fixed year to year. For example, during the years 1989–1995, Jiaojiao migrated upward between June 14 and June 26, a span of only 12 days. The upward migrations of Ruixue, Nüxia, and Dahuo also displayed little variation (≤6 days) by year. Similarly, most migrations to lower elevations varied little within individual; for example, the initiation of downward migrations for Xiwang, Dahuo, Huayang, and Baoma varied by less than 5 days annually. This suggests that differences in migration dates were largely a function of individual behavior.

Maternal status (e.g., pregnant or caring for cubs) was another factor affecting the timing of descent. For example, Jiaojiao in 1989, 1992, 1994, and 1996, Ruixue in 1993, and Momo in 1993 all displayed early descents in years in which they produced cubs.

112   Chapter 5

Of our 22 radio-marked pandas, 21 displayed a clear migratory pattern, occupying lower elevations in winter and higher elevations in summer. By contrast, in Wolong, Schaller et al. (1985) documented that pandas spent most of their time in *Sinarundinaria fangiana* bamboo at 2,600–3,200 m, with only a few descending to 1,600–2,300 m to feed on newly emerged bamboo shoots in mixed *Fargesia robusta* groves for a month or so during spring, after which they returned to the *S. fangiana* groves. In the Min Mountains of Sichuan, Schaller (1994) documented a male radio-marked panda with seasonal vertical movements similar to those in the Qinling, except that there, most pandas lived in *Fargesia scabrida* bamboo groves above 2,000 m early in the bamboo growing season.

Distance of Seasonal Migrations

Although we consider the boundary between winter and summer panda habitat to be roughly the 2,000-m elevation contour, distances between these seasonal ranges varied among individual pandas. For example, distances were relatively large for males Dahuo (3,265–6,498 m) and Huayang (1,355–5,390 m) and much smaller for Jiaojiao (0–3,357) and Dabai (0–2,063 m; Figure 5.4).

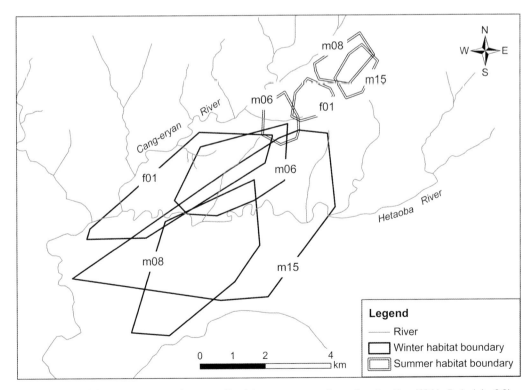

FIGURE 5.4. Minimum convex polygon method home ranges of pandas Jiaojiao (f01), Dabai (m06), Dahuo (m08), and Huayang (m15), showing spatial juxtaposition of summer and winter home ranges, Qinling study area.

Two examples illustrate these seasonal vertical movements. Beginning on June 21, 1991, we followed adult male Dahuo as he moved from a *Bashania* bamboo grove at about 1,450 m, eventually reaching an area of *Fargesia* bamboo at 2,950 m in mid-July. This was the largest single elevational movement by any of our radio-marked pandas; the total distance between his summer and winter ranges was 6.5 km. In late June 1994, female Xiwang, who was a bit less than 2 years old at the time, was in a *Bashania* bamboo grove at 1,850 m. However, by early July she had moved upward to a *Fargesia* grove at about 2,000 m. She returned to the *Bashania* grove on August 1, with the distance between her winter and summer ranges being less than 2 km.

During the early 1980s, there was a mass flowering of *Fargesia* in the southern slopes of the Qinling below 2,350 m. This meant that pandas moving up to summer range had no choice but to pass through an extensive band of habitat that was devoid of bamboo. This forced pandas to move longer distances than normal. By 1987 these *Fargesia* bamboo groves had begun to recover. By 1989 bamboo stems had achieved heights of 150–180 cm, and by 1990 the entire bamboo forest had recovered such that the geographic distribution of *Fargesia* around the 2,000-m elevation band had become adjacent to the *Bashania* bamboo groves. Thus, pandas migrating during summer no longer needed to travel so far.

Duration of Time Spent in Summer Range

Because the initiation of summer migration varied among individuals, the amount of time spent in summer range also varied. Excluding individuals who chose high-elevation areas for parturition or who engaged in unusual, extended movements, time spent in summer range varied from 28 to 116 days (Table 5.2).

We detected no difference between sexes in time spent in summer range. Males spent an average of 79 days (standard error [SE] = 21 days, $n = 21$) in their summer range; females averaged 68 days (SE = 32 days, $n = 21$). These means were not significantly different because of the substantial variation within each sex. For example, we observed Jiaojiao in her summer range during 7 years. She spent a mean of 40 days (SE = 10 days) there, which differed from the 14 summer range durations observed for other females. The time spent in the summer range by Jiaojiao's daughter Xiwang was similar to that of Jiaojiao.

Adult females reduced their time spent in summer range during years in which they produced cubs. The mean number of days spent in summer range by Jiaojiao and Ruixue during years in which they had cubs was less ($\bar{x} = 32.8$ days, SE = 4.9 days, $n = 4$) than in other years ($\bar{x} = 56.5$ days, SE = 18.7 days, $n = 6$; $P = 0.0285$). This reflects the needs of parturient pandas to return to the winter range earlier in order to look for suitable denning locations. For example, on June 21, 1992, Jiaojiao led her cub Huzi from winter range up to summer range, but the two stayed there only 29 days, returning on July 20 to a denning area at 1,950 m (Jiaojiao gave birth to Xiwang there on August 15). On June 16, 1995, Jiaojiao led her second male cub Xiaosan up to a *Fargesia* bamboo forest at 2,300 m, where they roamed for 2 months, descending to a *Bashania* bamboo forest at less than 2,000 m on August 16. On June 10, 1993, Ruixue ascended to her summer range but returned to *Bashania* bamboo at 1,800–1,900 m only 28 days later (on July 8) and gave birth to her cub Shigen on August 21. In contrast, in 1994, Ruixue moved up to her summer range on June 6 and did not return until August 26, some 80 days later.

Factors Influencing Seasonal Migrations

TEMPERATURE

In March 1989, we began recording temperature at our research station at 1,500 m, twice daily (0800 and 2000 hours) until August 1991, a period of 30 months. Official meteorological standards call for more frequent monitoring, but logistical constraints limited us to twice-daily measurements. We present the mean elevations of radio-marked pandas as well as mean monthly temperatures during 1989–1991 in Table 5.3.

During these 3 years (1989–1991), temperatures in the high-elevation areas favored by pandas remained very stable from May to September. Means were 11.1°C (SE =1.2°C) in 1989, 11.8°C (SE = 1.2°C) in 1990, and 10.1°C (SE = 1.2°C) in 1991 (this last mean is through August only). We consider the

Table 5.3.

MEAN ELEVATIONS (M) OF RADIO-MARKED PANDAS AND MEAN TEMPERATURE (°C) AT THE QINLING PANDA RESEARCH STATION, JANUARY 1989 TO AUGUST 1991. TEMPERATURES WERE EXTRAPOLATED FROM LOWER-ELEVATION STATIONS ASSUMING A DROP OF 0.6°C PER 100-M INCREASE IN ELEVATION.

Month	1989		1990		1991	
	Elevation	Temperature	Elevation	Temperature	Elevation	Temperature
January	1,600	−4.5	1,625	−3.3	1,640	−3.4
February	1,620		1,645	−3.0	1,657	−1.1
March	1,700	0.9	1,725	2.1	1,751	2.0
April	1,680	6.7	1,695	6.1	1,687	5.8
May	1,670	10.6	1,656	10.7	1,750	8.4
June	2,125	10.5	2,175	11.5	2,267	10.3
July	2,552	12.0	2,600	12.3	2,454	11.0
August	2,275	12.7	2,263	13.7	2,350	10.6
September	2,130	9.8	1,950	10.9		
October	1,800	6.4	1,831	6.0		
November	1,800	0.5	1,775	3.1		
December	1,750	−1.4	1,767	−3.1		

temperatures in May through September in these high-elevation areas to be optimal for pandas, whereas from October to the following April temperatures are colder than what pandas prefer.

In early October each year, snow begins to fall in *Fargesia* bamboo groves above 2,000 m, and by late December streams and gullies are dry. Mean temperatures in these bamboo groves at 2,400 m during the winters of 1989–1991 were 2.6°C in October, −1.9°C in November, −6.1°C in December, −8.4°C in January, −6.6°C in February, −2.4°C in March, and 1.9°C in April. Colder temperatures may be especially significant for pregnant females, who give birth in August or September and who seek warmth for their very altricial cubs. Six natal dens in *Bashania* bamboo groves at 1,700–1,950 m that we investigated during 1986–1991 had interior temperatures about 2°C warmer than the ambient air temperatures.

During the coldest month of January, pandas were active as low as about 1,350 m in the lowest portion of *Bashania* bamboo groves. Below this elevation, essentially all land was appropriated for agricultural use or otherwise degraded. These human-caused constraints have given pandas no choice but to occupy areas that are cooler than optimal. That said, our observations suggested that pandas made use of microclimates created by the complex topography to find habitats in which temperatures were somewhat warmer than the surrounding area.

Examples of a few pandas that continued to occupy high-elevation *Fargesia* bamboo grove even in winter suggested that they could tolerate colder temperatures, but downward migrations certainly provided pandas with a more hospitable environment. Pandas may move to higher elevations as summer approaches to find cooler climes even though the temperatures at lower elevations are still well within their tolerance (e.g., mean July daily high temperatures of 21.6°C at 1,500 m and about 30°C even in direct sun at noon). The typical temperature of about 11°C from May to September in summer range enables pandas to avoid temperature extremes and thus conserve energy.

FORAGE

The palatability of bamboo also followed changes in temperature. In early September 1988, we noted the onset of frost among *Fargesia* groves at 2,200 m. Leaves of *Fargesia* bamboo had turned a grayish yellow. That year, the bamboo shoots had already grown to over 2 m in height but still had not produced branches and leaves. In early October, the north slopes at 2,400 m began to get covered by snow, leaves of *Fargesia* bamboo began to curl up, terminal leaves had turned purple, and those in midsections had turned yellow. By mid- to late October, a series of snowfalls had occurred, the temperature at 2,600 m had hovered around zero, creeks had begun to freeze, the ground had become hard, and *Fargesia* bamboo leaves began to dry and curl. Daily means during the coldest times at 2,500 m fell to −15°C, streams were completely frozen, snow on southerly slopes had accumulated to a depth of about 40 cm (and to >60 cm on northerly slopes), and bamboo leaves were curled tightly, like withered grass.

In April 1989, following winter's passage, leaves of *Fargesia* had gradually unfurled, but about 50% were withered, and the other 50% were partly dry. Most leaves had littered the ground by late May, after which new leaves began growing from the terminal buds and joints. Although *Fargesia* bamboo stalks are protected by a thick waxy layer, they tend to become desiccated in April.

Any panda remaining in *Fargesia* habitat during winter would be faced with frozen streams, making it difficult to meet their requirement for water, as well as unpalatable, dried forage, and their mobility would be limited by accumulated snow. Through chemical analysis, we found that Qinling pandas subsisting on *Fargesia* obtained only about 4,750 kcal/day (slightly more than their subsistence requirement of 3,130 kcal/day; see chapter 11). These energetic constraints would strain pandas living full-time in these environments.

In *Bashania* forests below 2,000 m winter does not arrive until after mid-November. In mid-November 1989, mean temperature was 1.2°C at 1,500 m, and the first snow of the season fell on November 13. In 1990, the mean temperature in late November fell from 5.6°C to 1.7°C, and the first snow fell on November 26. At this time, deciduous leaves had begun falling but remained green. Even at the coldest time of year when the snowline had descended to 1,350 m, water seeped through the cracks of streams' frozen surfaces, and sunshine on clear days was capable of melting the accumulated snow. Leaves of *Bashania*, although not as moist and lustrous as during summer, remained fleshy and uncurled. As a result, *Bashania* forests provided a much more suitable habitat for pandas in winter than *Fargesia* forests. Thus, *Bashania* was the primary food item for pandas from mid-September through mid-April, providing them with >6,000 kcal/day.

Signs of spring appeared gradually, beginning at the lowest elevations. February 8, 1991, saw that season's first spring rain shower, and we noted new grass putting forth shoots at 1,300 m on February 23, as well as increased activity of small fish in mountain streams. We observed the first bamboo shoots of the year poking up from the ground at 1,400 m on March 25, at which time the ground had largely thawed and ice and snow had mostly melted. *Bashania* bamboo, which had turned dark gray following the long winter, resumed its fresh green appearance after a few spring rains. From March through mid-April, as the temperature warmed, pandas began to move upward from the river bottoms to the upper tributaries on the mountain ridges (Figure 5.1). With the continued rise in temperature and the sprouting of new leaves on trees and shrubs in mid-April, the mountain slopes at 1,300–1,900 m became filled with blooming flowers, and spring was very much in the air. At this time, *Bashania* bamboo shoots began to grow thicker and taller. In 1990, we recorded the growth of bamboo shoots: on April 16 bamboo shoots by the roadside at 1,400–1,500 m had grown to 10–20 cm in height; on April 22 bamboo shoots on a mountain slope at 1,400 m had grown to this same height, and by April 25 they had reached 20–39 cm.

Pandas began foraging on *Bashania* shoots in late April and followed this food source as it gradually moved upward in elevation. For example, Jiaojiao and Daxiong displayed the following patterns in 1990:

- April 25: Daxiong was at 1,390 m, eating bamboo shoots 20–39 cm high.
- April 27: Jiaojiao was at 1,450 m, eating bamboo shoots 18–40 cm high.
- May 12: Jiaojiao was at 1,500 m, eating bamboo shoots 27–44 cm high.
- May 14: Daxiong was at 1,520 m, eating bamboo shoots 32–48 cm high.
- May 24: Jiaojiao was at 1,800 m, eating bamboo shoots 28–64 cm high.
- May 28 to June 2: Jiaojiao was at 1,900 m, eating bamboo shoots 36–80 cm high.
- June 8: Daxiong was at 1,600 m, eating bamboo shoots 42–120 cm high.

From early June, *Bashania* bamboo growing below 1,900 m had generally grown to >1 m in height and had become increasingly woody and hard. At the same time, *Fargesia* bamboo had just begun to bud.

- May 29: *Fargesia* shoots at 2,200 m were 10–25 cm.
- May 31: *Fargesia* shoots at 2,500 m had just barely emerged.
- June 7: *Fargesia* shoots at 2,200 m were 30–50 cm high.
- June 20: The previous year's *Fargesia* shoots had just begun to sprout new leaves.

In June, when pandas ascended toward *Fargesia* groves, the previous years' growth had just begun thawing and had become palatable. Our observations indicated that pandas reaching these groves primarily ate previous years' growth of bamboo from mid-June through July. After August, these *Fargesia* bamboo plants grew new branches and leaves, and pandas gradually turned to these new shoots as their main forage.

Evidently, pandas time their movements to coincide with the phenology of these two bamboo species. Nutritional analyses showed that shoots at this stage of development are the most digestible parts of the bamboo plant (digestibility for pandas of about 40%; Schaller et al., 1985) and that pandas in the Qinling can obtain >5,100 kcal/day from them (Pan et al., 1988). Our inference is that the palatability of these bamboos is one of the factors underlying the upward migration among pandas.

Daily Movements

We used two methods to quantify how pandas moved about on a daily basis. During June 1987, during our radio tracking and monitoring of the male Shanshan, we recorded all of his locations, allowing us to directly map

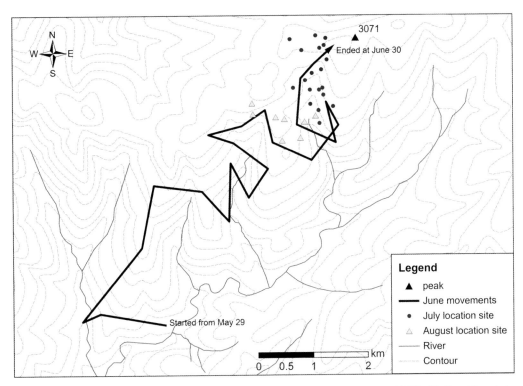

N W E S

3071
Ended at June 30

Started from May 29

Legend
▲ peak
━━ June movements
• July location site
△ August location site
── River
---- Contour

km
0 0.5 1 2

FIGURE 5.5. Movements of male panda Shanshan from May 29 to June 30, 1987, Qinling Mountain study area.

his continuous route of travel and to calculate distances traveled and elevations of each location (Figure 5.5). We also collected a large volume of telemetry data during the course of the study and built a simple program to calculate the distances and times between every radiolocation of each radio-marked animal. This produced a total of 75,551 straight-line distances between locations of individual pandas at time intervals of 1 to 30 days.

We use four individual pandas to illustrate variation in daily movements:

- March 1989: The old female Yanghe failed to come into estrus; distances between her locations were quite short (Figure 5.6A).
- April 1989: The old male Huayang spent the entire month traveling along either side of the Hetaoba River feeding on *Bashania* bamboo (Figure 5.6B).
- June 1989: Adult female Jiaojiao spent the early part of the month in Shuidong Gorge eating new *Bashania* shoots, traveling only short distances. In mid-June she quickly moved up in elevation and began traveling more widely. Within a few days, however, she entered a grove of *Fargesia* to seek the previous year's bamboo shoots and returned to her pattern of short daily movements (Figure 5.6C).

Panda Movements and Activity 119

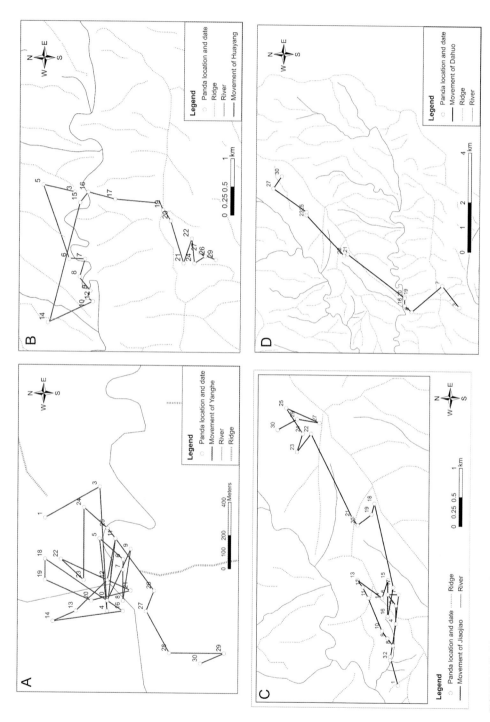

FIGURE 5.6. Movements of (A) Yanghe in March 1989, (B) Huayang in April 1989, (C) Jiaojiao in June 1989, and (D) Dahuo in June 1991. Numbers near each location (circles) indicate the date.

- June 1991: Adult male Dahuo moved from an elevation of 1,450 to 2,900 m in a period of 1 week (June 20–27) and climbed from 2,600 to 3,000 m in a single day (Figure 5.6D).

We selected six individual pandas to conduct complete analyses, Jiaojiao, Huzi, Dabai, Ruixue, Baoma, and Huayang, because sample sizes for each were relatively large and they represented both males (Huzi, Dabai, Huayang) and females (Jiaojiao, Ruixue, Baoma) and included both prime-aged and older (Huayang) individuals. We examined mean and maximum straight-line distances and distributions of straight-line distances. We binned straight-line distances between locations into 50-m increments, tallying the number in each bin for varying time intervals (Figure 5.7).

Males tended to move farther than females during the same time duration. This is evident when movement data by time step are combined by sex (Table 5.4). The mean distance traveled by males within a single day (506 m) was greater than for females (333 m); averaged across both sexes, daily movement was 411 m.

Dispersal

An important question is whether subadult pandas disperse from their natal home ranges prior to establishing their own home ranges and to mating, as is often seen in other mammals (Bertram, 1976). A subadult female, Shuilan, provided us a good opportunity to study this question. We fitted her with a radio collar on February 1, 1991, in Shuidong Gorge and estimated that she was 3.5 years old at the time, just entering reproductive maturity. Shortly afterward, on February 19, 1991, our radio tracking indicated that she left Shuidong Gorge and entered Tudi Gorge, and on February 27 we found her in the Diaoba River, a linear distance of 6.5 km from her original capture location (Figure 5.8a).

She then continued moving to the northwest and disappeared from the study area. We did not make contact with her again until June 7, 1991, when we found her in a *Fargesia* bamboo grove in the upper Xigou area. This was a 16-km straight-line distance from the area she had used the previous February. Shortly afterward, she moved to Huorenping and the fifth gully in Xiaoping, where she spent much of the summer. She descended in August and returned to Shuidong Gorge, the area where we had captured her, in mid-September (Figure 5.8a). At this time, she appeared to be emaciated, and numerous injuries were evident; she died in December 1991. During her disappearance, we searched nearly every corner of the study area but never heard her radio signal; we are thus quite sure she was elsewhere during that time. On the basis of where we last documented her and where she reappeared, our best guess is that she spent March through May, the estrous period for female pandas, in the nearby Dajiang and Xiaojian Valleys or perhaps farther

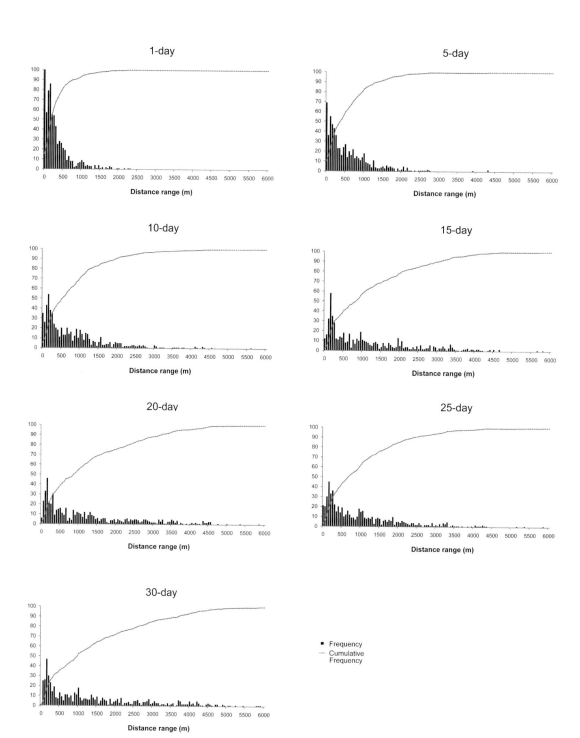

FIGURE 5.7. Straight-line distances for variable time intervals of Jiaojiao during radio tracking from March 1989 to August 1997. We indicate the number of occurrences for each distance class and the cumulative frequency of those movements. The distance classes on the x axis are 50-m increments.

Table 5.4.

MEAN ± STANDARD DEVIATION (SD) STRAIGHT-LINE DISTANCES BETWEEN LOCATIONS OF SIX RADIO-MARKED GIANT PANDAS (THREE MALES AND THREE FEMALES) IN THE QINLING STUDY AREA BY TIME INTERVAL BETWEEN LOCATIONS.

Time interval (days)	Male		Female		All	
	Mean distance (m)	SD (m)	Mean distance (m)	SD (m)	Mean distance (m)	SD (m)
1	506	578	333	375	411	48
2	596	595	426	448	521	534
3	729	690	496	562	601	634
4	839	794	549	632	681	723
5	904	836	602	678	742	770
6	971	885	633	738	788	826
7	1,029	910	690	765	845	850
8	1,055	900	762	876	894	899
9	1,130	992	786	876	943	947
10	1,168	1,022	843	951	990	997
11	1,243	1,067	857	956	1,035	1,027
12	1,252	1,056	904	1,009	1,067	1,046
13	1,312	1,101	918	1,035	1,104	1,085
14	1,351	1,161	969	1,078	1,148	1,133
15	1,390	1,178	1,055	1,118	1,216	1,159
16	1,414	1,231	1,063	1,164	1,226	1,208
17	1,492	1,297	1,043	1,149	1,248	1,239
18	1,512	1,335	1,146	1,235	1,324	1,298
19	1,552	1,353	1,177	1,253	1,351	1,314
20	1,527	1,348	1,201	1,301	1,348	1,333
21	1,570	1,368	1,196	1,275	1,370	1,332
22	1,640	1,466	1,218	1,299	1,416	1,396
23	1,607	1,401	1,259	1,343	1,424	1,382
24	1,703	1,509	1,258	1,342	1,464	1,439
25	1,710	1,467	1,265	1,348	1,476	1,423
26	1,759	1,533	1,306	1,363	1,518	1,462
27	1,766	1,511	1,366	1,412	1,552	1,472
28	1,789	1,564	1,407	1,461	1,593	1,524
29	1,852	1,613	1,403	1,465	1,610	1,551
30	1,861	1,627	1,460	1,532	1,653	1,591

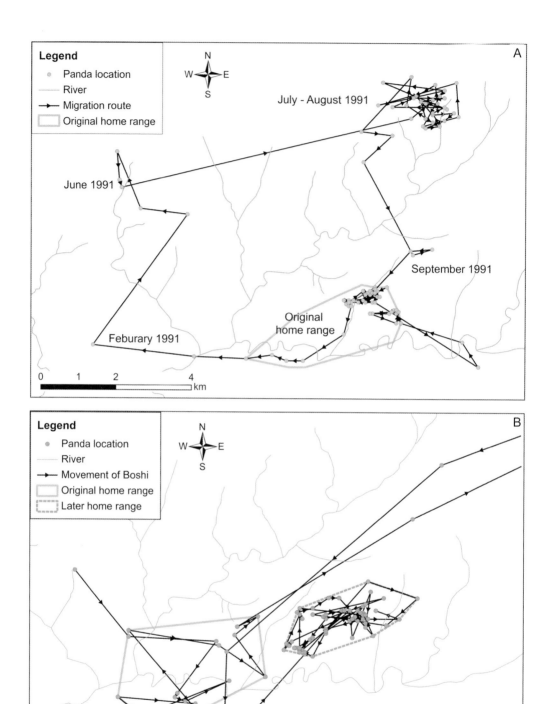

FIGURE 5.8. (a) Movements of subadult female Shuilan from February 1991, when first captured, to mid-September 1991. Shown are her original home range (solid polygon), and estimated travel routes (arrows). Individual radiolocations are marked with a gray dot. (b) Long-distance movements of subadult female Boshi, 1993–1994.

away. We cannot be sure if her long-distance movement was an aberration or a common occurrence for young females. Nor do we know whether she was born in Shuidong Gorge, where we captured her. Regardless, her movements were greater than any of our other radio-marked animals (Table 4.3).

In spring 1993, we fitted a collar on another subadult female, who we named Boshi, estimated to be 2.5 years old. She disappeared from her original area in June 1993, but our radio tracking revealed that she traveled east toward the area around Huangtong Ridge in Foping County (Figure 5.8b). In November 1994, she returned to the Xinglongling's south slope and established a home range in Chaijiawan, where she bred in April 1995.

These two examples of long-distance movements were both by relatively young females. Shuilan moved 34 km over an 8-month period; Boshi moved a straight-line distance of 24 km over a period of 17 months. Such lengthy movements, however, were unusual.

Activity Patterns

We elucidated the diurnal patterns of giant panda activity through continuous (24-hour) radio tracking. Radio collars were equipped with motion sensors that caused the radio signal to change pulse rate from 55 beats/minute when resting to 90 beats/minute when moving. This enabled us to determine whether pandas were active or resting each time we listened to their signal.

We conducted continuous radio tracking at regular intervals (e.g., once or twice monthly), as well as during particularly important periods of the pandas' life history, such as during estrous or around parturition. During each 24-hour tracking session we recorded at 15-minute intervals whether the pulse rate indicated that the animal was active or resting. We also noted whether there had been a change from the previous activity state because the activity sensor had a 1-minute delay when reacting to a change in state from active to inactive but did not have a similar delay when changing from inactive to active.

We conducted a total of 124 continuous 24-hour radio-tracking bouts on 18 animals (containing a total of over 30,000 individual activity indications) from March 1989 to November 1996. We used methods earlier applied in the Wolong research (Schaller et al., 1985) in the 1980s to calculate the proportion of time active as well as to contrast activity patterns by season, time of day, and gender.

We present the proportion of time active for eight radio-collared female and nine radio-collared male giant pandas (Table 5.5). Overall, pandas were active 49% of the time, and the proportion of time active did not differ between males ($\bar{x}$ = 50%) and females ($\bar{x}$ = 48.5%; P = 0.22).

Although gender did not explain differences in the proportion of time active, differences among individuals were notable. In searching for explanations for these differences, we examined the relationship between activity

Table 5.5.

PERCENT OF TIME ACTIVE FOR EIGHT FEMALE AND NINE MALE GIANT PANDAS, QINLING STUDY AREA.

Females		Males	
Name	Time active (%)	Name	Time active (%)
Jiaojiao	49.4	Huzi	61.0
Xiwang	54.9	Xiaosan[b]	13.8
Xiaosi[a]	18.4	Dabai	44.3
Nüxia	39.6	Daxiong	41.5
Ruixue	41.2	Xiaohuo	69.4
Baoma	59.4	Jiaoshou	54.4
Yanghe	58.7	Huayang	57.6
Momo	21.7	Xinxing	38.9
		Moming	72.3

[a] Three-month-old cub.
[b] Activity sensor failed.

and home range area. Females Nüxia and Ruixue both had small home ranges and a low proportion of time active. After removing some seemingly anomalous individuals, the correlation between home range size and proportion of time spent active was rather weak ($r = 0.20$). When we examined only those pandas with home range sizes of 15 km² or less, we found a stronger relationship ($r = 0.80$, $n = 11$).

The mean proportion of time active by hour of the day was similar for both sexes (Figure 5.9): activity peaks tended to occur at 1300–1700 hours and troughs at 0300–0600 hours. This differed from the pattern in Wolong, where pandas were most active in early morning and again in late afternoon. We observed variation in patterns among individuals, but none had a crepuscular pattern.

We considered time periods from 0700 to 1845 hours as daytime and 1900 to 0645 hours as nighttime. We calculated the proportion of time active for 10 individual pandas (5 males, 5 females) during daytime and nighttime (Table 5.6). Daytime activity exceeded nighttime activity for all 10 individuals, in most cases substantially; however, modest differences between daytime and nighttime were observed for the female Nüxia and the male Daxiong. Overall, the mean proportion of time active during daytime was 55.3%, compared to only 42.8% at night. Among males, the proportion time active during daytime was 57.2% versus 43.7% at night; females were active 54.6% during daytime versus 42.5% at night.

We also examined activity by four ecological seasons, namely, the mating period (March–April), the spring period when pandas forage on bamboo shoots (May–June), the summer period, when they migrate and give birth

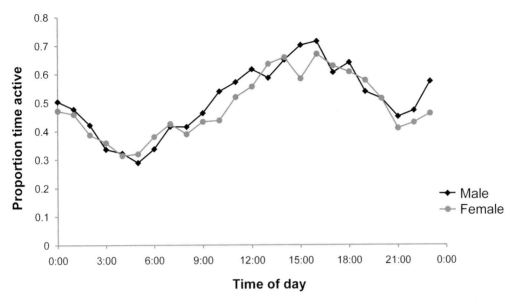

Time of day

FIGURE 5.9. Proportion of time active by time of day for male (diamonds) and female (circles) radio-marked pandas, Qinling study area.

Table 5.6.

PROPORTION OF TIME SPENT ACTIVE DURING NIGHTTIME AND DAYTIME PERIODS FOR NINE INDIVIDUAL PANDAS, QINLING STUDY AREA. MALES ARE INDICATED WITH AN ASTERISK (*).

	Nighttime	Daytime	Whole day
Jiaojiao	0.42	0.56	0.49
Xiwang	0.47	0.64	0.55
Nüxia	0.38	0.41	0.40
Ruixue	0.36	0.46	0.41
Yanghe	0.51	0.66	0.59
Huzi*	0.45	0.77	0.61
Dabai*	0.33	0.60	0.44
Daxiong*	0.41	0.42	0.41
Jiaoshou*	0.49	0.59	0.54
Huayang*	0.53	0.62	0.58

(July–September), and winter (October–February), when pandas are engaged in normal feeding and movement behaviors. Regardless of gender, activity tended to peak in spring, when pandas were foraging on fresh *Bashania* bamboo shoots; the second most active season was the mating period. Activity reached a nadir in summer.

Panda Movements and Activity 127

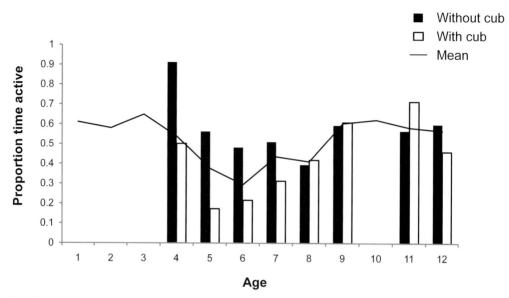

FIGURE 5.10. Proportion of time active for adult female giant pandas by age (in years), showing years in which they raised cubs (open bars) and years in which they did not raise cubs (solid bars).

When cubs were quite young (August and September), the activity of female pandas was low. As the cubs grew and their strength and ability to evade predators increased, the activity of adult females gradually increased. We compared the proportion of time active throughout the year between females raising cubs versus those without cubs (Figure 5.10).

Behavioral Patterns in Panda Society

The five adult males we tracked all contested with each other for mating privileges during the time when females were in estrus. The five adult females experienced cycles lasting 2–3 years, consisting of estrous, mating, parturition, and raising cubs.

We also focused particular attention on the behavior of younger pandas, from their birth (or at least when very young) through their maturity. Males and females differed greatly. Two subadult females made long-distance movements just as they were becoming sexually mature. A third subadult, Xiwang, embarked on a 17-month-long journey, after which she returned to an area quite close to where she had been born, where she went into estrous and mated. In contrast, none of the three subadult males displayed any long-distance movements. Rather, while they were still too young to breed, they made occasional movements to the courtship arena, where they observed the activities of adults. Huzi and Jiaoshou became sexually mature in 1994 and remained in their natal areas.

Activity of Qinling versus Wolong Pandas

In Wolong, pandas were found to spend at least 55% of their time foraging and 41% of their time resting (Schaller et al., 1985). In the Qinling, pandas were active 49% of the time, which included 13% moving in search of food and only 36% actually eating. This was less time spent actually consuming food than at Wolong and also less than omnivorous brown bears (*Ursus arctos*) in North America, which were found to spend 50% of their time actually consuming forage (Gebhard, 1982).

Why did we find such a large discrepancy between Qinling and Wolong pandas in time spent searching for food? Although nutritional differences exist between the bamboo species in the two areas, both can supply pandas with sufficient quantities of leaves, shoots, and stalks. In general, bamboo leaves have six times the nutritional value of stalks, and shoots are twice as nutritious as stalks (Pan et al., 1988). In the Qinling, pandas spend 10 months per year eating the very highly nutritious leaves of bamboo and only 2 months eating the slightly less nutritious bamboo shoots; they eat very little of the bamboo stalk, which is the least nutritious part. In contrast, Wolong pandas eat bamboo leaves for only 6 months of the year, spend 1 month eating primarily bamboo shoots, and for 5 months must eat nutritionally inferior bamboo stalks. These differences would suggest that for Wolong pandas to obtain enough energy, they need to consume more bamboo daily than their counterparts in the Qinling.

The daily activity patterns of pandas from the two areas support this hypothesis. Qinling pandas displayed a single daily peak, being most active between 13:00 and 17:00 hours. Pandas in Wolong, in contrast, displayed two daily peaks of activity, being most active at 05:00 and again at 17:00 hours. Scientists have concluded that because pandas have such low digestive rates and rapid rates of gut passage, they need to eat frequently in order to maintain a full digestive tract (Schaller et al., 1985). Qinling pandas can obtain sufficient nutrition from the bamboo leaves and shoots that they consume during the late afternoon to obviate the need for eating again right way and thus can spend more of their early morning hours resting than pandas in Wolong.

References

Bertram, B. C. R. 1976. Kin Selection in Lions and Evolution. In *Growing Points in Ethology*, edited by P. P. G. Bateson and R. A. Hinde, pp. 281–301. Cambridge: Cambridge University Press.

Gebhard, J. G. 1982. Annual Activities and Behavior of a Grizzly Bear (*Ursus arctos*) Family in Northern Alaska. MS thesis, University of Alaska Fairbanks.

Pan, W. S., Z. S. Gao, Z. Lü, Z. K. Xia, M. D. Zhang, L. L. Ma, G. L. Meng, X. Y. She, X. Z. Liu, H. T. Cui, and F. X. Chen. 1988. 秦岭大熊猫的自然庇护所 [The giant panda's natural refuge in the Qinling Mountains]. Beijing: Peking University Press.

Schaller, G. B. 1994. *The Last Panda*. Chicago: University of Chicago Press.

Schaller, G. B., J. Hu, W. Pan, and J. Zhu. 1985. *The Giant Pandas of Wolong*. Chicago: Chicago University Press.

Chapter 6

Abundance of Giant Pandas

Summary: Previous research has been hampered by the lack of a well-explored and logistically feasible method to estimate the abundance of free-ranging panda populations. We used three methods to obtain estimates of panda abundance and density within our 15-km^2 study area. In spring 1986 and again in spring 1995, we counted pandas directly on strip transects, yielding an average estimate of 28 pandas. During spring 1993, 1994, and 1995, we used mark-recapture (resight) of radio-collared pandas to estimate the presence of 17 adult individuals at breeding congregations; combined with an estimate of the population age structure, this produced a total estimate of 34 pandas. From 1989 to 1997, total counts of identifiable individuals and knowledge of some that had died resulted in an estimate of at least 27 individuals. These three population estimates were all fairly consistent, indicating the presence of about 30 individuals within the research area, or 2 pandas/km^2. We used data from the intensive study area to derive rough estimates of panda abundance within the entire Qinling Mountains. On the basis of the work of previous investigators as well as our own research, we propose an integrated approach for estimating panda abundance that utilizes distances between individual panda signs as well differences in the lengths of fecal bamboo fragments. These approaches have been tested and improved through practical field work and were later adopted for use by the Third National Giant Panda Survey. This method has become the current standard for surveying and monitoring giant pandas.

Introduction

Counting giant pandas is quite a challenge. Pandas live in habitats in which they are very easily concealed; even experienced hunters rarely catch a glimpse of them. While walking through the dense bamboo forests and shrublands that giant pandas call home, visibility is severely limited, at times to just a few meters in each direction. As well, it is almost impossible to move about without making noise, allowing pandas, who have excellent hearing, to flee well before they can be seen. Radio telemetry enabled us to locate pandas that we had collared and then slowly approach them. Some individuals were relatively easy to approach closely, but most individuals were not, and they normally disappeared before we got within a few dozen meters.

Another difficulty is the vast area and complex topography with difficult access. In the Qinling, panda habitat extends up to alpine peaks and subalpine mountain slopes. Much panda habitat is accessible only via crude roads cut through dense forests by logging operations, hunters, and medicinal plant gatherers or, in some cases, only by narrow trails created by the animals themselves. We had no choice but to travel on foot, which was difficult in the Qinling, with its great elevational gradient and high density of vegetation. Even a long day's hike would not cover very much ground.

Nevertheless, we sought to find a way to assess panda numbers because this knowledge is important for conservation. It is currently impossible for us to establish nature reserves in all places currently inhabited by giant pandas; we can only select a few of the more important areas. Equipped with an understanding of panda abundance and distribution, we can determine which areas are in urgent need of protection and how large these protected areas should be.

Do pandas exist as a single large, contiguous population, or have they already been fragmented into numerous small and isolated populations? Is it possible to reestablish connectivity among the smaller populations through increased protective efforts? As the economy develops and new roads are planned, should we not first gain an understanding of panda abundance and distribution to avoid further fragmenting their habitat? Smaller populations have much higher probabilities of extinction than larger ones. Consequently, conservation measures will differ according to the size of the population in question.

Is there a method capable of estimating the abundance of giant pandas in the Qinling? Is there a method capable of estimating the number of pandas in all of China? What degree of accuracy can we expect from such results? How many giant pandas are there in the Qinling? These are questions we address in this chapter.

Historical Efforts to Understand Panda Abundance and Distribution

From the first giant panda population survey in the 1960s to the present there have been more than 10 panda surveys varying in scope, including two nationwide surveys, one in the 1970s and the other in the 1980s. During 1974–1977, the Sichuan Endangered Species Survey Group, supervised by the Sichuan Forestry Bureau, conducted a systematic survey of specified taxa, including the giant panda and golden monkey. Particularly during late April 1975 and early May 1976, about 3,000 people participated in this large-scale panda survey, which included all three provinces with pandas (Sichuan, Shaanxi, and Gansu). Numerous teams trudged through panda habitat, making direct observations and tallying the number of feces encountered. Surveying a solitary species living in densely vegetated habitats in remote mountains is far from easy, and the effort produced only very crude numbers: 150–200 in Shaanxi, 100 in Gansu, and 800 in Sichuan, for a total of 1,050 to 1,100 animals (Schaller et al., 1985). The Shaanxi portion of this survey involved 475 people, who also tallied the number of golden monkeys and takin across the five counties of Foping, Taibai, Yang, Zhouzhi, and Ningshaan (Pan et al., 1988). This survey was also the first real survey of panda abundance in the Qinling. As indicated in this survey's final report, pandas in Shaanxi were found to be distributed only in the western Qinling along either side of its main crest, in parts of Ningshaan, Foping, Yang, Taibai, and Zhouzhi Counties. The main areas of distribution were in the subalpine belt along the southern slopes of the central Qinling. On northerly slopes, pandas were found only in Zhouzhi County. The entire distribution fell within $107°25'–108°35'$E longitude and $33°32'–33°58'$N latitude. The report included a listing of the areal extent of bamboo forest, the density of pandas, and the number of pandas in each county (Table 6.1; Pan et al., 1988).

Between November 1980 and May 1981, Yong Yange and colleagues used line transect and tracking methods to estimate the distribution, numbers, activity, and reproduction within an approximately 100-km^2 region around Sanguanmiao in the center of Foping Nature Reserve. They estimated the number of pandas in the survey area to be about 40 (Yong, 1981).

In 1986, the wildlife protection station of the Shaanxi Forestry Bureau conducted another survey of bamboo forest coverage and panda numbers within a nine-county area in the central Qinling's southern slopes. This produced estimates of 1,400 km^2 of bamboo, with pandas living in about half that area. The estimated numbers of pandas for each county were 82 in Foping, 79 in Yang, 33 in Taibai, 19 in Zhouzhi, 26 in Ningshaan, and 2 in Liuba, for a total of 241 individuals. In addition, giant pandas were also said to exist in Zhashui, Zhen'an, and Feng Counties, but no numbers were put forward (Pan et al., 1988).

Table 6.1.

ABUNDANCE AND DISTRIBUTION OF GIANT PANDAS AND BAMBOO IN SHAANXI PROVINCE, ACCORDING TO THE 1974 SURVEY (GIANT PANDA SURVEY TEAM, 1991).

Statistic	County					Total
	Yangxian	Foping	Ningshaan	Zhouzhi	Taibai	
Area of bamboo (km^2)	174	228	126	54	82	664
Pandas	78	102	16	15	26	237
Panda density (number/km^2)	0.45	0.45	0.13	0.28	0.33	0.36

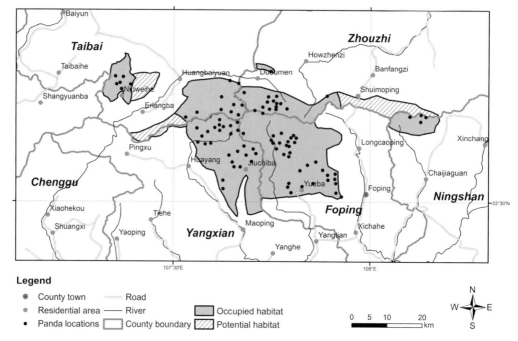

Legend

- County town —— Road
- Residential area —— River
- Panda locations ☐ County boundary — Occupied habitat / Potential habitat

FIGURE 6.1. Distribution of pandas and habitat in Shaanxi Province (data source: Forestry Ministry of the People's Republic of China and WWF, 1989).

In February 1985, the Chinese Forestry Bureau signed an agreement with the World Wildlife Fund (WWF) called the "Protection and Management Plan for Pandas and Their Habitat in China." Between 1985 and 1988, survey teams consisting of Chinese and foreign experts and technicians conducted general surveys in areas of panda distribution in Sichuan, Gansu, and Shaanxi. In Shaanxi, they spent 6 months in the field (September–December 1987 and May–June 1988) and found pandas in 1,037 km² of habitat within Foping, Yang, Taibai, Zhouzhi, and Ningshaan Counties (with an additional 98 km² having potential to support pandas; Figure 6.1). They estimated 109 ± 23 individuals (Table 6.2; Forestry Ministry of the People's Republic of China and WWF, 1989).

During the spring seasons of 1985–1987, our small research team attempted to calculate the density of pandas throughout the bamboo forests within the known distribution in the Qinling. We estimated a population of approximately 220–240 individuals by extrapolating previously estimated density. From the results obtained during the surveys conducted from November 1984 to March 1987, we believe that the current distribution of giant pandas on the southern slopes of the Qinling extends from Ningshaan County in the east to Liuba County in the west and from Taibai and Zhouzhi Counties in the north to Foping and Yang Counties in the south, an area of about 1,650 km². Within this area, we identified areas of concentrated

134   Chapter 6

Table 6.2.

ABUNDANCE AND DISTRIBUTION OF GIANT PANDAS IN SHAANXI PROVINCE FROM THE JOINT CHINESE FORESTRY BUREAU–WWF SURVEYS OF THE MID-1980S (FORESTRY MINISTRY OF THE PEOPLE'S REPUBLIC OF CHINA AND WWF, 1989).

County	Area of panda habitat (km^2)			Number of giant pandas	
	Occupied area	Potential area	Total	Integrated method	Calculated from density
Foping	412.00	0	412.00	39	39 ± 8
Yang	160.23	35.86	196.09	31	31 ± 7
Taibai	253.28	56.82	310.10	19	19 ± 4
Zhouzhi	158.60	5.10	163.70	16	16 ± 3
Ningshaan	52.56	0	52.65	4	4 ± 1
Total	**1,036.76**	**97.78**	**1,134.54**	**109**	**109 ± 23**

density centered on Landianzi ridge, where four counties straddling the main spine of the Xinglongling Range (Yang, Foping, Taibai, and Zhouzhi) come together. This is the heart of giant panda habitat in the Qinling, with an area of 346 km^2 and a population of about 132 pandas (Pan et al., 1988).

An assessment of historical records on the abundance and geographic distribution of pandas in the Qinling suggests that pandas are known from the six counties of Foping, Yang, Zhouzhi, Taibai, Ningshaan, and Liuba, with a habitat of between 1,100 and 1,700 km^2. Previous estimates put the numbers at as few as 109 individuals (Forestry Ministry of the People's Republic of China and WWF, 1989) to over 240 individuals (Pan et al., 1988).

Panda Abundance within the Study Area

Pandas in the Qinling are primarily distributed within four separate patches of habitat (Figure 6.2). Our understanding is that the largest piece of habitat is that near where the four counties of Yang, Foping, Taibai, and Zhouzhi intersect; our intensive study area is in the southwestern portion of this habitat patch, mostly in Yang County. Our field research station was initially established in a flat area of cypress trees on the Hetaoba River but later moved to an area of fir trees on the Cangeryan River. The varying intensity of use of different portions of the study area by our radio-collared pandas is shown in the bottom left in Figure 6.2. The outer polygon connects the outermost location points from 21 pandas. Within this 92 km^2 area, in addition to pandas we captured and followed with telemetry, there were others that we knew of but never captured. Thus, the density of location points of collared pandas does not represent the density of all pandas.

The individual pandas that we became familiar with all lived close to our field camp in the valleys centered around Shuidong Gulch. The bottom right plot in Figure 6.2 shows the spatial juxtaposition of these valleys. These

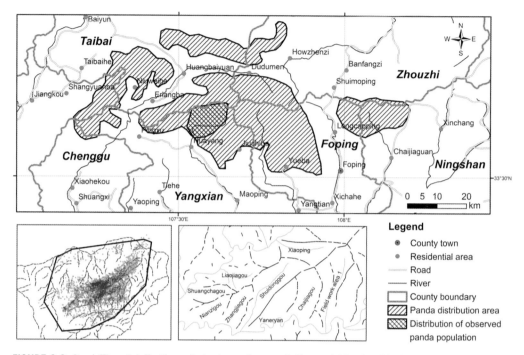

FIGURE 6.2. (top) The distribution of giant panda populations within the Qinling Mountains. The hatched area in the central population was the study area. (bottom left) The distribution of radiolocations for collared individuals within the study area. (bottom right) The location names described in this chapter.

mountain valley gullies are at a relatively low elevation and provide habitat where pandas live most of the year. Additionally, the important activities of estrus and breeding, parturition, and raising young cubs also occurred in these patches of habitat. Situated at elevations of 1,350–2,100 m, these habitat patches contain large groves of *Bashania* bamboo, providing abundant forage for pandas. In the spring, courtship was concentrated on several of these main ridges: those between Liaojiao, Shuidong, and Xiaoping Valleys and those demarking the Shuidong, Yan'eryan, and Chaijiawan Valleys. Caves used for natal dens were also found in the Shuidong, Liaojiao, and Xiaoping Valleys. This area of concentrated use by our collared pandas encompassed about 15 km². We were able to gain a reasonably clear understanding of panda abundance within this small area, using multiple approaches.

ABSOLUTE COUNTS

As of the end of 1996, with the aid of our radio-marked pandas as well as other approaches for discriminating among unmarked individuals, we tallied 30 individual pandas. On the basis of our knowledge of each individual's status (alive or dead), we devised a classification system to combine them into a population estimate.

Category 1 contains individuals that we knew with certainty to be alive. There were 13 animals in this category: Jiaojiao, Huzi, Ruixue, Xiaohuo, Dabai, Xiwang, Boshi, Nüxia, 0940, Momo, Momo's cub, Xiaosan, and Xiaosi.

Category 2 consists of individuals for whom we had detailed information at some point, but we were not entirely sure whether they were still alive. There were 10 animals in this category: Qingqing, Yanghe, Chunbao, Xiaof, Chaima, Xiaobudian, Xinxing, 1290, Keke, and Shiying. Some of these animals had been captured and collared, but because of battery failure, dropped collars, or other reasons, we had lost radio contact with them. Some were offspring of pandas that we had captured and collared that had since dispersed from their mothers, making it impossible for us to follow them.

Category 3 includes individuals for whom we had detailed information and were known to have died. There were seven pandas in this category: Shanshan, Huayang, Daxiong, Shuilan, Dahuo, Jiaoshou, and Guiye (five of which were adult males).

Category 4 represents unmarked animals that we observed within the study area, but we were unsure whether or not they were still alive. We often observed unmarked pandas, particularly during the estrous season of March and April. During this time of year, adult males congregated around estrous females in search of mating opportunities. This type of aggregation typically lasted for several days, enabling us to observe numerous pandas simultaneously, including both marked and unmarked individuals.

Additionally, there were surely unmarked pandas that we never observed. Cubs and subadults are more difficult to observe than adults. However, in some cases, we were able to infer their existence from the behavior of the mother. Female pandas generally reproduce every 2 years, going into estrus in spring and parturition at the end of summer. If a panda fails to conceive or if her cub dies during its first year, she may reproduce the subsequent year, or she may wait another year.

During the 12 years from 1986 to 1997, we followed a total of 30 pandas, including cubs, subadults, adults, and old (senescent) individuals. Of these, seven died and nine were born during the course of the study.

We have summarized pandas we observed in the study area during the 4 years from 1993 to 1996. Included are individuals we recognized and distinguished in the field (category 1 animals), as well as some individuals that we observed only in earlier years (category 2; Table 6.3). These two categories constituted the portion of the total population that was differentiable. In addition, during the mating season we were generally able to observe a considerable number of unmarked pandas (category 4). In some cases, we observed two or more unmarked individuals simultaneously, so we knew they were different individuals and could count them all toward the total. For individuals whom we observed at different times and at different locations and for whom there remained a possibility of duplicate counts, we

Abundance of Giant Pandas   137

Table 6.3.

THE NUMBER OF PANDAS WE DOCUMENTED IN THE QINLING STUDY AREA DURING 1993–1996.

Category	1993	1994	1995	1996
Radio-collared				
Observed	17	15	13	14
Alive but not observed	5	8	8	8
Observed, not collared				
Visually distinguishable	3	3	3	2
Visually indistinguishable	14	6	10	4
Inferred				
Known breeding females	8	2	4	3
Cubs inferred from known breeding females	2	2	3	4
Total (lower bound–upper bound)	**27–38**	**28–31**	**27–34**	**27–30**

calculated an upper bound to the population by treating each observation as a unique individual. We calculated a lower bound as the number of individuals identified with certainty. In addition, there were individuals that we never observed but could be logically inferred, primarily newly born cubs up to about 1 year of age. During this 4-year period, the minimum numbers of pandas we accounted for were 27, 28, 27, and 27 animals (Table 6.3). These consistent numbers indicate that the population within the study area was essentially stable and consisted of no fewer than 27 individuals. This method is relatively accurate but difficult to implement within a short time period or across a large area.

TRANSECT METHOD

We used direct observations of pandas obtained on fixed-width transects that we conducted twice, once in 1986 and then again in 1995 (Table 6.4). We used 40-m-wide transects, assuming that survey personnel were able to detect any pandas within 20 m on either side of their travel route. In practical application, field workers were at times able to detect animals farther than this distance, and at times, the effective detection width was narrower.

Table 6.4.

ESTIMATED NUMBER OF GIANT PANDAS IN RESEARCH AREA ACCORDING TO THE TRANSECT SURVEY METHOD IN 1986 AND 1995.

Time	Length of survey routes (km)	Area surveyed (km^2)	Number of pandas seen	Panda density (number/km^2)	Number of pandas in the research area
March 1986	91.3	3.62	7	1.93	29.0
April 1995	29.0	1.16	2	1.72	25.9

Because the transect length in the 1995 survey was relatively short and the number of pandas observed very few, our results are rather imprecise. This outcome is an inherent weakness of the transect survey method. The specific choice of a route may affect whether a panda is within the strip or not, and with such a small sample size, the difference between observing two pandas (as we did) or three pandas would lead to a 50% difference in the population estimate. In the March 1986 survey, the transect was fairly long, and we observed more individuals, so our results were more precise. If we assume that the population within the study area was static, then we can ignore any effects of time and thus calculate a weighted average of the two surveys (taking into account the differing transect lengths), resulting in an estimated density of 1.88 pandas/km². Extrapolating this to the study area (approximately 15 km²) yielded a total of about 28 pandas.

MARK-RECAPTURE (RESIGHT) ESTIMATE

During estrus, Qinling pandas gather together on certain ridges to contest and mate. This behavior provided us an opportunity to observe several pandas within a single day. By distinguishing those that we had captured and marked, we could use mark-recapture (resight) methodology to estimate the abundance of pandas within our study area.

This approach requires satisfying three conditions: (1) Marked animals were distributed uniformly within the surveyed population, and their chance of being observed was equal to that of unmarked animals. (2) No immigration or emigration occurred during the survey period. (3) No births or deaths occurred within the survey period.

We consider condition 1 to be reasonably well met. We used radio collars as marks and believe that wearing collars had no discernible effect on the activity of the pandas or their chance of being seen in a mating congregation. Because the time period within which we conducted these surveys was short, usually less than 2 months, we believe that conditions 2 and 3 were also relatively well satisfied. During the mating season, pandas congregated in order to contest for breeding opportunities and often could be found by their loud roars. We used both these vocalizations and radio signals to locate these congregation areas and begin our observations. Our observations of animals participating in these mating congregations allowed us to estimate the number of adults, and then, using our estimate of the population age structure, we were able to infer the total population size.

Mark-recapture estimates were made in three consecutive breeding seasons:

- March 24–31, 1993: We observed mating behavior at Shuidong Gorge and Chaijiawan.

Table 6.5.

NUMBER OF ADULT PANDAS IN THE QINLING STUDY AREA, 1993–1995, AS CALCULATED USING A SIMPLE MARK-RECAPTURE APPROACH.

Survey	Observations of all pandas during mating aggregations surveys	Observations of marked pandas during surveys	Marks in population at this time	Calculated number of adults[a]
Spring 1993	25	15	11	18.3
Spring 1994	47	40	11	12.9
Spring 1995	25	15	12	20.0

[a] The mean for all surveys is 17.1 (+3.7 standard deviations).

- March 19–25, 1994: We conducted comparatively detailed and continuous observations of the mating activities on a ridge near the junction of Shuidong Gorge and Chaijiawan for a period of 7 days.
- March 11–April 11, 1995: We observed mating behavior at Shuidong, Chaijiawan, and at Yan'eryan.

Using marked and unmarked animals participating in the mating aggregation, we used the following equation to calculate the total number of adults within the study area: total number of adults = total number of marked (collared) adult animals × number of animal observations/number of observations of marked animals. This yielded an estimate of 17 adult pandas (Table 6.5).

We estimated that the population consists of roughly 50% adults of breeding age, about 3% individuals past reproductive age, about 24% subadults, and 24% juveniles. Thus, we extrapolated 17 adults to an estimated 34 total pandas in the study area.

All three methods produced fairly similar results of about 30 pandas in the study area. The study area was about 15 km², suggesting a density of 2 pandas/km². This density applies only to our study area and only during the period of winter residency. This region probably contains one of the highest densities of giant pandas in the Qinling.

Panda Abundance within the Qinling

Using our understanding of the panda population within our study area, we attempted to assess the abundance of pandas throughout the Qinling Range. Human activity has restricted panda occupancy to those regions near the mountain ridges; these are surrounded in all directions by agricultural fields. The present distribution of panda populations in the Qinling is fragmented into four isolated mountain islands (Plate 3), each of which constitutes a local population: Xinglongling (occupying an area of 960 km²),

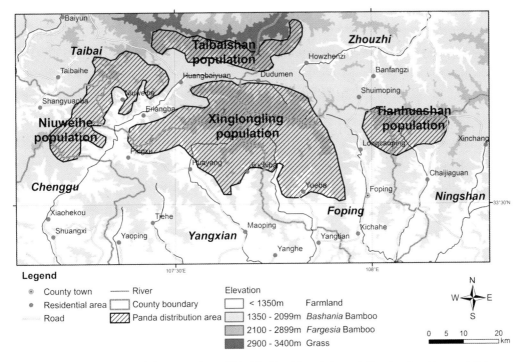

FIGURE 6.3. Distribution of giant pandas within Qinling Mountains relative to elevation and the distribution of bamboo species in the region.

Niuweihe River (320 km^2), Taibai Shan (200 km^2), and Tianhuashan (190 km^2; Figure 6.3).

As described earlier, our study area had a relatively high density of pandas. This area was primarily winter panda habitat, but pandas make seasonal movements in the Qinling, moving to higher altitudes in the summer, where they forage on *Fargesia* bamboo. Thus, the true density over the course of the year is considerably lower than the 2 pandas/km^2 that we documented for just the winter period. Over the course of the year, we found that 20 individuals occupied about 58 km^2, for a density of 0.34 pandas/km^2. Alternately, we could assume that at least two pandas live within an average panda home range (11.5 km^2), given no evidence that female giant pandas exclude each other (i.e., there is a high degree of home range overlap), yielding a minimum density of 0.17 individuals/km^2. With these as upper and lower bounds of panda density in the Qinling, we estimated abundance within the four local populations, which summed to 291–576 pandas (Table 6.6).

We caution against interpreting these estimates as precise. Whereas the Xinglongling local population occupies a relatively large and homogeneous area, the three smaller local populations occupy much smaller areas that have long and narrow shapes that are subject to edge effects from surrounding

Abundance of Giant Pandas   141

Table 6.6.

ESTIMATES OF THE NUMBER OF GIANT PANDAS IN THE QINLING MOUNTAINS AS EXTRAPOLATED FROM DENSITY ESTIMATES MADE UNDER TWO DIFFERING ASSUMPTIONS.

		Number of pandas	
Statistic	Area of local population (km^2)	Low-density assumption	High-density assumption
Home range size (km^2)		5.74	2.90
Density (panda/km^2)		0.17	0.34
Local population			
Xinglongling	960	167	331
Niuwehe	320	56	110
Taibai Shan	200	35	69
Tianhuashan	190	33	66
Local population total	1,670	291	576

human influence. Because we remain uncertain about the patterns of panda occupancy in these local populations, our extrapolations of density are quite speculative. We provide them here merely as reference for future research.

Earlier, the joint Chinese Ministry of Forestry (MOF)–WWF survey (1985–1988) for Shaanxi developed an estimate of density, from which they inferred an abundance of 109 ± 23 individuals. At a smaller geographic scale, their estimates were 31 individuals for Yang County, 14 for Maoping Township, and 17 for Huayang Town (Forestry Ministry of the People's Republic of China and WWF, 1989). These counts are clearly a bit low. The aggregate minimum convex polygon home range of the 21 collared pandas in our study area was 92 km^2, but the MOF-WWF survey documented only 12 pandas (as deduced from sign) within this polygon (making the usual assumption that each evidence of panda occupancy represented only a single individual). Our study area occupied only a small portion of the extent of panda occupancy within Yang County's Huayang Town, but our estimate of panda abundance there was greater than the MOF-WWF estimate for all of Yang County. We suspect the reasons for the discrepancy lie in the methods and inadequate sampling of the earlier survey. If we assume that the underestimation in the MOF-WWF survey in our study area was similar distribution-wide, we can make a simple extrapolation that they documented 12 of the 21 pandas actually present (i.e., biased low by >40%); thus, their survey result of 109 individuals would be corrected to no fewer than 191 pandas.

Objectives of Abundance Estimation

Estimates of population abundance are useful for understanding current circumstances in relation to conservation policies and thereby predicting future prospects. Because of the particularities of habitat configuration and variation in the quality of habitats, panda density varies spatially. Naturally occurring geographic barriers, such as high mountains and wide rivers, also disrupt the continuous nature of panda occupancy. Human activities, such as agriculture in valley bottoms and highways, also constrain panda distribution.

Conducting a large-scale investigation of panda abundance (e.g., national in scope) would likely require at least 3 years. During this time, the panda population would be subject to the processes of births and deaths, as well as dynamics of exchange among local populations. These fluctuations would inevitably affect survey results. Within smaller sections of the distribution, surveys could be completed within rather short times, allowing one to assume stability. A national-scale survey would, in actuality, be a blending of numerous smaller-scale results, each of which was made at slightly different times.

Survey methodology and implementation are also likely to have a large impact on the ultimate accuracy of the population estimate. The accuracy of the estimate may be influenced by survey methods; the season of the survey; the number of field workers involved and their levels of skill, motivation, and sense of responsibility; time spent in the field; specific routes selected for the survey; weather conditions; and other unpredictable factors. Thus, although one's goal may be to obtain a result with high accuracy, in practice this goal may simply not be attainable. However, if the results of a survey satisfy the minimum needs of the work that follows, then that survey can be judged a success.

A variety of field methods have been employed in the various attempts to count pandas. Most of these methods relied on investigators moving throughout an area of panda distribution and recording pandas observed or their sign and extrapolating abundance from these data. Although methods were diverse, all faced the same fundamental difficulty: the huge gap between the enormous area to be surveyed and the resources marshaled to survey it. Thus, results had low accuracy, and it was difficult to compare results of one survey with another.

We experimented with various methods in our attempt to understand panda abundance in our Qinling study area. Because our study area was small relative to the time and manpower we had at our disposal and because we continued our work over a series of years, we ultimately became acquainted with almost every individual panda in the population. Thus, we were reasonably sure that we obtained an accurate estimate of the number of pandas in the study area. However, this method is not feasible over larger areas or shorter time frames.

A good survey method should be able to obtain results that are sufficiently accurate to satisfy the needs of the subsequent work. A subsidiary consideration is that to the extent possible, survey methods should use manpower and resources efficiently and avoid wasting precious research time, energy, and money. A suitable research method should be integrated into the normal work conducted within nature reserves, allowing population and distribution data to be accumulated by reserve staff. Thus, survey methods should be simple and easy to carry out and possible for typical reserve staff to master. Further, data should be easy to analyze and results should be quickly obtainable.

Putting Survey Results in Perspective

The abundance of wild giant pandas is of great interest to government officials, nongovernmental organizations, scientists, and the general public. Population status and trends reflect conservation achievements. During the time period spanning the Second National Giant Panda Survey (1986–1988), the 1988 survey in Sichuan's Pingwu County, the pilot study in Sichuan's Qingchuan County in 1999, and the Third National Giant Panda Survey beginning in 2000, China was in the midst of reform and opening up. Economic development was rushing forward at a headlong pace; many new systems were sprouting up, developing, and being perfected. Environmental issues, wildlife conservation, and sustainable use of resources gradually entered into the lives of the general populace and became issues of concern.

However, economic development and resource consumption are closely related, particularly for a developing country. From the perspectives of local governments and communities in and near giant panda habitat, the quickest route to economic development came from timber harvesting. This meant that the earlier approach of selective logging was abandoned in favor of unplanned clear-cutting. Large swaths of primary forest disappeared within a short time period. The overwhelming majority of these lost forests were giant panda habitat, and pandas that lived in them were exposed to the pressures of reduced and fragmented habitat.

At the same time, efforts and measures aimed at nature protection were increasing, and a new group of nature reserves for giant pandas was established, increasing the area of panda habitat coming under such protection. The training and background of nature reserve staff improved, as did the quality of work conducted within the reserves. With forest resources nearly depleted by the old style of forestry enterprises, everyone was looking for a new way out, with many transforming themselves from loggers into tree planters, from exploiters of natural resources to protectors of those same resources.

However, with economic development still being paramount, investments in long-term and potential future benefits were easily overlooked. Governments and industry typically focused on projects with short-term and obvious profits. In contrast, the benefits engendered by nature conservation were long term and not always obvious, superficially requiring expenditures with no direct material benefit, and thus failed to obtain the attention and investment they deserved. Only after huge floods affected the middle and lower Yangtze in 1998 did a consciousness arise that these immense losses had been caused by insufficient conservation of forests upstream.

Estimation of abundance does not answer the larger questions pertaining to long-term persistence of pandas. Under natural conditions, animal populations fluctuate according to internal dynamics (e.g., age structure effects) and environmental effects (e.g., mass flowering of bamboo). Human factors can also affect giant panda distribution and numbers. Logging can lead to the loss or degradation of giant panda habitat; pandas originally living in a logged area will move to surrounding habitats that are somewhat better, with the result that panda numbers in relatively good habitat may increase.

Often, we cannot judge whether a true population change has occurred by simply looking at the results of two surveys. We may find that where surveys seem to indicate that pandas have increased, the underlying reason is that survey effort increased. To some degree, counts and population estimates are positively correlated with the sample sizes used to obtain them, so if samples are relatively low, it may be that some areas of panda distribution have been overlooked.

In assessing data on giant panda abundance, we should not be concerned only with whether or not panda numbers have increased; we should also consider the overall geographic distribution, as well as where density is high or low and the reasons for this variation. We need to consider whether the habitat is conducive to long-term persistence, what the current threats to the population are, and whether there are key areas requiring particular conservation attention.

Integrated Methods for Surveying Panda Abundance and Habitat

During the past few years, we suggested a suite of methods to estimate abundance on the basis of particular aspects of the biology and ecology of pandas. We employed these methods during portions of the Third National Giant Panda Survey in Pingwu County in 1998 and Qingchuan County in 1999 (both in Sichuan) while simultaneously checking and improving them. Following revision, these methods were adopted as the standard for inventory and monitoring for the rest of the national survey.

USING DISTANCES AMONG SIGNS TO DIFFERENTIATE INDIVIDUALS

Every panda occupies a home range, has a definite area of activity, and can only move a limited distance in a given period of time. The 5,976 locations we obtained from 23 collared pandas produced a total of 1,456,560 possible straight-line distances between locations, with the longest being 12.7 km. From this, we plotted the distribution of all possible distances between locations of the same individual and found that less than 5% exceeded 6 km. Thus, even

Table 6.7.

DISTRIBUTION OF DISTANCES BETWEEN RADIO RELOCATIONS OF INDIVIDUAL PANDAS BY SPECIFIC TIME DURATIONS. NOTE THAT N IS THE SAMPLE SIZE OF PAIRS OF LOCATION POINTS USED IN CALCULATION; THE MAXIMUM DISTANCES ARE THE GREATEST DISTANCE BETWEEN ANY TWO LOCATIONS OF THE SAME INDIVIDUAL OBTAINED WITHIN THE SPECIFIED TIME DURATION.

Duration between locations (days)	n	Maximum distance (km)	Percent in distance category							
			>1 km	>1.5 km	>2 km	>3 km	>4 km	>4.5 km	>5 km	>7 km
1	2,197	4.85	10.0	3.7	1.5	0.5	0.2	0.2	0	0
3	6,792	6.35	14.0	5.1	2.0	0.6	0.2	0.2	0	0
7	15,092	7.35	19.6	8.3	3.8	1.1	0.4	0.2	0.1	0
All	197,307	7.70	51.4	33.2	22.5	11.4	5.1	2.8	1.2	0

without considering habitat features, two signs of panda presence farther than 6 km apart are most likely to have come from different individuals.

We also examined how far individual pandas moved during short periods. We found that excluding the few days during which pandas migrate between their summer and winter ranges, pandas generally do not move a great deal (Table 6.7). Data from 19 of the 23 pandas were used to produce Table 6.7 because they could be identified as coming from either the summer or winter range. Among 2,197 pairs of location points no more than a single day apart, the greatest straight-line distance was 4.8 km. During survey work, we can generally judge the rough age of the spoor and in this way more reliably assess the probability of two spoors a certain distance apart being from the same individual. There should exist a threshold distance beyond which it is not possible for any individual to leave spoor within a given time period. That is, we can produce distributions of the distances between locations at 1-, 3-, and 7-day intervals and apply a threshold value (e.g., 95% probability) using Table 6.7, beyond which spoor most likely comes from separate individuals. The values we generated using this approach were Day $1_{95\%}$ = 1.5 km, Day $3_{95\%}$ = 2 km, Day $7_{95\%}$ = 2 km, and $DS_{95\%}$ = 4.5 km (for 1, 3, 7, and no specific number of days, respectively).

DISCRIMINATING INDIVIDUALS BASED ON BAMBOO STEM FRAGMENT LENGTHS IN SCATS

When foraging on bamboo, giant pandas bite off the stems, section by section, but do not chew them completely or break the stems into small fragments. Thus, after these bamboo stem fragments pass through the giant panda's digestive track and are expelled, they largely maintain their original shape and length. Within panda feces, each such piece is called a bamboo stem fragment (BSF; Mainka et al., 1994). It is generally recognized that individual pandas produce BSFs of consistent lengths.

In their 1985 work, Hu Jinchu, George Schaller, and others considered that the length of bamboo stem fragments could be related to approximate age groups of giant pandas. They categorized pandas in their Wolong study area into three different BSF length groups: older cubs (1.25–1.5 years), 2.7 cm; young subadults (2.5–3.5 years), 3.5 cm; and older subadults and adults, 3.8–4.1 cm. However, they noted that discrimination could not be extended to the level of an individual (Schaller et al., 1985). These investigators identified pandas producing mean BSF average lengths of <30 mm as young cubs (to 1.5 years old), 30–36 mm as animals aged 1.5–5.5 years, 35–40 mm as 5.5- to 20-year-old adults, and 40–45 mm as 20- to 27-year-old adults (Hu, 1987).

Giant panda foraging consists of a series of highly mechanical motions. Because the nutritional value of bamboo for pandas is low, pandas must ingest copious amounts of it to fulfill their requirements. Assuming pandas excrete around 100 scats/day, each scat contains 60 stem fragments, and one bite results in 3 fragments, pandas must bite bamboo stems into segments some 2,000 times/day, or about 60,000 times/month. In the Qinling, pandas forage on bamboo stems only during March and April, but in Sichuan the season during which pandas eat bamboo stems is even longer. Thus, the number of bamboo stem fragments created by pandas each year may reach into the millions.

Such a high-level repetitive behavior might well result in individual characteristic behavior; that is, individual pandas could develop characteristic bite sizes and hence fragment lengths. Although one would expect segment lengths to vary with growth as the individual panda grows and the skull size increases with somatic growth, such variation would be quite small during the relatively short time period occupied by a survey; we would thus expect fragment lengths produced by any given panda to remain relatively constant within the survey period.

We believe that neither the skull size nor age class make BSF length individualistic, but rather the high level of repetition and individual habits of each individual. When pandas eat bamboo stems, they place them into their months from the side, using their premolars or molars to bite them into small pieces. Because this behavior is done repeatedly, the length of the bamboo stem entering the mouth becomes consistent. Pandas do not insert the bamboo into their mouths deeply enough to reach the teeth on the other side; thus, the bamboo segment length does not necessarily equal the width of the space between teeth on either side of the jaw. Although it is logical that individuals with larger skulls will generally produce longer BSF lengths, no research establishes a direct correlation between BSF length and skull size. Additionally, skull sizes differ among animals of the same age

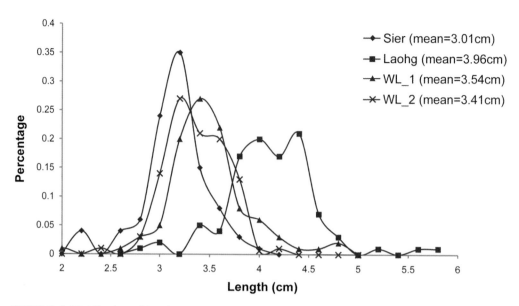

FIGURE 6.4. Distribution of bamboo stem fragment lengths collected from four different fecal piles from giant pandas in 1998: WL_1 and WL_2 were collected from a single individual in the Wanglang Nature Reserve, Sier was collected in the Si'er Nature Reserve, and Laohg was collected in Laohe Gorge.

group, and skeletal growth ceases once animals reach maturity, so using BSF length to determine age is not appropriate.

Using the distribution of BSF lengths from the 1998 survey in Pingwu County, we created curves labeled WL_1 and WL_2 in Figure 6.4 representing fragments from two separate fecal piles of a single individual in Wanglang Nature Reserve. The mean lengths of the fragments in the two piles were 3.41 and 3.54 cm. The curve labeled Sier represents fragments collected in the Si'er Nature Reserve that had a mean segment length of 3.01 cm. The curve labeled Laohg shows fragments collected in Laohe Gorge, with a mean segment length of 3.96 cm. Because the three areas (Wanglang, Si'er, and Laohe Gorge) were separated from each other by fairly large distances, fecal piles found in each area were unlikely to have come from the same individual. Note that the distribution of BSF lengths from a single individual tends toward a single mode or perhaps multiple modal values. The shapes of these peaks are probably related to the posture assumed by the individual while eating, whether the left or right molars are used to break the stem, exactly which teeth are used, and the thickness of the bamboo stems. That said, the peaks from the two fecal piles of the same individual were very close to each other, whereas the three individuals displayed rather larger differences. Thus, BSF data may be a useful criterion to determine whether or not feces came from a single individual.

148   Chapter 6

Table 6.8.

MEAN BSF LENGTHS FROM FECAL PILES OF AN INDIVIDUAL GIANT PANDA, WANGLANG NATURE RESERVE, SICHUAN. PLOT MEANS (MEAN OF FECAL PILE MEAN ± STANDARD DEVIATION) ARE SHOWN AT THE BOTTOM OF EACH COLUMN. VALUES ARE ALL IN CM.

Mean BSF of fecal piles within 5 separate plots				
$n = 4$	$n = 12$	$n = 7$	$n = 4$	$n = 18$
3.53	3.39	3.31	3.53	3.24
3.64	3.43	3.51	3.57	3.30
3.54	3.32	3.50	3.52	3.22
3.36	3.39	3.45	3.52	3.17
	3.35	3.46		3.25
	3.34	3.41		3.24
	3.39	3.42		3.21
	3.28			3.20
	3.23			3.20
	3.44			3.16
	3.40			3.21
	3.38			3.30
				3.22
				3.18
				3.18
				3.27
				3.15
				3.22
3.52 (+0.12)	3.36 (+0.06)	3.44 (+0.07)	3.54 (+.02)	3.22 (+0.04)

We collected fragments from fecal piles belonging to the same individual in the Wanglang Nature Reserve in Sichuan's Pingwu County and measured the mean BSF lengths in each pile (Table 6.8). We calculated the residual differences between the mean BSF lengths and the grand mean (0.008 cm) and the standard deviation of these residuals (0.057 cm). Thus, we calculated that the mean BSF lengths of an individual panda fecal pile correspond to a normal distribution (0, 0.057). From this, we can calculate a 95% confidence interval (i.e., 2 standard deviations) as the mean ± 0.15 cm and a 99% confidence interval (i.e., 2.5 standard deviations) as the mean ± 0.19 cm. That is, if two piles of panda feces are collected in the wild and we wish to know if they came from a single individual, we can measure BSF lengths for both piles and calculate their respective means. If the means differ by 0.15 cm or more, we can say that they have at most a 5% probability of coming from the same individual, and if the means differ by 0.19 cm or more, the

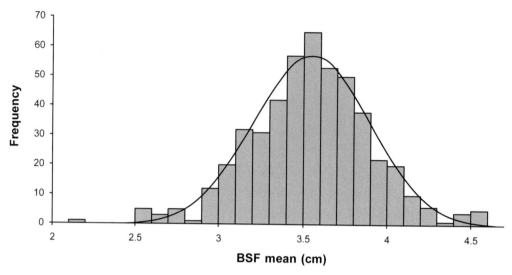

FIGURE 6.5. Distribution of mean BSF values from giant panda feces collected in the Pingwu survey in 1998. A normal curve is fitted to these data.

probability of them belonging to the same individual panda is less than 1%. With such low probabilities, we can judge these as coming from different individuals. In practice, we could raise the threshold to 0.2 cm and thus increase the assurance that a single animal was not erroneously classified as two individuals, although the higher the threshold value is set, the weaker the discriminating power of the method.

During the 1998 Pingwu survey, we accumulated and measured 483 fecal samples, calculating for each sample group the mean BSF length. The smallest BSF was 2.08 cm, the largest was 4.50 cm, and the mean was 3.45 cm (standard deviation = 0.36; Figure 6.5). The range in means was 2.42 cm, and 95% of the mean BSF length values fell within a span of 1.4 cm (i.e., 2.7 to 4.1 cm). In this case, if we used a threshold of 0.2 cm to separate scats from different pandas, we would distinguish only 7–12 individuals. If the threshold was 0.15 cm, 9–16 individuals could be distinguished.

The ability to identify unique individuals from BSF length alone is limited; this would be so even if we could find every single panda dropping. However, we can combine these data with the distances between scats as well as their degree of freshness and in this way greatly increase our power to discriminate individuals. In fact, distances between scats and degree of freshness can distinguish far more individuals than BSF.

Implementation of Field Survey Methods

The initial task in surveying pandas is to determine the geographic scope of the survey. Here we primarily refer to existing written material, relevant material from the forest industry, data from informal or pilot surveys, and interviews with knowledgeable people. We attempt to locate areas requiring particular focus, using local maps, topographic maps, vegetation maps, and land use maps. Areas meriting particular survey focus include possible travel corridors between disjunct areas of distribution, as well as any places subject to specific research, such as where bamboo is flowering or suspected panda courting arenas.

Following the determination of geographic coverage, sample sizes are allocated to zones. For the Third National Giant Panda Survey, a sample line was allocated for each 200 ha within "key research zones" and for each 600 ha in other areas. These zones were plotted on 1:50,000 or 1:100,000 topographic maps. Field staff then selected travel routes within these zones considering field conditions, but ensuring a variety of panda habitat types (e.g., gullies, mountain slopes, ridges, and other topographic features) that typified the entire zone (Chinese Ministry of Forestry Wildlife Protection Office).

In practice, we did not insist that field staff strictly follow the survey lines that had been plotted on maps. Topographic constraints and other issues of field logistics conspired to make it impossible to follow the preplanned survey routes exactly. Additionally, some preplanned sample lines were drawn in areas with essentially no probability of encountering pandas, whereas alternative sites were found to have higher probabilities of encounter. Thus, field staff were given considerable latitude in selecting travel routes as field conditions dictated. The overriding objective was to maximize the sample size of panda scats, bedding sites, and other indications of presence. Field staff were instructed to plot their actual travel routes on topographic maps, as well as the locations of each panda sign encountered, using GPS and topographic maps.

For any pandas visually observed, staff were instructed to document the number of animals, size, gender, and behavior (e.g., foraging or resting). For scats, staff documented the number of piles and, for each pile, the degree of freshness, degree of mastication, and contents. For fresh scats containing bamboo stems, if the sample was collected, then the field form included a specimen reference number. For each such scat, field staff collected and measured 100 BSF lengths; if the scat consisted of fewer than 100 BSF, all fragments were collected and measured. Old scats were not collected and were only recorded for purposes of documenting panda distribution. Small scat piles only needed to be counted, measured, and documented. Scats were measured on the day collected (or, at worst, the second day after) using calipers and were recorded on the data form (Chinese Ministry of Forestry Wildlife Protection Office).

Analytical Methods

Field data form the foundation for analysis; sufficient field data can support a variety of analyses. For example, from field surveys, we can calculate abundance, ascertain changes in geographic distribution, and assess habitat quality. Here we describe the essentials of analytical methods for abundance and distribution.

ANALYSIS OF ABUNDANCE

We determined the abundance of giant pandas through an integrated analysis that made use of the distances between panda signs and discrimination based on BSF lengths. Thus, both the exact locations of each panda scat and the length of bamboo fragments found in them took on great importance.

First, we marked each of the locations on the appropriate map, next to which we indicated its mean BSF length. Then, we picked a starting point, for example, a panda scat located on the periphery of the spatial distribution of all of them. From that scat we drew a circle to encompass other scats that constituted a cluster of potentially the same animal based on distance from the starting scat. Scats within this cluster were evaluated by the BSF length method, and any that were assessed as coming from a different individual were excluded. Those remaining were considered as coming from a single individual. Then, the location of the next scat was taken as the center of a new circle, and the same procedure was followed. The very last point evaluated in this manner was the number of pandas. During the first step, when determining the geographic scope for differentiation, one can overlay certain geographic characteristics, such as agricultural fields, roads, high mountain ridges, or impassable valleys, that function as obvious barriers to panda movements. This helps to reduce the scope of the area within which further differentiation is needed and thus increases the precision of differentiation. With the aid of geographic information system software and computer programs, it is possible to automate this process.

ANALYSIS OF DISTRIBUTION

From the records of panda sign obtained during a survey, one can readily produce a coarse distribution map. To create a more precise distribution map, one can build on these data by adding topography, vegetation, bamboo groves, and inferences from analyses of panda habitat use.

Once a distribution map has been produced, one can quite objectively discern where habitat has become isolated or fragmented, how far fragments are from one another, and the prospects for reconnection. Further, one can consider which habitat patches are large and therefore key to the future persistence of the panda population and which populations are most at risk from human threats. When we add this to information on panda abundance and analyses of population viability, we have the basis for understanding which areas have the potential to ensure persistence and where the most severe threats are located. This information allows us then to target various conservation strategies.

References

Chinese Ministry of Forestry Wildlife Protection Office. 全国第三次大熊猫调查工作手册 [The 3rd National Giant Panda Survey work handbook]. Internal material.

Forestry Ministry of the People's Republic of China and WWF. 1989. 中国大熊猫及其栖息地综合考察报告 [Comprehensive research report on giant pandas and their habitats in China]. Internal report. Chengdu, China.

Giant Panda Survey Team. 1991. 大熊猫数量调查方法的探讨 [A discussion about population survey methods for giant pandas]. In 大熊猫繁殖与疾病研究 [Reproduction and disease analyses of giant pandas], edited by Wenhe Feng and Anju Zhang, pp. 1–6. Chengdu, China: Sichuan Publishing House of Science and Technology.

Hu, J. C. 1987. 从野外大熊猫的粪便估计年龄及其种群年龄结构的研究 [A study on the age and population composition of the giant panda by judging droppings in the wild]. *Acta Theriologica Sinica* 7:81–84.

Mainka, S. A., Y. Huang, and J. Yang. 1994. 五一棚野生大熊猫和其他动物观察 [Observation on wild giant pandas and other wildlife in Wuyipeng, Wolong Nature Reserve, Sichuan, China]. In 成都国际大熊猫保护学术研讨会论文集 [International Workshop on Giant Panda Conservation and Research, Chengdu], pp. 139–143. Chengdu, China: Sichuan Publishing House of Science and Technology.

Pan, W. S., Z. S. Gao, Z. Lü, Z. K. Xia, M. D. Zhang, L. L. Ma, G. L. Meng, X. Y. She, X. Z. Liu, H. T. Cui, and F. X. Chen. 1988. 秦岭大熊猫的自然庇护所 [The giant panda's natural refuge in the Qinling Mountains]. Beijing: Peking University Press.

Schaller, G. B., J. C. Hu, W. S. Pan, and J. Zhu. 1985. *The Giant Pandas of Wolong*. Chicago: Chicago University Press.

Yong, Y. G. 1981. 佛坪大熊猫的初步观察 [Preliminary observations of giant pandas in Foping]. *Chinese Journal of Wildlife* 4:10–16.

Chapter 7

Population Dynamics of Giant Pandas

Summary: We recognized four distinct ecological age groups for giant pandas in the Qinling: cubs (0–1.5 years), subadults (1.5–4.5 years), reproductive adults (4.5+ years), and postreproductive adults (>17 years). Yearly survival rates were about 90% for cubs and subadults, and 93% for reproductive adults; however, among our radio-collared sample, the only adults that died were males. Adult females produced, on average, one cub every 2 years. The average female born in our study area contributed an average of 1.5 new females to the population during her lifetime, indicating a growing population. Projection of the population using a Leslie matrix yielded an estimated yearly growth rate of 4.1%. Even adding stochasticity and density dependence, both of which increase risks of population decline, the population of more than 200 pandas in the Qinling has a high probability of persistence. However, the additional deaths of just two pandas per year, from poaching or incidental capture in traps set for other species, could set the population on a negative trajectory. We are hopeful that the population will increase and then remain stable when it reaches carrying capacity, assuming no major changes in the natural or social environment. To conserve this population, we emphasize protecting existing resources and conducting long-term and continuous population monitoring and research.

Introduction

Temporal and spatial variations in abundance are referred to as population dynamics. Research on the dynamics of a population may help us understand the effects of habitat on the population's persistence and trajectory. Excluding effects of human activities, populations also fluctuate because of changes in population structure. By population structure, we mainly refer to the sex ratio and age composition, both of which affect vital rates. In speaking of vital rates, we refer to birth and death rates, as well as immigration and emigration. Our research on Qinling pandas covered a span of over 10 years, during which we came to know over 30 individuals.

Information on these animals formed the basis for our analyses of vital rates and population dynamics.

Population Structure

Age Groupings

Determining ages of pandas in the wild remains an unresolved problem. There have been attempts to derive specific ages from cementum annuli (annual rings formed on teeth), as well as to categorize pandas into general age classes based on patterns of molar wear and fusion of skull bones (Wei et al., 1988). However, these methods all require collecting large numbers of wild panda skulls. Thus, these methods were not practical for us, studying living pandas. Other researchers have suggested that panda age may be estimated from the quantity and characteristics of their feces (Hu, 1987), but we found no reliable, empirical support for this approach.

Our field observations provided us opportunities to observe wild pandas directly and to consider their physical appearances and behaviors as well as their social interactions with other individuals. Thus, we relied upon these external, easily observable characteristics to develop rough age categorizations. On the basis of their physical characteristics and behavior, we developed four ecological age classes for pandas and used these to classify each of our studied individual pandas.

Cubs (0–1.5 years)

During this stage, cubs live with their mothers, who provide them with food and protection. Cubs generally remain in dens up to the age of 6 months. After that time, cubs typically climb trees in response to perceived danger. Young cubs are of such small body size that it is relatively straightforward to differentiate them from older pandas. Cubs generally leave their mothers to live independently at 1.5 to 2 years of age, although the timing may be slightly later in some cases, depending on the mother's level of experience in raising cubs or the specific circumstances surrounding her subsequent mating.

Subadults (1.5–4.5 years)

Although differences in appearance between subadults and adults are not striking, we learned to use differences in behavior, feeding, and other attributes related to their activity to distinguish the subadult age class, particularly around estrus when these differences were most pronounced.

Mature Adults (4.5+ years)

Males during this stage of life engage in intense struggles for mating privileges during the mating season; females enter a reproductive cycle of

2–3 years consisting of estrus, mating, parturition, and rearing young. By the time they reach 4.5 years, females are capable of reproduction. The breeding system of giant pandas is one in which males contest for mating privileges with females. At 4.5 years old, males begin participating in these struggles, although they are rarely successful at this age because of their small stature and lack of experience. The oldest male we observed breeding was 16.5 years old; zoos have recorded females breeding up to the age of 20.

Postreproductive Adults

Some adult pandas, owing to advanced age, injury, or illness, no longer reproduce. We had one such animal in our radio-collared sample: the male Huayang, who we followed for 23 months from his first capture in December 1988 until his death in November 1990. Although this time period included two mating seasons (1989 and 1990), he was never observed to participate in courtship or mating; we thus judged him to be postreproductive. Wild populations likely have very few animals in this age group. The rigors of nature are such that animals reaching this age group are usually in very poor condition and generally die before long.

SEX AND AGES OF STUDY SUBJECTS

Here we briefly overview the sex, age, and other relevant details regarding each of our study animals, ordered by the time when we first knew of them:

- Qingqing: Subadult female, captured in June 1985 and released about 2 weeks later.
- Shanshan: Adult male, captured near Shanshuping on May 17, 1987; his carcass was discovered in early March 1988 and examined on March 7.
- Huayang: Postreproduction male, captured on December 29, 1988, at Baiyangping and died in November 1990.
- Yanghe: Adult female, captured near Hetaoba on February 10, 1989; she shed her collar in March 1990.
- Jiaojiao: Adult female; captured in Shuidonggou on March 11, 1989. She produced four cubs during the course of the study. As of March 1997, she was still under observation and in good health.
- Chunbao: A cub produced by female Baoma, born in August 1989; sex uncertain. This animal was last observed on February 19, 1991, when we found the den and footprints of Baoma and her cub and heard them moving.
- Huzi: Male cub of Jiaojiao, born August 1989 in Shuidonggou. In March 1994, Huzi participated actively in the mating season, but he had low status among the male pandas. In March 1996, when Nüxia was in estrus near the ridge separating Shuidong Gorge from Yanerya, we observed Huzi fight, gain a position of advantage, and mate with

Population Dynamics of Giant Pandas   157

her. He evidently shed his collar in April 1996, after which time we lost contact with him.

- Xiaof: Subadult female, captured on December 23, 1989 in Shuidong Gorge; she shed her collar in April 1990.
- Daxiong: Adult male, captured in Shuidong Gorge on December 14, 1989; his carcass was found on April 16, 1991, in Shuidong Gorge, 2–3 days after his death.
- Baoma: Adult female, captured in March 1990 in Chaijiawan. At the time, she had a 6-month-old cub. On March 27, 1993, we observed a radio-collared panda traveling with Dabai in Chaijiawan but were unable to obtain a signal from the collar; from the external appearance, we inferred that this must have been Baoma.
- Xiaobudian: Offspring of Ruixue, sex unknown. Born August 1990, last seen on March 22, 1992, with its mother on the right slope of S3 at Shuidonggou.
- Shuilan: Subadult female, captured on February 2, 1991, in Shuidong Gorge. We found her carcass on December 9, 1991, in a small ravine in Chaijiawan, 3–4 days after her death.
- Xinxing: Adult male, captured February 1991 at Tudigou. We found his collar on October 2, 1993, near the ridge separating Tudigou and Xigou. We had located him at this spot 3 months earlier and speculate that he lost his collar at about that time.
- Dahuo: Adult male, captured on February 2, 1991, at Shuidonggou. We found his carcass in December 1994 at Shifogou.
- Ruixue: Adult female, captured on December 31, 1991, at Shuidong Gorge. She was lactating at the time; we lost contact with her in May 1995, probably because of battery failure.
- 1290: Adult male, captured on January 31, 1992, at the head of a ravine in the area in which we first began our field work (which we called Yizuoyequ). We lost his signal in November 1992.
- Keke: Adult female, brown and white in color, captured on February 14, 1992. She was lactating at the time; the last date that we received a signal from her was May 2, 1993.
- Xiaohuo: Adult male, captured on March 13, 1992, at Shuidong Gorge. As of November 1996, his collar was still operating.
- Dabai: Adult male, also captured on March 13, 1992, at Shuidong Gorge. As of November 1996, his collar was still operating.
- Xiwang: Female cub of Jiaojiao, born in August 1992 at Shuidong Gorge. Our most recent observation of her was in March 1997 in a small ravine on the south side of Liaojiagou.
- Jiaoshou: Adult male, captured on February 17, 1993, at Shuidong Gorge; he died in February 1996 near the mouth of Shuidong Gorge.

- Boshi: Subadult female, captured on February 19, 1993, at Shanshuping. She was observed to be in estrus in April 1995 at Chaijiawan. As of November 1996, her collar was still operating.
- Nüxia: Adult female, captured on February 24, 1993, at Shuidong Gorge. As of March 1997, her collar was still operating.
- 0940: Adult male, captured on March 31, 1993, at Shuidong Gorge. As of November 1996, his collar was still operating.
- Momo: Adult female, captured at the same time as 0940 on March 31, 1993, at Shuidong Gorge. As of March 1997, her collar was still operating.
- Shigen: Cub of Ruixue, for whom we could not determine gender. This animal was first observed on August 21, 1993, at Shuidong Gorge soon after birth. We also observed small footprints that appeared consistent with a cub on November 3, 1993, in a cave not far from where Ruixue was located, in Shuidong Gorge.
- Guiye: Female cub of Momo. On August 26, 1993, we observed that Momo had given birth to twins at Dishuiya. Although one cub had already died, the other survived initially, and we named it Guiye. On November 5, 1993, as we approached Momo near the ridge separating Shuidong Gorge from Chaijiawan, we heard vocalizations from a cub that we took to be Guiye. However, in March 1994, Momo again went into estrus and gave birth in August of that year. We thus infer that Guiye died between November 1993 and March 1994. Taking the midpoint of these times, we estimate that she died in January 1994.
- Cub 064: Male cub of Momo; born at Xiaoping in August 1994. Our last observation of him was in March 1996, traveling with Momo in Shuidong Gorge.
- Xiaosan: Male cub of Jiaojiao; born in August 1994 at Shuidong Gorge. The last signal from him was received on March 24, 1997.
- Xiaosi: Female cub of Jiaojiao; born in August 1996 at Shuidong Gorge. Our last observation of her was on November 11, 1996.

SEX AND AGE STRUCTURE OF STUDY POPULATION

Within our study population we observed 9 cubs (3 male, 3 female, 3 unknown sex), 9 subadults (3 male, 5 female, 1 unknown), 19 reproductive adults (10 male, 9 females), and 1 postreproductive male. The observed sex ratio was very close to 1:1.

Survival

To estimate survival, we followed the approach described by White and Garrott (1990), based on deaths observed in animal-months of monitoring radio-collared pandas:

$$S_m = 1 - (D/T_m),$$
$$\text{Var}(S_m) = [S_m \times (1 - S_m)]/T_m,$$
$$S_y = S_m^{12},$$

where D is the number of deaths during the study period, T_m is animal-months (1 animal monitored for 1 month = 1 animal-month), S_m is the monthly survival rate, S_y is the yearly survival rate, and $\text{Var}(S_m)$ is the variance of monthly survival rate. We present survival estimates by sex and ecological age (Table 7.1) and the number of months of monitoring each individual in the sample at each age (Table 7.2). Of note is that we documented no mortalities among adult females, nor did we document deaths among male cubs or subadults. With sexes combined, annual survival was 90% for cubs, 91% for subadults, 93% for adults, and only 59% for postreproductive adults.

Reproductive Rate

During 1989–1998, we gathered detailed data on estrus and reproduction of five adult females in the Qinling giant panda population:

- Jiaojiao gave birth to Huzi in August 1989, Xiwang in August 1992, Xiaosan in August 1994, and Xiaosi in August 1996. In August 1998, she gave birth to a fifth cub.
- Baoma gave birth to Chunbao in August 1989 but did not go into estrus in 1990 or 1991. We do not know whether she went into estrus in spring 1992. In March 1993, we observed her in estrus.
- Ruixue gave birth to Xiaobudian in 1990; she was known to enter estrus again in spring 1993 and gave birth to Shigen later in 1993. We observed her in estrus again in spring 1995.
- Nüxia went into estrus in 1993 and may have given birth to a cub that year. We observed her in estrus in spring 1995 and again in spring 1996. We deduced from her radio signal in fall of 1996 that she had already given birth to a cub, although her estrus of 1995 may not have resulted in a cub.
- Momo gave birth to Guiye in August 1993, but Guiye did not survive to the following spring. Momo went into estrus again in spring 1994 and gave birth to a male cub in August of that year. In spring 1996, we again observed her in estrus.

Because we are somewhat uncertain about Baoma's reproductive history, we have excluded her from calculations of the reproductive rate calculations. We documented 10 reproductive cycles: 2 cycles of 1 year, 6 cycles of 2 years, and 2 cycles of 3 years. Thus, females in the Qinling produced, on average, 1 cub every 2 years (counting unsuccessful litters).

Table 7.1.

ESTIMATES OF SURVIVAL AMONG MALE AND FEMALE PANDAS OF EACH ECOLOGICAL STAGE, QINLING STUDY AREA, 1985–1997. A SINGLE MALE WAS OBSERVED IN THE ADULT POSTREPRODUCTIVE (AP) STAGE. A DASH (—) INDICATES NOT APPLICABLE BECAUSE NO DEATHS OCCURRED.

Ecological stage	All				Female				Male			
	Cub	Subadult	Adult Reproductive	AP	Cub	Subadult	Adult Reproductive	Cub	Subadult	Adult Reproductive	AP
Observation time (months)	119	126	659	23	26	75	330	54	50	329	23
Individuals	9	9	19	1	3	5	9	3	3	10	1
Mortalities	1	1	4	1	1	1	0	0	0	4	1
Survival rate											
Monthly	0.992	0.992	0.994	0.957	0.962	0.987	1.000	1.000	1.000	0.988	0.957
Standard deviation	0.008	0.008	0.003	0.043	0.038	0.013	—	—	—	0.006	0.042
Annual	0.904	0.909	0.930	0.587	0.625	0.851	1.000	1.000	1.000	0.863	0.587

Table 7.2.

HISTORY OF TRACKING FOR EACH PANDA INDIVIDUAL CONSIDERED IN THE ANALYSIS, INDICATING SEX, INITIAL MONTH OF TRACKING, FINAL MONTH OF TRACKING, MONTHS IN EACH OF THE FOUR POSSIBLE ECOLOGICAL STAGES, TOTAL MONTHS TRACKED, AND STATUS AT THE CONCLUSION OF THE STUDY. U = UNKNOWN; L = LIVING; D = DEAD.

Panda name	Sex	Tracking period		Life Stage					Months in sample	Status at end of study
		Start	End	Cub	Subadult	Adult Reproductive	Adult Postreproductive			
Qingqing	F	June 1985	June 1985		1				1	L
Shanshan	M	May 1987	March 1988			10			10	D
Huayang	M	December 1988	November 1990				23		23	D
Yanghe	F	February 1989	March 1990			13			13	L
Jiaojiao	F	March 1989	March 1993			96			96	L
Chunbao	U	August 1989	February 1991	18					18	L
Huzi	M	August 1989	April 1996	18	36	26			80	L
Xiaof	F	December 1989	April 1990		4				4	L
Daxiong	M	December 1989	April 1991			16			16	D
Baoma	F	March 1990	March 1993			36			36	L
Xiaobudian	U	August 1990	March 1992	18	1				19	L
Shuilan	F	February 1991	December 1991		10				10	D
Xinxing	M	February 1991	July 1993			29			29	L
Dahuo	M	February 1991	December 1994			46			46	D
Ruixue	F	December 1991	May 1996			53			53	L
1290	M	January 1992	November 1992			10			10	L
Keke	F	February 1992	May 1993			13			13	L
Xiaohuo	M	March 1992	November 1996			56			56	L
Dabai	M	March 1992	November 1996			56			56	L
Xiwang	F	August 1992	March 1997	18	36	1			55	L

162   Chapter 7

Name	Sex							
Jiaoshou	M	February 1993	February 1996			36	36	D
Boshi	F	February 1993	November 1996	24		21	45	L
Nüxia	F	February 1993	March 1997			49	49	L
0940	M	March 1993	November 1996			44	44	L
Morro	F	March 1993	March 1997			48	48	L
Shigen	U	August 1993	November 1993		3		3	L
Guiye	F	August 1993	January 1994		5		5	D
Cub 064	M	August 1994	March 1996	1	18		19	L
Sun	M	August 1994	March 1997	13	18		31	L
Xiaosi	F	August 1996	November 1996		3		3	L

Population Dynamics

ESTIMATING POPULATION GROWTH

A fairly simple method of gauging population dynamics over the short term is to calculate the average number of cubs produced by the average female, considering both reproductive rates and survival. If the average female that is born produces more than one female offspring in her lifetime, then the population should increase. Using the rate of survival and reproduction estimated from our data, we calculated a net reproductive rate of 1.53 (female cubs per female), suggesting that the population within our study area was increasing at the time of our study.

We also entered the vital rates estimated from the Qinling panda population into a Leslie matrix (Ma, 1996). This yielded an annual population growth rate of 4.1%. The stable age distribution predicted by the Leslie matrix compared well with the age structure that we actually documented through field surveys (Table 7.3), providing support for the reliability of the results of this model. With this rate of increase, the population would double in about 17 years.

The accuracy of a population projection depends both on the accuracy of the vital rates entered into it and on the correspondence between the model assumptions and reality. Sample size is a vexing problem in population dynamics research on large, elusive animals like giant pandas, with slow reproduction and high survival. Conceivably, the birth or death of a single individual within the studied population could change the perceived population trend. Thus, deterministic models, like the Leslie matrix, may be inappropriate for projecting population trend into the future because they do not consider uncertainty in parameter estimation and random events that can affect small populations.

EFFECTS OF UNCERTAINTY AND STOCHASTICITY

We used a simulation model (Sun, 1992) that reflected the life history of giant pandas (Figure 7.1) to investigate the effects of random fluctuations

Table 7.3.

THE STABLE AGE DISTRIBUTION IMPLIED BY THE VITAL RATES WE DOCUMENTED IN THE QINLING PANDA POPULATION AS APPLIED TO A LESLIE MATRIX FORMULATION AND ESTIMATES OF THE STANDING AGE STRUCTURE WE OBTAINED THROUGH DIRECT OBSERVATION, 1985–1997.				
Statistic	Cubs	Subadults	Reproductive Adults	Postreproductive Adults
Stable age distribution (%)	24.2	26.0	48.3	1.5
Observed composition (%)	24.3	18.9	54.1	2.7
N	9	7	20	1

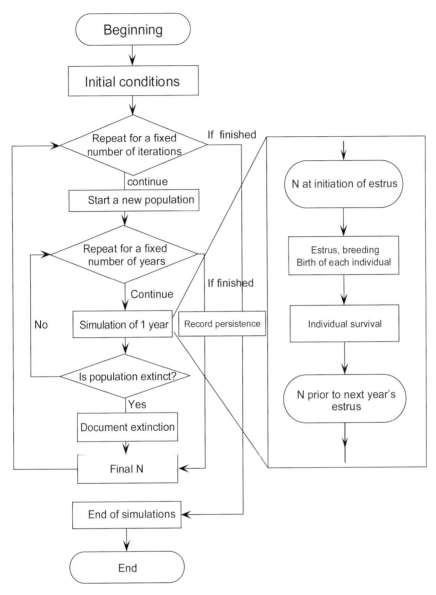

FIGURE 7.1. Flowchart of the stochastic simulation model of panda population.

in vital rates on population persistence. We represented individual animals within a computer program, each of which was assigned a sex, age, and social status. We began each simulation run with a particular population size, age structure, age-specific survival rates, sex ratio at birth, social relationships, and environmental factors. Each year, some individuals within the population mated, gave birth, and survived to the next year. As the simulation progressed, we queried the population for specific data. We were mainly concerned with extinction probability, so we continually queried

Table 7.4.

EXTINCTION PROBABILITIES FOR SIMULATED PANDA POPULATIONS OF INITIAL SIZE 10, 20, 25, 28, 30, 40, 50, AND 70 OVER 200 YEARS.

Statistic	Initial population size							
	10	20	25	28	30	40	50	70
Extinction probability within 200 years (%)	41.2	13.2	7.6	6.5	4.1	1	0.2	0

population size. When extinction of the simulated population occurred, we recorded the year of extinction. Otherwise, we continued the projection for the desired number of years. We repeated this simulation through a large number of iterations.

As our research progresses, we can gradually add data into the model to investigate effects of environmental, genetic, and geographic factors, as well as social structure. At present, however, our understanding of these factors is insufficient to merit including them in the model.

Our primary focus in this simulation was on how stochastic factors affected the probability of extinction for various initial population sizes from 10 to 70 for periods of up to 200 years. For each initial population size we conducted 1,000 iterations. Populations of smaller initial size were more sensitive to stochastic factors and thus were more likely to go extinct (Table 7.4). With at least 30 pandas and the vital rates we observed (i.e., no major changes in habitat in the future), the probability of population extirpation was <5%.

CONSIDERING DENSITY-DEPENDENT AND -INDEPENDENT FACTORS

Population growth rates slow as density increases because animals compete for food and space; thus, the vital rates of an increasing population will change through time with increasing density. We incorporated a form of density dependence in our modeling. We assumed that individual survival rates declined with increasing population density, although we included density effects only when the population exceeded its carrying capacity ($N > K$). This reduced population persistence because it set a cap on population size (Figure 7.2).

We also added density-independent factors to our model by altering survival rates by a certain factor and calculated extinction probabilities incorporating additional density-independent factors under an array of initial population sizes (Figure 7.3). Poaching and trapping can be viewed as density independent. Our analysis suggests that if the Qinling contains a population of about 200 pandas, mortality from poaching and trapping must be kept to less than 1% annually to assure long-term persistence. This means that the

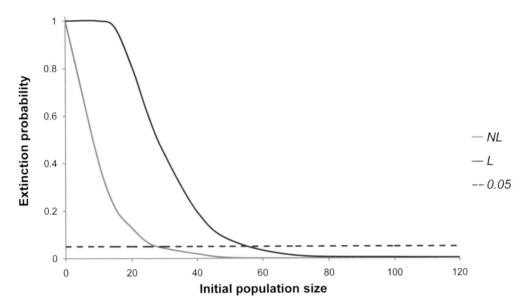

FIGURE 7.2. Probability of extinction of a hypothetical panda population given varying initial size. Line NL represents probability in the absence of a carrying capacity and a density-dependent function relating population size to it. L represents extinction probability with this function added. The dashed line represents a 5% probability of extinction.

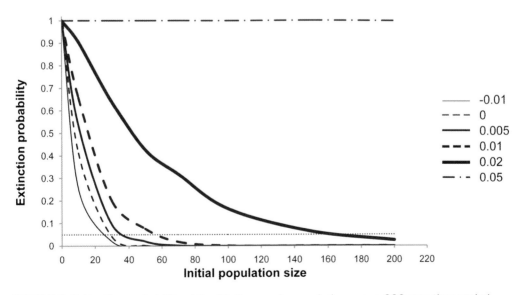

FIGURE 7.3. Extinction probability of the Qinling panda population over a 200-year time period at various initial population sizes (x axis) for various levels of added mortality from density-independent causes.

number of pandas killed by poaching or trapping within the Qinling should not exceed two pandas per year.

Trajectory of the Qinling Giant Panda Population

Our analysis thus far has provided an approximate understanding of population trend of giant pandas in the Qinling. From this, we believe there is hope for Qinling pandas: our data indicate that they are probably increasing. We cannot say definitively that the population was increasing, because our model did not include stochastic events (which can have a large effect with small populations); however, the probability that the population was increasing was greater than the probability that it was in decline. Also, with over 200 pandas in the Qinling, our models suggested that the chance of extinction is very low (Figures 7.2, 7.3). However, if this population is near carrying capacity, then it will soon reach a point when it can no longer increase. Thus, we think the most probable future for the Qinling panda population is one of general stability characterized by fluctuations. As long as both their habitat and the social environment they require do not change greatly, pandas should persist in the Qinling.

Our analyses are based on vital rates obtained over 10 years of intensive research. Nevertheless, our results remain more ambiguous than we would have liked. Although we have begun to understand the dynamics of the Qinling panda population, we feel that providing precise long-term prognostications is neither scientific nor necessary.

Our understanding of the Qinling panda population stems from our study of a sample of individuals in one portion of this mountain range. Our study area is situated centrally within the largest local panda population of the Qinling, backed up against the main Xinglongling ridge, and surrounded on the other three sides by high mountain ridges. Additionally, our studies have been of relatively short duration on the scale of panda population dynamics. It is possible that this period represented nothing more than a small upward bump within a longer downward process. It is also possible that our study area represents a source population with a high growth rate, from which surplus pandas migrate and settle in nearby sink populations, with lower-quality habitat and lower potential for growth (Meffe and Carroll, 1997; Figure 7.4).

How to Better Conserve the Qinling Giant Panda Population

From existing data, we can, at best, obtain an approximate understanding of the population's capacity for growth, but we cannot produce an accurate forecast. We need to understand not only the population's current status but how it is likely to change in the future. This information will be important in crafting appropriate conservation policies for the effective protection of wild giant pandas.

Through continuous monitoring we can obtain an understanding of how the population is changing spatially and temporally with changing

168   Chapter 7

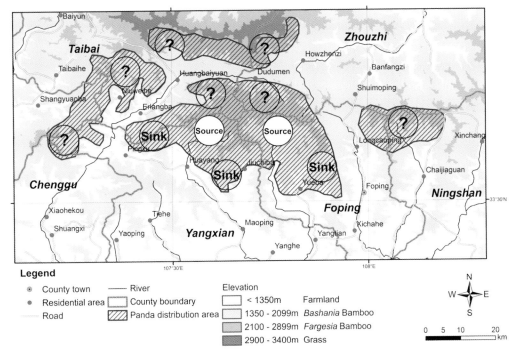

FIGURE 7.4. Distribution of giant pandas within the Qinling Mountains relative to elevation and the distribution of bamboo species in the region. We indicate our sense of whether each region serves as a source or sink population for the region.

conditions. By continuous monitoring, we mean surveys conducted at fixed times, during which information is collected on abundance, population structure, and habitat conditions.

To an extent, our 10 years of work have yielded results that fell short of expectations. We had originally hoped to produce accurate estimates of population parameters and a clear understanding of population dynamics. We have come to realize that truly understanding the population dynamics of such a long-lived species will require many more years of arduous fieldwork.

We conclude that if there are no major changes in habitat and if there are no large looming genetic issues, the Qinling population of giant pandas should persist. We believe that it would be better to invest more energy into continuing research on this species than to adopt conservation measures without a fuller understanding of the conditions on the ground.

References

Hu, J. C. 1987. 从野外大熊猫的粪便估计年龄及其种群年龄结构的研究 [A study on the age and population composition of the giant panda by judging droppings in the wild]. *Acta Theriologica Sinica* 7:81–84.

Ma, Z. E. 1996. 种群生态学的数学建模与研究 [Mathematical modeling and research of population ecology]. Hefei, China: Anhui Education Publishing House.

Meffe, G. K., and C. R. Carroll. 1997. *Principles of Conservation Biology*. Sunderland, MA: Sinauer.

Sun, R. Y. 1992. 动物生态学原理 [Principles of animal ecology]. Beijing: Beijing Normal University Publishing Group.

Wei, F. W., G. Z. Xu, J. C. Hu, and B. Li. 1988. 野生大熊猫的年龄鉴定 [The age determination for giant panda]. *Acta Theriologica Sinica* 8:161–165.

White, G. C., and R. A. Garrott. 1990. *Analysis of Wildlife Radio-Tracking Data*. San Diego: Academic Press.

Chapter 8

The Mating System of the Giant Panda

Summary: We observed both female ($n = 2$) and male ($n = 1$) giant pandas reach sexual maturity at age 4.5. From 1985 to 1996, we observed a total of 21 courtship bouts among the giant pandas. The mating season occurred from early March (March 7) to mid-April (April 11). In 1993, the mating season lasted for at least 33 days. We divided courtship bouts into two types depending on the number of males involved, monogamous ($n = 3$) and polyandrous ($n = 15$). In three bouts, we could not determine the number of males. During the mating season, males establish a hierarchy through a mechanism we do not yet understand, with the highest-ranking male positioning himself nearest to the estrous female. However, the female appears to exercise some choice of mating partners. During the 15 polyandrous mating events we observed, we documented injuries to 4 males. We characterize the mating system of Qinling giant pandas as promiscuous. A 1995 study of scent-marked trees showed that out of 22 species of trees present on the mountain ridge, 16 species were scent marked by pandas. The most frequently marked species were *Pinus armandii*, *Quercus spinosa*, *Populus davidiana*, and *Quercus aliena* var. *acuteserrata*. Pandas tended to mark trees that were distributed in close proximity to one another, as well as trees with relatively large diameter at breast height and rough bark.

Introduction

In spring, the snow-covered valleys of the Qinling resound with the calls of the giant pandas, at times resonant and melodious, at times low and harsh. Successful mating and conception are necessary conditions for the survival of a species, and the giant panda is no exception. In this chapter, we analyze the age of sexual maturity, the mating season, and mate-seeking and mating behavior of Qinling giant pandas. We also present our investigation of panda message stations, scent-marked trees, and discuss the mating system generally. This information will not only enhance our understanding of the natural history of giant pandas and, in particular, of their mating behavior in the wild but also help in the maintenance of healthy populations of giant pandas in captivity. In order to maintain healthy populations of giant pandas in captivity, it is not sufficient to have enough individuals; they must also exhibit normal sexual behavior and be able to reproduce normally without external assistance. The information we provide on giant panda behavior under natural conditions will help improve management of individuals in captivity and allow their behavior to more closely approach its natural state.

Age of Sexual Maturity

Following birth, the period of growth and development is one in which all physiological functions of an individual animal gradually mature. Only when individuals of both sexes have reached physiological maturity (i.e., the stage at which females produce mature egg cells and males produce sperm) can they engage in reproduction and produce offspring.

The normally solitary giant panda changes its behavior during the mating season, when males and females gather together and interact through sounds, chemical messages, and active behaviors. In other words, pandas have the ability, via specific behavioral signals, to communicate to the opposite sex that they are sexually mature and receptive to mating. Following such communication, individuals of both sexes can commence courtship and, ultimately, mating. Only individuals who exhibit these behavioral changes and take part in courtship can be considered to have achieved behavioral sexual maturity.

In our study we could only indirectly determined the state of an individual's physical sexual maturity from its behavior. Unless otherwise noted, sexual maturity as used hereafter refers to physiological sexual maturity expressed through behavior, by observing the giant panda's behavior during mating season. The age of sexual maturity refers to the earliest age at which we observed free-ranging giant pandas displaying reproductive behavior. In order to determine the age of sexual maturity for giant pandas, we must first determine their true age; of course, this is difficult when researching the ecology and behavior of many wildlife species. In cases in which we already know the date of birth (at least to the nearest month and year) of an individual, we can accurately calculate its age. Alas, in other cases, we were forced to rely on other methods to estimate age indirectly.

Seasonal variation in the abundance of food can also result in enamel being deposited on teeth in a layered configuration. It might be possible to use seasonal deposits of enamel on teeth (cementum annuli) to determine age in seasonal environments (Klevezal' and Kleĭnenberg, 1969; Dimmick and Pelton, 1994). Earlier work had indicated its potential for use with giant pandas (Wei et al., 1988; Xu and Wei, 1988), and thus, we conducted some experiments with this method. During April 1991 in the Qinling, a male panda named Daxiong died. We used this opportunity to section his lower right third incisor. At the same time, we conducted separate analyses of incisors from a captive panda with the same known age and a black bear, counting the annuli on these animals' tooth cross sections. The results of our investigation showed that although annuli were clearly visible on the panda's tooth, there was no obvious correlation between the number of annuli and the age of the panda. Therefore, we concluded that further research is needed before we can determine if this method can be used to accurately age wild pandas and, if so, which tooth (incisor, canine, premolar, or molar) should be used to give the most accurate results.

We decided to integrate a number of other physical characteristics, such as condition of teeth, body shape, coloration of fur, and characteristics of the animal's gait, to age giant pandas in the wild. Among these, the most important were tooth characteristics. Teeth of younger individuals were generally a clean, white color and, because they had been eating bamboo for a shorter time than older individuals, displayed only light wear, particularly on their molars. In contrast, teeth of older individuals were increasingly yellowed, and molars displayed increasing wear. Adult pandas are generally larger than juveniles, with a steady gait. In comparison, subadults and juveniles are smaller in size and livelier in their gait. Pandas in their prime generally have a sturdy build, with neat fur and lustrous coloration, whereas pandas past prime age appear to be in poorer condition, with ragged fur and faded coloration. In particular, the fur at the edge of the upper lip of older animals is long and ragged. For free-ranging pandas, we were only able to infer age from an integrated consideration of these characteristics.

FEMALES

We were able to assess the age of female sexual maturity from the behavioral changes of two females, Xiwang and Boshi. Xiwang was born on August 15, 1992, in our study area. We were able to follow and document her physical and behavioral development from birth.

In spring 1994, when Xiwang was 1 year and 7 months old, she left her mother, Jiaojiao, and began to live independently. In spring 1995, Xiwang was 2 years and 7 months old. On March 22 of that year, as we observed Xiwang walking in Shuidonggou valley, we observed her carefully sniffing the scent markings left by other pandas on the trees, sometimes stretching out her tongue to lick them. At this time, her vulva displayed none of the red swelling characteristics of female pandas in estrus (Zeng et al., 1990). In spring 1996, Xiwang was 3 years and 7 months old. During 2 weeks of continuous observation, we observed no indication that Xiwang was in estrus.

We were unable to conduct observations of Xiwang in spring 1997. However, in early August of that year, she descended before the other female pandas from the high-elevation area to a lower-elevation area on the mountain and restricted her activity to a small area. These are characteristics displayed by female pandas prior to giving birth (see chapter 9). After approaching Xiwang, we observed a scar on her back in the shoulder area that had since healed over. We inferred that this scar was the result of a bite from a male panda during mating. Thus, Xiwang went into estrus and successfully mated in the spring of 1997 when she was 4.5 years of age.

On February 18, 1993, we fitted the female Boshi with a radio collar. At that time, Boshi was still following her mother everywhere. From her physical characteristics, she appeared to be a large juvenile. According to our observations, mother pandas typically leave their offspring when they are 1.5 to 2.5 years of age, at which time they enter their next reproductive cycle (see Chapter 9). At this time, Boshi was probably between 1.5 and 2.5 years of age. Two years later, on April 11, 1995, while observing Boshi in a ravine 5 km from the location where we had radio-collared her, we observed her to be in estrus. There were at least three adult male pandas in the vicinity, engaged in combat. At the time, Boshi's age was either 3.5 or 4.5 years, and we considered this her first estrus.

Observations on these two pandas show that female giant pandas in the wild probably go into estrus and mate between 3.5 and 4.5 years of age. Given that some females, such as Xiwang, had their first estrus after the age of 3.5 years, we consider the age of sexual maturity to be 4.5 years.

On March 11, 1989, we anesthetized a female named Jiaojiao. At the time she was quite small, only 120 cm in length, and weighed 60–65 kg. Her teeth were straight and white with very little wear, so we deduced that she was a subadult at that time. However, 5 months later, on August 15, she gave birth to a male cub who we named Huzi. Beginning in 1990, Jiaojiao grew in size steadily. When we measured her again in 1995, she was 150–153 cm in length. She kept Huzi under her care for 2.5 years. However, the succeeding three cubs she raised all left her after only 18 months each. Considering this information, we inferred that Huzi was her first cub and that she had raised him for an extended period, possibly to compensate for her lack of maternal experience. In summary, when Jiaojiao went into estrus and mated in the spring of 1989, we believe that she was probably between 3.5 and 4.5 years of age.

MALES

To date, only one wild-born male giant panda has provided us data on the age of sexual maturity, Jiaojiao's first cub, Huzi. Huzi was born on August 15, 1989. In the spring of 1992, his mother Jiaojiao came into estrus, at which time Huzi (who was then 2.5 years old) left Jiaojiao and began to live independently. In spring 1993, when Huzi was 3.5 years old, he encountered the female Nüxia. This was around the time that we trapped Nüxia to radio-collar her. Huzi discovered Nüxia in the trap, whereupon we noted that he paced around the trap in circles continuously while emitting a bleating sound. This sound is a type of call used to express nonaggression or amicability (Schaller et al., 1985); in addition, it is always used by males during mating season when approaching a female in estrus. But 2 weeks later, when the female Momo went into estrus, Huzi simply lay motionless in a nearby tree; he seemed to be sleeping. The male closest to Momo gave a warning call to other males in the vicinity but otherwise completely ignored Huzi. Only when Momo gave off a sharp cry did Huzi get up and look down at her. Later, Huzi climbed down the tree and left the mating area. Although at this time Huzi appeared very strong, about 160 cm in length, with a body conformation very similar to other adult males, his testicles were much smaller in volume than the other males, suggesting to us that he had still not achieved sexual maturity.

In the spring of 1994, when Huzi was 4.5 years old, his behavior began to change in an obvious way from the previous year. He altered his behavior from being primarily arboreal to being more terrestrial; at the same time, his role changed from that of observer of courtship to that of participant, as we documented Huzi actively scent marking trees. In mid-March, we observed him chase an uncollared panda up a tree, after which he waited below. During the period March 19–25, when Momo was in estrus, Huzi was a constant and active participant. He continually bleated and often tried to approach Momo.

Although Huzi had grown to about 160 cm in length by spring of 1993 and by 1994 he was larger than some other adult male pandas, he failed to achieve a dominant position in contests with them. On March 24, we observed the male Xiaohuo, who was smaller than Huzi, chasing him from behind; Huzi ran down the mountain from the ridge where he had been to lower slopes far from the estrous females. Several minutes later, we noted Huzi again bleating as he climbed back up the mountain. The next day, we again observed Huzi scent marking as he walked along a ridge. At this time, his testicles had become very large (we were able to see them protruding clearly as we followed behind him). These behavioral and physiological characteristics typify male pandas during the mating season. Therefore, we believe that Huzi, who at the time was 4.5 years old, had achieved sexual maturity in March 1994. However, for various reasons, he evidently missed his opportunity to mate with any of the estrous females. In sum, we conclude that free-ranging male giant pandas achieve sexual maturity at 4.5 years, just as females do.

Generation Length of the Free-Ranging Giant Panda

We considered the time span from birth to first reproduction as a generation. As we have concluded above, wild giant pandas achieve sexual maturity at 4.5 years. At this age, a female panda can become pregnant and, in the fall of the same year, give birth to her first offspring (at which time she would be approximately 5). Thus, according to our data, the generation length of female giant pandas in the wild is 5 years.

According to our observations, although the male Huzi had already reached sexual maturity in 1994, it was not until the mating season of 1996 (6.5 years old) that he achieved a dominant position within the mating hierarchy. Male and female pandas in captivity also often have different generation lengths (Xie and Gipps, 1997).

Mating Season

SEASONALITY OF ESTRUS

From March 1985 to April 1996, we observed 21 panda courtship events (because the estrous period for females may last several days, we have treated each estrous period as a single courtship event). We have listed each courtship event we observed, including the number of male and female pandas involved, in Table 8.1. Here we must point out that we could only positively identify radio-marked individuals and then only if we obtained a radio signal. When uncollared individuals were involved, in order to avoid double-counting them, we treated the maximum number of uncollared individuals seen at any one time as the true number at that courtship scene. In reality, this represents the minimum number of uncollared individuals participating in any given courtship event. Only 2 (9.5%) of the 21 courtship events (Table 8.1, events 6 and 18) involved only uncollared pandas; at least one collared male or female was involved in the remaining 19 instances.

We documented courtship events from as early as March 7 to as late as April 11, a duration of 36 days (Table 8.1). In the years 1992 to 1996, the observed durations of courtship events were as follows:

- 1992: 17 days (March 13 to 29, $n = 2$)
- 1993: 15 days (March 19 to April 2, $n = 8$)
- 1994: 18 days (March 24 to April 10, $n = 2$)
- 1995: 32 days (March 11 to April 11, $n = 4$)
- 1996: 18 days (March 7 to 24, $n = 2$)

We list the distribution of courtship events for each 2-week period during March and April 1985–1996 in Figure 8.1. In captivity, the mating season observed during some of the same years is much longer than we report here. Working at the Giant Panda Protection and Research Center in Wolong, Sichuan, during 1991–1993, other researchers (Zhang et al., 1994) have summarized a total of 13 natural mating events involving the five females, Dongdong, Jiajia, Tangtang, Taotao, and Jiasi. Of these, over half occurred in May or June. Of note is that in 1995, the earliest of the five events occurred on February 27, whereas the last two occurred on June 28 and 29, a time span of 121 days. In 1996, also at the Wolong Center, another

Table 8.1.

COURTSHIP EVENTS INVOLVING AGGREGATIONS OF MALE AND FEMALE GIANT PANDAS, 1985–1996, QINLING STUDY AREA. A DASH (—) INDICATES UNDETERMINED.

Event	Date	Location	Female	Identifiable males	Male closest to female	Number of males involved	Primary Observers
1	March 11, 1985	Foping Daoliushuigou	Unknown	None	—	4	Pan Wenshi, Lü Zhi, Yong Yange
2	March 11, 1989	Shuidonggou	Jiaojiao	None	—	>3	Guo Jianwei, Xiang Bangfa
3	March 20, 1991	Liaojiagou	Jiaojiao	Xinxing	—	2	Pan Wenshi, Lü Zhi
4	March 13, 1992	Shuidonggou	Jiaojiao	Xinxing, Dabai, Xiaohuo	—	4	Pan Wenshi, Xiang Dinggan, Mao Xiaorong
5	March 29, 1992	Guojiawan	Unknown	Dabai, Xiaohuo, 129	—	3	Xiang Dinggan
6	March 19, 1993	Yizuoyequ	Unknown		—	—	Xiang Dinggan
7	March 19, 1993	Shuidongg	Keke	Dabai, Dahuo	—	2	Mao Xiaorong, Xaing Dinggan
8	March 25, 1993	Shuidonggou	Unknown	Dabai, Jiaoshou	Dabai	2	Pan Wenshi, Zhu Xiaojian
9	March 24–26, 1993	Shuidonggou	Ruixue	Xiaohuo	—	1	Pan Wenshi, Zhu Xiaojian
10	March 27–28, 1993	Chaijiawan	Baoma	Dabai, Xiaohuo	Dabai	2	Xiang Dinggan
11	March 29, 1993	Zhongxinmiaopu	Unknown	Dahuo	—	—	Pan Wenshi, Zhu Xiaojian
12	March 30–31, 1993	Shuidonggou	Momo	0940, Dabai, Jiaoshou	Dabai	2	Pan Wenshi, Zhu Xiaojian
13	April 2, 1993	Shuidonggou	Nüxia	Dabai	—	1	Zhu Xiaojian, Xiang Dinggan
14	March 24, 1994	Shuidonggou	Momo	Dabai, Xiaohuo, Huzi	Unknown	5	Pan Wenshi, Wang Dajun
15	April 10, 1994	Shuidonggou	Jiaojiao	Dahuo, Jiaoshou, 0940	Dahuo	4	Wang Dajun
16	March 11, 1995	Shuidonggou	Nüxia	Huzi, Dabai	Unknown	3	Xu Zhaohui, Xiang Bangfa
17	March 19, 1995	Shuidonggou	Ruixue	None	—	1	Wang Dajun, Zhu Xiaojian
18	March 23, 1995	Shuangchagou	Unknown	None	Unknown	2	Wang Dajun, Zhu Xiaojian
19	April 11, 1995	Chaijiawan	Boshi	Dabai, 0940	—	>3	Wang Hao, Fa Dali
20	March 7, 1996	Shuidonggou	Momo	Huzi	—	—	Wang Dajun, Wang Hao, Lü Zhi
21	March 23–24, 1996	Yan'erya	Nüxia	Huzi, Dabai	Huzi	4	Wang Dajun, Wang Hao, Zhang Lei

The Mating System of the Giant Panda   177

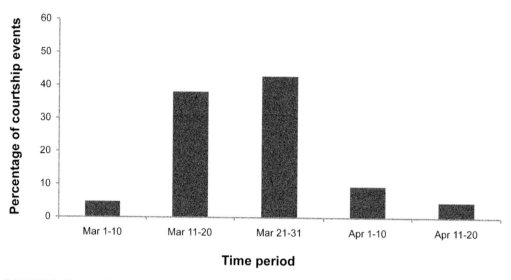

FIGURE 8.1. Timing of giant panda courtship events, 1986–1996, Qinling study area.

female came into estrus in June and gave birth in November. By contrast, of the courtship events we observed in the Qinling during 1985–1996, 18 (86%) occurred in March; the other three occurred in April.

WEATHER CONDITIONS DURING THE MATING SEASON

During the spring mating season, our practice was to spend entire days on mountain ridges or in valleys searching, listening, and observing in order to document as many courtship events as possible. We had come to learn that it was easier to find and observe courtship bouts just after the first spring rains. We recorded meteorological data during the mating season of 1993 and quantified the weather conditions during times when mating activity was and was not recorded. We documented a total of eight courtship events in 1993 (Table 8.1), of which half occurred during rainy weather, although rain fell on only 22.6% of the days, suggesting that mating activity was more likely to occur during rainy weather. Additionally, we documented weather conditions within March (early, middle, and late periods), calculating the percentage of each weather condition by period.

It is widely known that giant pandas leave scent marks on tree trunks, rocks, and other rough surfaces (Schaller et al., 1985). Fresh scent marks left by female pandas have a somewhat sour odor, whereas the urine and scent marks of male pandas have strong and weak odors, respectively (some observers have characterized the odor as musky). While searching for mating areas, we noticed that we were able to detect scent-marking odors at relatively large distances during humid conditions but only at much smaller distances during clear and dry weather. It is possible that our observations

178   Chapter 8

are relevant to the correlation between panda mating activity and the arrival of the rainy season, but additional detailed research is needed.

Communication (Scent-Marked Trees)

Giant pandas live generally solitary lives. They depend on various sorts of visual, auditory, and olfactory cues to exchange information with other individuals at various times during the year (Schaller et al., 1985). Visual cues include information transmitted directly by body posture or through direct contact with one another, as well as signs pandas leave in their environment, such as bite marks left on plants from feeding activity, claw and bite marks on tree trunks, scent marks, feces, and urine. Auditory information exchange includes information transmitted between individuals through sounds. Olfactory information is the exchange of information between individuals through the chemical components of feces, urine, and scent marking.

During the mating season, giant pandas produce large amounts of olfactory information that persists in the environment, including anal gland secretions and urine. Giant pandas frequently leave scent marks on the trunks of trees we call scent trees, which are concentrated on mountain ridges. We conducted a survey of scent trees used by giant pandas in the study area during spring 1995. In the course of our investigation, we discovered that the locations where giant pandas gather, chase each other, and mate are close to areas with high densities of scent trees. Therefore, we considered these ridges to be scent passages or information corridors. It is possible these information corridors differ in their vegetation composition from other areas within panda habitat. We conducted a vegetation survey within these information corridors to gain further insight into the mating behavior of giant pandas.

RESEARCH METHODS

Our research area was located in the southern slopes of the middle section of the Qinling Mountain Range in Huayang District, Yang County, Shaanxi Province, at elevations of 1,400 to 2,468 m. The boundaries of the area were formed by the Hetaoba River to the south, the Cha'eryan River to the northwest, Xiaoping to the north, and Yizuoyequ to the east, an area of 14 km^2 (Figure 8.2). Topographically, the area is characterized by being higher in elevation in the north than the south, with gentle slopes, complex topography, and eight major gullies: Liaojiagou, Shuangchagou, Zhangjiawan, Shuidonggou, Xiaoping Sanhaogou, Yan'erya, Chaijiawan, and Yizuouyequ. Shuidonggou was at the heart of this area, where no fewer than approximately 30 giant pandas were active, including 13 pandas we had fitted with radio-tracking collars. Our study of scent marking took place on three of the major ridges within the research area: the ridge between Chaijiawan, Shuidonggou, and Yan'erya (ridge A), the ridge between Shuidonggou,

The Mating System of the Giant Panda   179

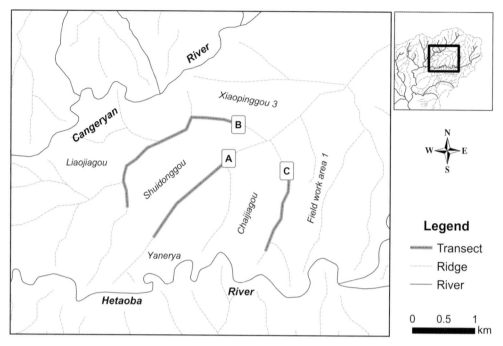

FIGURE 8.2. Schematic drawing of the study area in which we surveyed scent trees.

Liaojiagou, and Xiaoping Sanhaogou (ridge B), and the ridge between Chaijiawan and Yizuoyequ (ridge C). Elevations in the survey area varied from 1,870 to 2,050 m; the survey route was 4.9 km long. In addition, we also conducted a detailed habitat investigation on ridge A.

Through previous studies, we knew that scent trees exceeded 3 cm in diameter. Thus, we used the center quadrat method, limiting our recording to trees with diameter at breast height (DBH) >3 cm. Our survey was restricted to the main mountain ridge, ridge A, which we surveyed by following game trails along the ridge. Specifically, we used limited random sampling to select sample sites. We divided the full survey line into 20 sections and randomly selected a sampling site within each section or quadrat. In each quadrat, we determined the distance of the nearest tree to the sampling site, as well as its DBH (taken as the mean of two diameter measurements at perpendicular angles to one another). On the basis of these data, we calculated several forest metrics, including tree density, relative density, basal area, relative dominance, and importance values for each species.

Along each survey route we recorded tree species, DBH, and height of the center of scent mark from the ground (or, in the case of urine marking, its highest point) for trees that had fresh anal scent gland or urine marks. We also measured the area of some of the marks on the tree trunks. We used a χ^2 test of independence to examine giant panda selectivity for scent trees (having first lumped together infrequently encountered species) using

180   Chapter 8

the relative density of scent trees and the relative density of each tree species within the environment.

Survey Results

We used the central quadrat method to survey tree species in spring 1995 and found 22 tree species and 11 shrub species on ridge A (Figure 8.2; Appendix). All shrub species were deciduous and had yet to grow new leaves during the panda mating season (with the exception of the evergreens *Rhododendron* spp. and *Ilex pernyi*). Among trees, *Pinus armandii, Cyclobalanopsis glauca, Quercus spinosa,* and *Tsuga chinensis* were evergreens. Tree canopy density was approximately 40%–60% during the mating season, whereas the shrub canopy density layer was 0%. The bamboo *Bashania* was the dominant species in the lower tree layer, but its density was not high. Although in the area of densest growth the canopy density was as high as 80%, in most areas it was approximately 20% and was low as 10% on the ridge.

On ridge A, the mean intertree distance was 2.20 m, and the density of trees was 0.207 tree/m^2. In addition, *Pinus armandii* displayed the highest relative density and frequency, whereas *Quercus aliena* displayed the highest relative dominance. Summarizing these three indices, *Pinus armandii* had the greatest importance followed by *Quercus aliena*. The average DBH of all trees on ridge A was 15.4 cm (±12.0 cm; $n = 80$, range 3.8–89.5 cm).

On the three ridges, we documented scent marks on 18 different tree species. Among these, the primary species used as scent trees by the giant panda were *Pinus armandii, Quercus spinosa, Populus davidiana,* and *Quercus aliena* (Table 8.2). The mean DBH of scent trees on ridge A was 20.0 cm (±11.11 cm; $n = 69$, range 3–75 cm), somewhat greater than the DBH of trees in the surrounding environment.

Table 8.2.

FOR EACH DOMINANT SPECIES, THE NUMBER OF TREES DOCUMENTED WITHIN QUADRATS AND THE NUMBER WITH SCENT MARKS ON THREE RIDGES SURVEYED FOR SCENT TREES, QINLING STUDY AREA, SPRING 1995. NOTE THAT SOME TREES HAD <1 MARK.

Species	Ridge A		Ridge B		Ridge C		Total	
	Trees	Marks	Trees	Marks	Trees	Marks	Trees	Marks
Pinus armandii	14	17	16	19	6	6	36	42
Quercus spinosa	13	21	4	5	0	0	17	26
Populus davidiana	11	16	7	8	11	12	29	36
Cyclobalanopsis glauca	2	2	0	0	1	1	3	3
Quercus aliena	11	13	1	1	1	1	13	15
Other live trees	17	20	5	5	2	2	24	27
Dead snags	4	4	1	1	1	1	6	6
Total	**72**	**93**	**34**	**39**	**22**	**23**	**128**	**155**

Table 8.3.

RELATIVE DENSITY OF ALL TREES, RELATIVE DENSITY OF SCENT TREES, NUMBER OF SCENT TREES, AND SELECTION COEFFICIENTS FOR TREES FOUND IN THE SCENT TREE SURVEY AREA, SPRING 1995, QINLING MOUNTAINS.

Tree species	Relative density D_t (%)	Relative density of scent trees D_r (%)	Number of scent trees	Selectivity S^a
Quercus spinosa	16.25	18.84	13	1.16
Pinus armandii	23.75	17.39	12	0.73
Populus davidiana	12.5	15.94	11	1.28
Quercus aliena	7.5	15.94	11	2.13
Cyclobalanopsis glauca	11.25	1.45	1	0.13
Others live trees and snags	28.75	30.43	21	1.06

$^a S = D_t / D_r$.

We also found that pandas displayed variable intensities of selection for different species of scent trees. We used the ratio of the relative density of scent trees to the relative density of the species within the habitat to express the selectivity of giant pandas for a given tree species. In the Qinling, pandas exhibited positive selection coefficients for *Quercus spinosa, Populus davidiana,* and *Quercus aliena,* negative selection coefficients for *Pinus armandii* and *Cyclobalanopsis,* and no obvious selection for other tree species (Table 8.3).

ANALYSIS OF SURVEY RESULTS

Given that Qinling pandas exhibited variable selectivity for scent marking by tree species (Table 8.3), we recognize four main factors influencing selectivity:

1. Distribution pattern of trees. Among the 22 species of trees on ridge A, 16 were chosen by pandas as scent trees. Among these, the most common species were *Pinus armandii, Quercus spinosa, Populus davidiana,* and *Quercus aliena.* Although these species were all dominant locally, this dominance did not necessarily translate into selection by pandas. *Quercus aliena* was particularly concentrated along the middle and lower sections of ridge A, and there, pandas selected it for marking much more strongly than any other species. This suggests that the influence of a species' distribution pattern on the panda's scent tree selection should not be overlooked. In addition, *Populus davidiana* was quite concentrated on the upper third of ridge A. Of 11 *Populus davidiana* trees we documented on the ridge, 6 (54%) were scent trees.

2. DBH. The selectivity index for *Cyclobalanopsis glauca* was 0.13 (Table 8.3), indicating that pandas selected against the species for marking. However, on the middle portion of the ridge *C. glauca* was also characterized by small DBH, which we believe may help explain why pandas avoided it for marking. In contrast to the mean DBH of all trees on ridge A (20 ± 11.1 cm), with most trees having a DBH of 8.9–31.1 cm, only four of the nine *C. glauca*

had DBH > 8.9 cm. Since the sample points in the survey were randomly selected, we can assume that only 44% of the *Cyclobalanopsis* had a relatively high probability of being selected as scent trees.

On ridge A, although the mean DBH of scent trees was 20.0 cm (±11.1 cm; n = 69), the mean DBH of trees in the surrounding environment was 15.4 cm (±12.0 cm; n = 80). Comparatively speaking, the mean DBH of scent trees was larger. The mean DBH of scent trees on ridge B (20.0 ±10.3 cm; n = 34) was similar to ridge A, although scent trees on ridge C were smaller in DBH (16.0 ± 10.1 cm; n = 22). The density of scent trees was also lowest on ridge C.

3. Bark texture. As mentioned earlier, pandas showed positive selection for *Populus davidiana* and *Quercus aliena*. In addition to displaying a concentrated distribution in this area, these two species both also have very rough bark. This suggests that bark texture may also be a factor in panda's selection of scent trees, perhaps because odors can attach more easily and persist longer on coarse-textured than fine-textured bark. The fact that the bark of *C. glauca* is smooth may explain why pandas exhibit negative selection for it.

4. Tree odor. *Pinus armandii* is the dominant tree species in the area, and although it was the most frequently used for scent marking, pandas used few of these trees relative to their abundance. A reason for their avoidance may be obtained by comparing the Qinling with Wolong. In the Qinling, pandas exhibited no particular preference for scent marking on conifers, but in Wolong 94% of scent trees were conifers trees (Schaller et al., 1985). In the Qinling, *P. armandii* has a fairly strong odor, which can easily mask odors produced by the panda's scent marking. Thus, we infer that it may not be a desirable species for scent marking. At Wuyipeng in Wolong, the main conifer is the hemlock *Tsuga chinensis,* which has a relatively weak odor. In addition, we suspect that habitat differences also characterize the two regions. Wuyipeng has experienced little deforestation, with conifers dominating these forests, and scent trees are also largely conifers. However, in the Qinling, deforestation is more serious, and because of their higher-quality wood, conifer numbers have been reduced greatly. Our understanding of the factors influencing selection of scent trees is based on our direct observations and measurements in the field, but our assessment of odors was necessarily subjective and is subject to some possible biases that warrant further study.

Habitat Characteristics of Scent Tree Concentrations

A characteristic of early estrus is that pandas do a great deal of scent marking. Their anal gland secretions and urine serve as very important signals of physical condition and reproductive status. We have noted that areas with high concentrations of scent trees are generally located near areas where pandas congregate and contest breeding status with each other during the mating season.

We have discovered that panda scent trees are mainly distributed on mountain ridges, with almost none found on midslope or river valley locations. On the basis of our habitat and vegetation investigation on ridge A and also considering habitat information from ridges B and C, we produced an initial description of those areas in Qinling with concentrations of scent trees. Elevations of all three ridges (1,870 to 2,050 m) were similar to one another. All were also located on the Qinling's secondary plateau. Habitat conditions on ridge A were optimal for scent trees, and it was also surrounded by the largest surrounding concentration of pandas. Trees were less common on ridge C, which instead was characterized by dense bamboo (to the point where it was difficult to walk in many areas). We suspect that in such places the effectiveness of scent marks would be greatly reduced. Our 1995 survey thus leads us to following conclusion: In the Qinling, concentrations of panda scent markings occur where bamboo density is relatively low and tree canopy cover is 40%–60%.

FUNCTION OF SCENT MARKING: OBSERVATIONS OF ANIMAL BEHAVIOR

During the springs of 1994–1996, we focused our research attention on observing estrous behavior. We selected radio-collared individuals to follow, alternating among them at differing times in order to document their behavior at different stages of estrus. We generally followed an individual as it moved along an entire mountain ridge. Through our observations we discovered that both females and males were interested in scent marks during the mating season.

On March 26, 1995, we found Xiwang, who was 2 years and 7 months old at the time, on a ridge and began following her. She followed the ridge down the slope to a lower elevation, stopping at almost every scent tree and carefully sniffing. She stood on her hind legs when inspecting scent marks that were higher off the ground, leaning her front legs on the tree trunk to support herself. If scent marks were fresh, she lingered, sometimes also licking them with her tongue or biting the bark. Fresh scent marks seemed to have more attraction for her than older ones. We followed her continuously for 75 minutes, during which time Xiwang made no scent marks herself and ate only three bamboo stalks and neither urinated nor defecated. This was quite unusual behavior because pandas typically spend most of their time feeding.

During March 22–26, we had opportunities to observe Xiwang's vulva, which did not have the red and swollen appearance that would generally characterize a female panda during estrus (Zeng et al., 1990). Also, our previous analysis suggested that Xiwang had not yet achieved sexual maturity. On March 21, 1995, during the scent mark survey, we found fresh urine marks on a tree trunk, and the thick musky odor still present in the air indicated that this was the urine mark from a male. The following day, we observed an unmarked panda passing this location. When it passed the scent tree, it stopped and sniffed it carefully. Because the panda had its back to us, we were unable to see the movements of its snout. After sniffing, it turned, raised its right rear leg, and urinated on the tree. We do not know if it was trying to strengthen the odor of its own scent mark or was covering up the odor of another individual's mark.

On April 7, 1996, we saw Jiaojiao squatting in front of a larch tree, carefully sniffing a scent mark. At that time, Jiaojiao was accompanied by her cub Xiaosan, aged 1 year and 8 months. After wandering away from his mother's side a number of times, Xiaosan turned back and, using his body and head, nudged Jiaojiao, who paid no attention to Xiaosan's behavior.

Later on, in the fall of 1996 when we returned to the research area, we discovered that Xiaosan had by then left Jiaojiao and was living independently. On further investigation, we discovered that Jiaojiao had given birth to another cub. Thus, Jiaojiao must have gone into estrus and successfully mated shortly after we had observed her the previous spring. This then raises the question of whether there was a connection between interest in the scent marks and her subsequent estrus and mating.

Courtship and Mating

METHODS OF DETERMINING ESTRUS

Free-ranging pandas live in densely forested habitats. Although our field station was located on the mountain slope, to make detailed observations of pandas, we first had to determine where our radio-collared individuals were and then had to trudge through the mountains on small trails over distances of 10 or 20 km before we could get close to them. Even then, however, observing pandas was not easy. Normally, pandas would quietly slip away from us into dense bamboo as soon as we got close. During the mating season, however, pandas in estrus became somewhat less sensitive to our presence, allowing our long treks to be rewarded with short glimpses of their behavior.

We began our research into the behavior and habitat of free-ranging pandas in Qinling in 1985. However, we did not obtain our first observations of panda behavior during the mating season until 1989. As we gradually understood more about panda behavior and we slowly accumulated experience and interactions with free-ranging pandas, some individuals, gradually and to different degrees, adapted or habituated to our proximity. This provided us increased opportunities to observe them directly, so that by 1994 and 1995 we were actually able to observe giant pandas mating in the wild.

During estrus, pandas change their behavior in clear ways. Females exhibit decreased appetite and increase their scent marking and calling behavior. When actually approached by a male, a female will flatten her back and raise her tail when mounted by a male (Kleiman, 1983). In most cases we were unable to see all behaviors of both animals during copulation. However, it was clear that during mating season both sexes interacted with each other frequently: Males followed females, biting their backs or shoulders in attempted mating, and in general displayed an increase in direct physical contact (Kleiman, 1983). Therefore, we made determinations regarding which animals were in estrus (and, by extension, had reached sexual maturity) on the basis of their behavior. Thus, although we lacked the ability to make detailed observations and measurements of hormone levels common for captive animals (Zeng et al., 1990, 1992, 1994), we used these other signals to understand the reproductive condition of radio-marked females.

First, we used the aggregation of collared male and female individuals to judge whether or not they had entered estrus. Except for the time spent rearing cubs, female pandas are solitary. However, during the mating season, male and female individuals will gather together for several days. We classified these aggregations into three types of situations: (1) two or more radio-marked pandas present (possibly with additional unmarked individuals), (2) one radio-collared panda of either sex interacting with an unmarked pandas (or pandas), and (3) aggregations of unmarked pandas. In the first situation, we documented aggregations from radio signals. However, in the second two situations, we had no way to document that an animal was in estrus unless we managed to observe directly.

Each year during the mating season our procedure was to walk every day along the mountain ridges and valleys to locate each individual. When signals indicated an aggregation, we made every effort to approach these individuals to determine reproductive condition. Determining the distance between individuals from radio signals helped us more quickly understand relationships among radio-collared individuals. But most pandas were not radio marked, and uncollared pandas also showed up in mating aggregations. In any given courtship event, we observed the behavior of both collared and uncollared individuals.

Another method we used to find estrous pandas was to listen for and hone in on their mating calls. Pandas calls frequently during the mating season (Peters, 1982, 1985; Kleiman, 1983; Schaller et al., 1985), and their unique characteristics made them easy to detect. However, finding mating aggregations from panda calls was unpredictable, and we used this method primarily as an adjunct to radio tracking.

From 1989 to 1997, we observed a total of 21 aggregations and courtship events (Table 8.1). We categorized courtship into two types on the basis of the number of males involved: monogamous and polyandrous.

MONOGAMY

Of the 21 courtship bouts we observed, there were only 3 in which only a single male and a single female were present. These three instances were

(1) Ruixue and Xiaohuo, March 24–26, 1993 (Table 8.1, event 9), (2) Dabai and Nüxia, April 2, 1993 (Table 8.1, event 13), and (3) Ruixue and an uncollared male, March 18–20, 1995 (Table 8.1, event 17). Our information on the April 2 event involving the male Dabai is scanty: the female panda quickly climbed down from a tree, followed by Dabai, and both disappeared from view. Thus, we focus on the other two bouts.

1993 Courtship Bout of Ruixue and Xiaohuo

From March 18 to 20, 1993, telemetry locations indicated to us that the male Xiaohuo and the female Ruixue were in close proximity to each other; we confirmed that they were still close to each other on March 22 and 23. On March 24, we first heard the calls emanating from the two pandas engaged in combat. We headed in the direction of the sounds, and when we located them on the slope, Ruixue was leading Xiaohuo. She then sat down at a high point on the slope while he walked repeatedly back and forth below her, giving off moaning sounds, chomping and clapping his jaws, and appearing agitated. On several occasions, Xiaohuo attempted to approach Ruixue, but each time she retreated while making a chirping sound. Shortly afterward, both animals began foraging on nearby bamboo. Less than 5 minutes later, Ruixue stopped eating, climbed up a tree (~10–15 cm diameter), and settled down at the fork of a large branch. At this point, Xiaohuo walked to the foot of the tree, set his front paws on the trunk, and began climbing up. However, Ruixue gave off a shrill cry from her perch, stopping Xiaohuo in his tracks. Within 10 minutes, however, Ruixue became aware of us, descended from her perch, and ran off into deep forest, chased by Xiaohuo.

The next day, we followed radio signals to discover Ruixue and Xiaohuo together on a small ridge roughly 100 m higher than where they had been the previous day. When we arrived at <100 m distance, we began hearing roars, bleats, and jaw clapping. On observation, we discovered that Ruixue was sitting in the crotch of an oak (*Quercus spinosa*) tree (DBH about 20 cm), and Xiaohuo was below her. Both were moaning, roaring, bleating, and clapping their jaws, but it was hard to distinguish which sounds were made by which animal. As soon as Xiaohuo heard us, however, he rushed in our direction, making threatening moans, roars, and jaw claps. At this point, Ruixue descended her tree, ran away toward the valley, and climbed up another tree (a poplar this time), followed by Xiaohuo. By this time, rain had begun to fall, and a dense fog had rolled in, obscuring everything from view, so we made no attempt to follow them.

Later, as we were inspecting the scene of the courtship bout, we noticed both claw marks and the odor of scent marks on the oak tree in which Ruixue had earlier perched. On a neighboring pine tree, we noticed a bite mark about 30 cm long. We speculate that these were both made by Xiaohuo while Ruixue was up in the tree.

On the third day, both pandas continued to move toward higher elevations. When we found them, Ruixue was sitting in a tree, with Xiaohuo bleating away beneath her. They seemed oblivious to our presence. When we got close enough to see, we noted that Ruixue's vulva was wrinkled and pale pink, quite unlike the red, swollen genitalia of the female panda we would expect had she been in early or peak estrus. (Observations of female pandas in captivity have shown that in early estrus, the vulva begins to swell. By the peak of estrus, the vulva is swollen and a rosy red color; in late estrus, the vulva begins to shrink, its color gradually fading from rosy red to a yellowish, fleshy color [Zeng et al., 1990].) In addition, we noted bite marks on her back, which we guessed were made by a male during a mating bout. By the fourth day, the two pandas had separated, and Xiaohuo had moved to another valley. During the entire week, the two pandas had moved a straight-line distance of only approximately 350 m.

1995 Courtship Bout of Ruixue and an Uncollared Male Panda

In 1995, when Ruixue again entered estrus, there was also only one male panda nearby, and he was uncollared. We documented the two of them involved in a rather intense tussle.

On March 18, 1995, we used the sounds of their cries to locate the pair. We watched as Ruixue descended from a tree, whereupon the male climbed onto her back, holding her down. Ruixue then broke free of the male's grasp and climbed up another tree. When Ruixue again climbed down the tree to about a meter above the ground, the male panda came forward, bit her on the back, and dragged her back down. During this time, neither panda ceased vocalizing with groans and chirps. After 70 minutes, the female attempted to climb up the tree but was blocked from doing so by the male.

When the animals happened to move closer to us, we were able to hear a continuous bleating sound. Later, we observed the animals sitting, with the male behind Ruixue. Within a few minutes, Ruixue escaped from the male and climbed up a tree. Below the tree, the male continued vocalizing with honking sounds and jaw clapping, seemingly guarding her, heedless of our close proximity.

The following day (March 19), we documented a combat bout between Ruixue and a single, uncollared male. On the third day, Ruixue was still together with the uncollared male. Although we heard sounds of a struggle coming from where they were located, by the time we arrived, the two individuals were both lying on the ground not far from each other. When we got closer, Ruixue moved off, but the male remained behind. Within approximately an hour Ruixue's radio signal had become too weak to detect, suggesting that she had moved quite some distance away. We had no way to confirm that the male panda appearing on these three days was the same single individual, but we can be sure that a single uncollared male was following Ruixue on all three days.

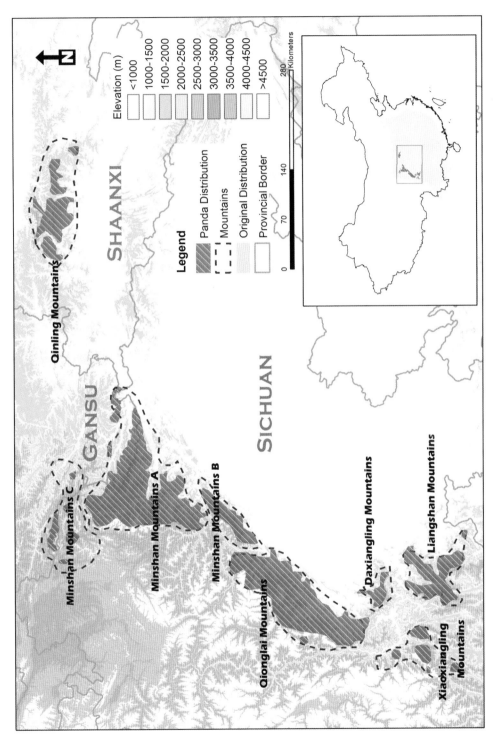

PLATE 1. Current (2012) and original (inset) distribution of giant pandas. For current distribution, we indicate the mountain ranges occupied and their province on the basis of information from the Third National Giant Panda Survey (China's Ministry of Forestry, 2006). The map was created by Wang Fang.

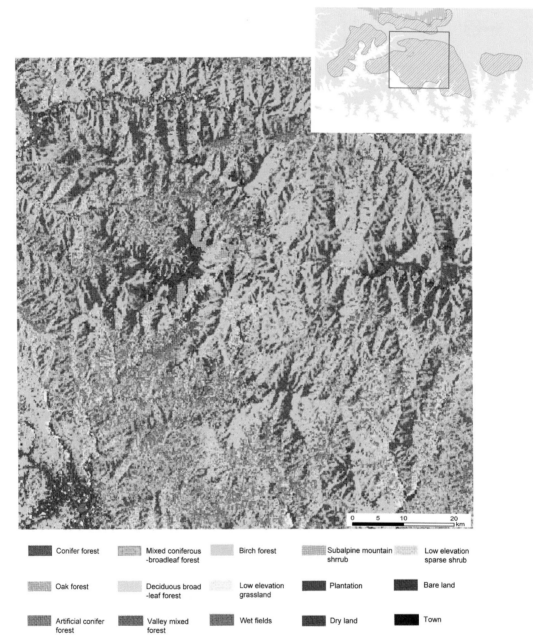

	Conifer forest		Mixed coniferous -broadleaf forest		Birch forest		Subalpine mountain shrrub		Low elevation sparse shrub
	Oak forest		Deciduous broad -leaf forest		Low elevation grassland		Plantation		Bare land
	Artificial conifer forest		Valley mixed forest		Wet fields		Dry land		Town

PLATE 2. Vegetation map of the Xinglongling area of the Qinling's southern slopes, as interpolated from satellite imagery (Landsat Thematic Mapper, August 15, 1994). The map was created by Fu Dali. The inset indicates the portion of the study area covered by the map.

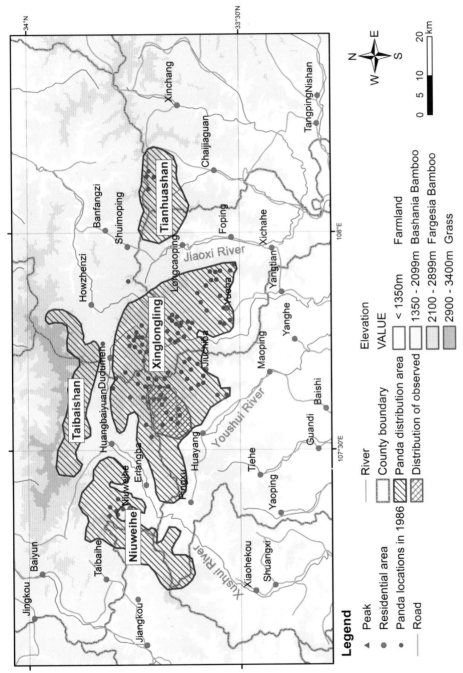

PLATE 3. Map of the study area surroundings in Shaanxi Province, China, showing the approximate area of panda distribution (red hatching), the study area in Shanshuping Township (green boundary), and individual panda sightings in 1986 (red circles). The four panda areas within the region are indicated with the names that are used in the text. The map was created by Wang Hao.

PHOTO 1. Jiaojiao and her cub in their den. Photo by Pan Wenshi.

PHOTO 2. The survival and well-being of Jiaojiao (pictured here with a cub) and her offspring weighed heavily on all of our hearts during the study. Jiaojiao produced five offspring during the period August 1989 through August 1998. Photo by Pan Wenshi.

PHOTO 3. Jiaojiao leads her 7-month-old female cub Xiwang through the snow-covered forest in March 1993. Photo by Lü Zhi.

PHOTO 4. Jiaojiao leads her male cub Xiaosan over a mountain stream in May 1995. Photo by Wang Dajun.

PHOTO 5. Momo, lifting her newborn cub Baobao with her mouth. This photo was taken on August 28, 1993, when we estimated the cub to be only 3 days old. Newborn panda cubs are only about 1/900 the mass of their mothers. Photo by Pan Wenshi.

PHOTO 6. (left) Jiaojiao's male offspring Huzi in May 1995, at the age of 5 years, 8 months. Although he had long before adopted a life independent of his mother, his home range overlapped Jiaojiao's home range by 75%. Photo by Pan Wenshi.

PHOTO 7. (below) By following its unique vocalization, we were able to locate Shigen, the offspring of Ruixue, which we estimated to be only about 36 hours old. It did not yet have the sense of sight or hearing, but its cry was loud and clear. Photo by Pan Wenshi.

PHOTO 8. The 9-month-old cub Huzi, resting on a pine tree above a steep gorge at 1,850 m. When this photo was taken, Huzi's mother, Jiaojiao, was foraging below him on new *Bashania* shoots. Photo by Pan Wenshi.

PHOTO 9. Jiaojiao holds her fourth newborn cub, Xiaosi, under her chin, as she typically did for many hours of the day. As with other ursids, giant pandas undergo a prolonged fast immediately following parturition. Photo by Pan Wenshi.

PHOTO 10. Hillsides below 1,350 m in the Qinling, such as the one pictured here, had long since been under human cultivation. Photo taken in April 1994. Photo by Pan Wenshi.

PHOTO 11. By August 1992, timber harvest had already progressed upward from the valley bottoms to the main crest of the Qinling. Photo by Pan Wenshi.

PHOTO 12. On March 2, 1994, we observed adult male Dabai rubbing his hindquarters on a tree, leaving behind a scent mark. Photo by Zhu Xiaojian.

PHOTO 13. Adult male Xinxing pursuing estrous female in Shuilan on March 7, 1994, in a mountain gorge at about 1,800 m. At this time of year, the hills resound with the vocalizations of pandas in search of mates. Photo by Pan Wenshi.

PHOTO 15. Adult female Ruixue (in tree) and adult male Xiaohuo (on ground) during the mating season. Our interpretation is that Ruixue was not yet ready to accept the male's overtures and thus took refuge in the tree. Xiaohuo guarded her and warned off other males by clapping his jaws and roaring. The photo was taken on March 10, 1995. Photo by Pan Wenshi.

PHOTO 14. Xiwang, at the age of 3 in February 1996, using her sense of smell to help her determine which panda left the scent mark she is investigating. Pandas frequently use olfaction to learn about their surroundings. Photo by Pan Wenshi.

PHOTO 16. In May 1994 at the age of 22 months, Xiwang (right) stayed very close to her mother, Jiaojiao (left). However, this duo broke up 2 months later when Jiaojiao gave birth to her third cub. Photo by Pan Wenshi.

PHOTO 17. Huzi (left) fighting with an unidentified adult male panda in April 1995, whom he subsequently drove off. At the age of 5 years and 9 months, Huzi patrolled his own home range, which overlapped that of his mother, and kept intruders away. Photo by Xiang Dingqian.

PHOTO 18. Jiaojiao sleeps while her fourth cub, Xiaosi, then 20 days old, suckles at her teat. The photo was taken on September 12, 1994. Photo by Pan Wenshi.

PHOTO 19. Graduate students Zhu Xiaojian (right) and Zhou Zhihua (left) atop a tree stand where they spent much time recording the behaviors of Jiaojiao and Xiwang at the den site. Photo by Pan Wenshi.

PHOTO 20. The research crew, accompanied by others, rescuing an elderly and ill panda so that it could receive medical attention. Photo by Xiang Dingqian.

PHOTO 21. When Jiaojiao left her den on September 25, 1994, we took the opportunity to examine her male cub Xiaosan closely. Photo by Pan Wenshi.

PHOTO 22. On March 26, 1985, our team was the first to document the phenomenon of a brown-and-white colored panda when we "rescued" this individual in the Qinling; we named her Dandan. In spring 1992, a second brown and white panda, whom we named Keke, was radio-collared. Photo by Pan Wenshi.

PHOTO 23. This structure, which had earlier been used by forestry workers, became the project's research base during 1986–1999. The photo shows some of the students of Professor Pan Wenshi setting out on a biodiversity survey on a tractor. Photo by Pan Wenshi.

PHOTO 24. The upper montane region, from 1,350 m to the highest peaks at 3,700 m, where no human dwellings occur, has become the last bastion of security for the giant pandas of the Qinling. Photo by Pan Wenshi.

PHOTO 25. In April 1995, the 6.5-year-old panda Huzi treated researchers as a normal part of his surroundings. Here he looks at researcher Wang Dajun while eating bamboo as Xiang Bangfa looks on. Photo by Pan Wenshi.

PHOTO 26. In March 1992, when Huzi, a subadult of 3 years, awoke from his nap in the tree, he found researcher Qin Dagong below him and descended to investigate. Photo by Pan Wenshi.

PHOTO 27. Researcher Zhu Xiaojian (left) offering a bamboo shoot to the 4 year old male Huzi, who had approached her (May 1993). Photo by Pan Wenshi.

PHOTO 28. Researchers Lü Zhi (foreground) and Pan Wenshi (background) checking the pulse of an evidently starving panda that had appeared at the door of a farmer's house in search of food on October 12, 1993. The animal was so weak that the researchers could handle it without immobilizing drugs. Photo by Xiang Dingqian.

PHOTO 29. Wang Dajun (far right) meeting with staff of Changqing Nature Reserve prior to a panda survey (2008). Photo by Xiang Dingqian.

PHOTO 30. The patrolling team of the Changqing Nature Reserve at one of their range stations (2008). Photo by Xiang Dingqian.

PHOTO 31. In March 2008, Lü Zhi (far left), Xu Zhihong (former president of Peking University, second from the left), and Wang Dajun (far right) dine at the restaurant owned by Wang Shuwen (second from the right), who used to be a staff member of the panda research team. Photo by Xiang Dingqian.

PHOTO 32. Huayang Town, located at the entrance of Changqing Nature Reserve, in 2012, it is now focused on ecotourism. Photo by Xiang Dingqian.

PHOTO 33. In April 2010, a wild giant panda near a ranger station in the Changqing Nature Reserve. Photo by Xiang Dingqian.

PHOTO 34. Deciduous forest in the Changqing Nature Reserve has an abundant bamboo understory (2008). Photo by Xiang Dingqian.

PHOTO 35. Pan Wenshi (right) revisited the Changqing Nature Reserve in 2010 and met his previous field assistant Xiang Bangfa (left). Photo by Xiang Dingqian.

POLYANDRY

Two or more males were present in 15 of the 21 aggregations or courtship events we observed. We observed mating behavior in two of these; in the other cases, we observed only combat behavior.

Observation of Polyandrous Mating 1

It rained at our research base from April 7 to 9, 1994. By April 10, the rain had stopped, but the mountains were enshrouded in dense fog. That day, we heard the sound of intense panda combat coming from a gully in Shuidong-gou. We could distinguish sounds from no less than five individuals: one gave low, resonant moans, two gave barks, one bleated, and another was moving in the distance. Although this fifth animal made no cries, we could hear the sounds of bamboo as it moved through it. Telemetry revealed that pandas in the vicinity included female Jiaojiao and males 0940, Jiaoshou, and Dahuo.

As Jiaojiao gave off a series of gentle bleats, an uncollared male (who was either Jiaoshou or 0940) followed her, also bleating, and proceeded to mount her. But Jiaojiao seemed uncooperative; this male mounted her three times without success. At this time, Dahuo charged over, roaring at the male mounting her and causing him to flee. Dahuo then proceeded to mount Jiaojiao, the animals bleating the whole time. Suddenly, an uncollared male (who we refer to below as unknown male) charged toward the mounted pair, pushing Dahuo off Jiaojiao's back. The two males bellowed at each other, but this fighting did not seem to result in serious injury. Dahuo then mounted Jiaojiao again, and the unknown male once again rushed up to stop him; this behavior was repeated four times before the unknown male ceased his harassment of Jiaojiao and Dahuo.

At approximately noon, we observed that Dahuo had climbed onto Jiaojiao's back and both animals were bleating. After 5 minutes, we noted that while Jiaojiao continued to emit a bleating sound, Dahuo's vocalization had changed to intermittent moans, and his abdomen jerked spasmodically. Within 4 minutes, the two had become rather still and ceased vocalizing, but Dahuo remained mounted on Jiaojiao's back. A minute later, the two individuals separated, whereupon Jiaojiao walked over to some nearby bamboo, while Dahuo sat on the ground, panting heavily, and then lay down as if very tired. He rested in this way for approximately 10 minutes.

Just as we were preparing to determine the identity of the other pandas in the vicinity, we heard Jiaojiao and Dahuo bleating again, and noticed that they had again come together. Jiaojiao was now alternately emitting two kinds of cries: toward Dahuo she gave gentle bleats that suggested receptivity, but toward the unknown male she gave undulating moans that seemed to express threat or intimidation. Dahuo then rushed and faced the unknown male, at which both males bellowed and groaned at each other but avoided physical contact. The unknown male then retreated, and Jiaojiao and Dahuo resumed feeding on bamboo while seated on the ground.

Half an hour later, Jiaojiao, who had been lying down, sat up and again began bleating to Dahuo. She then stood up, lowered her back and abdomen, raised her tail, and adopted a submissive posture suggesting receptivity to mating. At this, Dahuo threw down his bamboo, walked over to Jiaojiao, stood on his hind legs, and, using his forelegs, mounted her. However, this mounting did not last long, and Jiaojiao gave off several honks and once again sat down to eat. Several minutes later, we noted that once again they were in the copulatory position. Afterward, everything quieted down. Jiaojiao and Dahuo sat down to sleep amid the bamboo, with Dahuo snoring very loudly, neither animal showing an overt reaction to our proximity.

Observation of Polyandrous Mating 2

On March 23, 1995, one ridge away from our camp, we discovered another courtship bout. This bout was the closest to our camp we ever documented, and we could even hear panda vocalizations and observe activities right from the road. After climbing for about 100 meters we could see that on a slope across a gully, at about 1,600 m elevation, were three uncollared pandas. One panda (referred to hereafter as panda A) sat beneath a tree, while another (panda B) walked several meters along the slope, then turned back. The third (panda C) stood upslope from these two at some distance. While we observed this scene, we heard three types of vocalizations: bleats, sharp cries, and moos. Because of our distance, we were unable to distinguish which sounds were made by each individual.

We observed panda B gradually approaching panda A on the slope, until the two individuals stood face to face. As B approached, A stood up and stretched out its left front paw and swatted B, which remained upright. Panda B then stopped and sat down, raised its head and looked about briefly, then fell to the ground and began rolling. As this was occurring, we heard two vocalizations: bleats and sharp cries. We guessed that the bleats were made by B and the sharp cries by A. While B rolled, A sat on the ground with its head lowered and its back facing us. It then raised its head in the direction of B. We heard a rather soft shrill cry, at which point then B also stood up, turned around, and backed up toward A. At this time, we could still hear bleats and sharp cries. When B's hind-quarters reached A, A lowered its head and twice approached B's buttocks, then slowly turned around until its back faced B. At this, B gradually turned around. In this way, the two pandas shifted their positions such that A was now in front, facing B's hindquarters. B then turned around again, put its two front legs on A's back, and assumed the mounting posture. At this time, although we still heard bleating and sharp cries, the frequency of crying diminished, and the pitch of cry also lowered. Following this, as the two pandas in the copulatory position continued bleating, panda A gradually turned around, and B moved along with it, but the two individuals did not separate and continued to adopt a mating embrace. From these behaviors, we inferred that A was a female and that B and C were males.

After 10 minutes of observation, the third individual (C) walked down from the slope above toward the mounted pair. When he arrived to within 5 or 6 m, female A and male B separated, with A climbing up the nearest tree and B running 5 to 6 m down the slope, at which point he too climbed a tree; male C followed male B to the tree, where he waited. At this point, none of the animals were vocalizing. Male B sat motionless in his tree, as if resting. After 17 minutes, female A climbed down from the pine tree she had been in and walked up the slope, crossed the ridge, and disappeared from view. After this, first one and then the other male also departed from view.

Relations among Males during Mating Bouts

As we have seen, courtship bouts did not involve only individual males and females; often, multiple males were involved. From our observations we learned about interrelationships among these males.

ADULT AND SUBADULT MALES

From March 30 to 31, 1993, we documented a mating bout involving 0940 and Momo that occurred on Shuidonggou's east ridge. When we first arrived, Momo was beneath a tree and 0940 sat near her. At the same time, we noted that Huzi was sitting on a branch of the same tree, his head lowered. When Momo or 0940 vocalized loudly, Huzi would raise his head and look down at them, but at the other times, Huzi appeared to be asleep. The vocalizations of 0940 seemed intended to threaten any other individuals in the vicinity, but only twice did he climb up the tree trunk toward Huzi while bellowing at him, and each time he got only a meter or so up the trunk before retreating. Huzi, meanwhile, seemed to pay him little heed and continuing sitting on his branch. Within about a half hour, Huzi finally climbed down and walked away, which elicited no discernible reaction from Momo or 0940. (At this time, Huzi was 3 years and 7 months old and thus still a subadult.)

Relations among adult males involved in courtship situations are not quite so simple. As we discussed earlier in the section on age of sexual maturity, the next year, when Huzi reached 4.5 years, he played a completely different role during mating bouts, that of a prospective participant. By then, other pandas no longer ignored him as they had in 1993. Instead, they now displayed the rich and varied range of behaviors consistent with regarding him as an equal. We thus would characterize the relations we documented between adult and subadult male pandas during mating bouts as relatively restrained, and they likely involved no intense contesting or combat. We speculate that subadults are able to use some method of broadcasting their status (perhaps olfactory) and therefore avoid contests with older rivals.

RELATIONS AMONG ADULT MALES

On March 25, 1993, we observed Dabai (Table 8.1, event 8) sitting silently and motionless at the base of tree in which an uncollared female panda sat. Meanwhile, the male Jiaoshou was only about 5 m away in a bamboo grove, bleating but otherwise making no movements toward the tree in which the female sat.

During March 30 and 31, 1993, we observed the female Momo exhibiting signs of estrus (Table 8.1, event 12). We observed Momo sitting below a tree in which the subadult Huzi had perched and that the adult male 0940 was also situated close to her. Less than 10 m away, in a bamboo grove, we could dimly make out yet another panda, this one uncollared. Each time the uncollared animal emitted a bleat, 0940 responded with threatening roars, groans, and jaw claps. With each noise coming from where it seemed the uncollared panda was, 0940 immediately charged in its direction. This seemed to show that 0940, closest to Momo, was dominant in this situation, whereas others individuals, located slightly farther away, were subordinate.

On March 24, 1994, we again observed Momo displaying signs of estrus (Table 8.1, event 14). This time, five pandas were located within 10 m of one another and Momo (the radio-collared males Xiaohuo, Dabai, and Huzi, as well as two uncollared pandas), and all were vocalizing with roars and jaw claps. Before long, one of the unmarked males (unknown male 1) chased unknown male 2 into an adjacent bamboo grove. From our position, we could not see what was happening there, but we heard loud, high-pitched cries, sounding extremely miserable, as if an individual had been seriously injured. Upon approaching closely, we discovered unknown male 2 lying on his side, groaning loudly but seemingly completely unharmed, with unknown male 1 standing next to him, regarding him intently. Over our nearly 4 hours of observing these two individuals, we saw no additional injuries or bleeding. We did find some panda hair about 20 m from the courtship scene, which we guessed had been left during a fight, but we did not actually observe fighting between the pandas that might have produced bodily harm.

On March 23, 1996 (Table 8.1, event 21), when Nüxia came into estrus, Huzi was the male closest to her. At that time, Dabai and two uncollared males (unknown males A and B) were also nearby. Nüxia had just moved from one side of the ridge to the other, and Huzi had followed closely behind her, followed by Dabai, then unknown male A and unknown male B.

As Huzi sat on the ridge, Dabai, unknown male A, and unknown male B climbed up toward her, one after another. When they saw Huzi, they exchanged calls and then retreated. During our observations, Dabai climbed up the ridge twice; unknown male A charged up once, and unknown male B rushed up in front of Huzi and then retreated straight to the bottom of the slope, temporarily leaving the courtship scene.

On the basis of our observations, we postulate that when multiple male pandas fight for mating privileges with an estrous female, their positions within the dominance hierarchy can be determined by their closeness to or distance from the female panda. The more dominant male is nearest to the female because this male is able to block other males from approaching her.

In 1993, Dabai was the dominant male in two fights (events 8 and 12); however, in 1995, he was the loser in struggles with an uncollared individual (event 16); in 1996, his dominant position was superseded by Huzi (event 21).

In 1994, Huzi actively participated in mating events, but he was a timid explorer; as soon as a conflict occurred, he quickly fled, and only when the conflict had passed would he come forward again, vocalizing with bleats (event 14). However, in 1996, he had moved up to the dominant position (event 21). This progression provides evidence that the dominance rank of males can change with age.

We observed adult males participating in up to six mating-related fights within a year (Table 8.1). In 1993, the male Xiaohuo spent 3 days alone with the female Ruixue; we documented no other pandas in the vicinity during that time. However, on March 28, we heard a ferocious roar and saw Xiaohuo, breathing heavily, running down from the slope to one side of the gully from whence he proceeded to run up the slope from which we were observing him. It was clear that he had lost a fight with another individual. This type of situation also occurred with other male pandas. In 1993, the male Dabai appeared on March 25 along with the male Jiaoshou at the courting site of an uncollared female panda in estrus. As we watched the uncollared female sitting in a tree, Dabai sat below while Jiaoshou remained hidden beneath a clump of bamboo about 10 m away, bleating continuously. It was clear to us that Dabai was dominant to Jiaoshou. On March 28 and 30, Dabai was once again closest to the female in estrus. However, in two other courtship events, Dabai was far from the estrous female. Thus, it appears that the dominant male in any given courtship event does not necessarily maintain that position during every courtship bout within that year's mating season; rather, the dominance hierarchy is variable.

INJURIES TO MALES DURING THE MATING SEASON

Although we never directly observed battles between individuals resulting in fatal injuries, we did observe evidence of external injuries on some males. On March 13, 1992, we noted an obvious fresh bite wound on the right forelimb of Dabai. On March 28, 1993 (Table 8.1, event 10), while on a ridge in the Chaijiawan area, we heard furious cries coming from the opposite slope. Our radio tracking revealed that both Dabai and Xiaohuo were present on that slope. Two minutes later, the vocalizations ceased, after which Xiaohuo, breathing heavily, charged up to our ridge. We could make out a fresh wound on the left side of his nose; there was no blood, but we could clearly see flesh color, suggesting a wound that exposed muscle tissue. We deduced that he had been injured while fighting with Dabai. On March 24, 1996, we came across bloody footprints from a panda while we walked along a ridge. After careful observation we concluded that the blood had come from its paws. During the 1996 mating season, Huzi also had a scar on his back, but because it was nearly healed over, it was difficult for us to determine if the scar had been produced by fights with other males during the mating season or, alternatively, if he had already sustained the injury prior to the mating season.

Among the 15 courtship bouts that involved 2 or more males, we observed 4 that resulted in visible injuries. On March 31, 1993, we found panda hair some 20 m away from a courtship site and presumed that it had been produced in a fight; however, on that day we observed no bloody fights between pandas, nor did we see any individuals with external injuries. In addition, during the aforementioned courtship bout involving Jiaojiao, Dahuo, and several other individuals, we observed no serious fights among males. Physical contact between them was limited to one instance of an uncollared panda shoving the mating male off the female's back. Instead, to impress, threaten, or otherwise communicate, males depended primarily on vocalization.

To summarize, although on occasion males engaged in physical clashes and combat over estrous females, in the overwhelming majority of cases we observed, males relied on sound in their interactions with each other. That said, loud cries from pandas during the mating season did not necessarily represent fights, nor did they necessarily indicate that injuries had occurred.

Compared with the mild interactions among mating pandas we documented in the Qinling, the behaviors of mating pandas in Wolong were seemingly more intense. The literature records that during the two courtship bouts observed (one in 1981, the other in 1983), multiple males were involved in both (Schaller et al., 1985). Although in both situations observers heard loud cries from the pandas, only in the 1983 observation were blood marks observed on the forehead and ear of a large panda. At that time, the largest number of males recorded simultaneously at a courtship bout was five (although subsequently as many as seven males were documented at a courtship bout). Despite that, researchers in Wolong documented only a single fight among courting males.

Mating Systems

MALES

From the descriptions above, we have seen that courtship bouts are characterized by the presence of one female and one or more males. Table

8.1 provides our best attempt to enumerate the males and females observed during each bout. We hasten to point out that we could only identify individuals accurately when they were radio marked, and we also used radio telemetry to estimate distances among individuals present during a courtship bout. To avoid duplicating counts of uncollared individuals, we always used the maximum number of uncollared individuals observed at any one time.

We were able to confidently determine the number of males present in 18 courtship events (Table 8.1). Of these, only one male was present in three cases (16.7%). In the other 15 cases (83.3%) from two to five males were present in the vicinity of the female. At the same time, our summary illustrates that it was not uncommon for a male to appear at courtship bouts of more than just one estrous female. All four of our radio-collared males, Huzi, Dabai, Xiaohuo, and Jiaoshou, participated in two or more dominance-related fights during any given mating season. In particular, Dabai appeared at different courtship bouts on March 19, 25, 27, 28, and 30 at which he contested with collared and uncollared males for mating privileges.

Therefore, we suggest that the mating system among male giant pandas is best categorized as promiscuous. During the reproductive season, males compete with other males as much as possible in order to gain mating rights with adult females.

FEMALES

In the wild, our observations of giant pandas during estrus and mating have been extremely limited. However, even fragmentary information is very helpful and valuable as a supplement to research conducted in zoos and by other methods in making inference about the mating system of female pandas. Our studies used radio telemetry and audible sounds made by pandas to locate mating aggregations. As explained previously, on 15 instances we documented multiple males present at courtship sites. However, we could never determine if the female involved mated with multiple males or with just one.

We know that at Wolong, a female panda once mated with three different males within two days (Schaller et al., 1985). While making observations of an estrous female at the Beijing Zoo in 1995, we noted that she bleated continuously as she entered the peak estrous period. As soon as a male panda entered the cage, she approached him on her own initiative, assumed a mating posture (raised tail and lowered back), and backed toward him. However, she did not limit this type of behavior to just the one male, but rather repeated it with almost every male that entered the cage. If we assume that the peak of estrus in the wild also lasts for several days, females would have plenty of opportunity to mate with multiple males. Our observations have shown that after mating, males leave females to search for new females. At this time, other males would thus have opportunities to approach and mate with the female.

The Mating System of the Giant Panda   195

In 1995, a combination of observational and molecular evidence showed that an adult female panda could mate with multiple males during a given estrous period. We collected tissue samples from three juveniles born within the study area (Xiwang, Xiaosan, and Guiye), as well as from adult males we had radio collared. The results of our genetic analysis showed that none of the collared males had fathered any of the cubs (Lü et al., 2001). In 1992, we observed that the female Jiaojiao was close to the male Dabai and in 1993 that male 0940 was closest to the female Momo; we thus judged these to be the most dominant males and inferred that they had sired those females' cubs, Xiwang and Guiye. However, evidence later showed that we were wrong. In the spring of 1994, we observed Jiaojiao and Dahuo mating. This was our closest and most definitive observation of giant pandas mating in the wild. Although we observed Dahuo mounting Jiaojiao directly, and we actually observed that Dahuo ejaculated during copulation, Dahuo was not the father of Xiaosan. In this case, we were unable to determine the actual father of Xiaosan. Was it the male who had shoved Dahuo from Jiaojiao? Or might it have been another male we did not even know about? Regardless, we can be sure that either before or after Jiaojiao copulated with Dahuo, she must have done so with another male panda. In other words, female giant pandas in the wild have been found to mate with more than one male during a single estrus; that is, they have a promiscuous mating system.

Behavioral Patterns during Estrus

Through the observations of free-ranging pandas in Qinling during estrus, we suggest that the estrous period can be divided into early, middle, and late stages. During early estrus, males and females congregate in response to a variety of cues (such as scent marking and vocalizations). At this time, males generally pursue females, following them closely, vocalizing with nonthreatening bleats. However, females generally strive to escape this pursuit by climbing trees or simply running away, while growling, chirping, or making other threatening, aggressive calls. In order to get close to and control a female, a male may make physical contact with her and attempt to mount her, often biting the female's shoulder forcefully. This behavior can very easily injure her.

As the peak period of estrus arrives, females become willing to approach males. In captivity, we know that females will approach male pandas of their own volition during this period, assume mating postures such as flattened back and raised tail, and copulate with the male. Then the estrous period enters its second stage: mating. During mating, both sexes may vocalize with bleating sounds. From early estrus to the mating period, males determine their position in the hierarchy through a variety of measures. Their position, manifested by their proximity to the female, probably determines their priority in mating with the female. Under these conditions, the female is not simply the passive recipient of mating but takes an active role in selecting among competing males. However, we currently do not understand the mechanism of this selection.

Finally, in the latter part of estrus, the male leaves the female. We believe this departure represents the conclusion of a discrete mating event between a particular male and female. However, a male may also participate in other mating events as other females move through their estrous periods.

FEMALE MATE CHOICE

It was not until spring 1994, that in the foggy conditions during Jiaojiao's estrus, we were able to observe a unique behavior of the female panda: choice of a male mate. We had fitted the male Dahuo with a radio collar in 1991. He was a relatively large male, but he was missing a chunk of flesh between his upper jaw and nose, and we guessed he had been injured in a fight. When we observed Dahuo on April 10, 1994, he appeared to be quite old, his fur no longer lustrous, and he was quite dirty and thin. On December 17 of the same year, we discovered his carcass in a bamboo grove at 1,280 m elevation in Shifogou. On his back was a very old scar measuring about 9 × 8 cm, and another scar, round in shape and about 4 × 3.7 cm, was at the base of his right forepaw. He was emaciated and evidently malnourished. Upon dissection, we noted no abnormalities of internal organs. In his mouth were remnants of green bamboo leaves, but his digestive tract was empty. The necropsy suggested that he had died of starvation. On examination, we noted that Dahuo's teeth were seriously worn down. In his upper jaw, except for the posterior premolar and two other molars that still retained four to six cusps, all other molar and premolar cusps had been worn flat; some premolars and molars had been worn down to varying degrees, to pulp in some cases. His lower jaw displayed similar degrees of tooth wear. His right canine had broken off; his left first incisor and right third incisor were missing. All cusps from the third premolar to the last molar were worn flat, exposing the pulp, and the final molar on the right side had broken into two pieces, one large and one small. It was clear to us that Dahuo was quite old. In addition, we also examined his skull. The fissure between his occipital bones was almost completely joined; only between the nasal and frontal bones was any fissure still visible. There was also a relatively large fissure still remaining between his right and left nasal bones. A fissure remained between his jawbone and upper jaw, but it was quite solid. Viewed frontally, except for a small fissure between his left and right sphenoid bones, the other bones appeared to be completely fused. The condition of Dahuo's skull provided further evidence that he was quite old. As detailed earlier, we knew at least three different males were present on April 10, 1994, at the site of Jiaojiao's courtship; Dahuo, Jiaoshou or 0940, and an unknown male. Another radio-collared male attempted to mate with Jiaojiao, but she was not very cooperative. When Dahuo was copulating with Jiaojiao, she not only assumed the typical female panda mating posture of flattened back and raised tail but also emitted bleats. However, when the unknown male approached Jiaojiao, she vocalized with threatening sounds.

From this behavior it appears, at least during some stages of estrus, that females display obvious preferences among males. It is worth noting that among the males present, Dahuo seemed to be the oldest and was also in the worst physical condition. Although we know that his copulation with Jiaojiao did not produce live offspring, he evidently managed to hold a position of dominance in the face of two younger, healthier competitors. Thus, at least under these conditions, the male's physical condition did not appear to be the determining factor for female mate choice.

Appendix

Using the central quadrat method to survey tree species in spring 1995, we found 22 tree species and 11 shrub species on ridge A (Table 8.A.1).

Table 8.A.1.

LIST OF TREES AND SHRUBS ON RIDGE A WITHIN THE STUDY AREA THAT ARE DESCRIBED IN TEXT. SPECIES PREDOMINATELY FOUND IN SHRUB FORM ARE INDICATED WITH AN ASTERISK.

Local name	Common name	Scientific name
Frog maple	Père David's maple	*Acer davidii*
Pentagon maple		*Acer shenkanense*
Cowhide birch	Chinese red birch	*Betula albosinensis* var. *septentrionalis*
Iron birch		*Betula ceratoptera*
Wheat tip		*Carpinus cordata* var. *mollis*
Mountain chestnut		*Corylus tibetica*
Qinggang		*Cyclobalanopsis glauca*
Stone date		*Dendrobenthamia japonica*
Tea wax		*Fraxinus* sp.
Guangui	Japanese spicebush	*Lindera obtusiloba*
Ginger wood		*Litsea szechuanica*
Huashan pine	Chinese white pine	*Pinus armandii*
White poplar		*Populus davidiana*
Steel wood		*Quercus aliena* var. *acuteserrata*
Iron oak		*Quercus spinosa*
Spotted-leaf tree		*Quercus* sp.
Willow		*Salix* sp.
Shunzi wood		*Sorbus* sp.
Mountain locust tree	Mountain ash, rowan	*Sorbus discolor*
Beany tree		*Staphylea holocarpa*

(continued)

Table 8.A.1. (*continued*)

Local name	Common name	Scientific name
Basswood		*Tilia paucicostata*
Iron hemlock	Chinese hemlock	*Tsuga chinensis*
Immortal leaf*	Engler's Abelia	*Abelia engleriana*
Chicken bone*		*Deutzia vilmorinae*
Ba wood*	Winged spindle	*Euonymus alatus*
Flower on leaf*		*Helwingia japonica*
Rat thorn*	Perny holly	*Ilex pernyi*
Monkey fur*		*Ligustrum acutissimum*
Azalea 1*		*Rhododendron* sp.
Azalea 2*		*Rhododendron* sp.
Tonghuagan*	Chinese stachyurus	*Stachyurus chinensis*
Sticky rice strip*	Birchleaf viburnum	*Viburnum betulifolium*
Black fork*		*Viburnum glomeratum*

References

Dimmick, R. W., and M. R. Pelton. 1994. "Criteria of Sex and Age." In *Research and Management Techniques for Wildlife and Habitats*, 5th ed., edited by T. A. Bookhout, pp. 169–214. Bethesda, MD: Wildlife Society.

Kleiman, D. G. 1983. "Ethology and Reproduction of Captive Giant Pandas (*Ailuropoda melanoleuca*)." *Zeitschrift für Tierpsychologie* 62:1–46.

Klevezal', G. A., and S. E. Kleĭnenberg. 1969. *Age Determination of Mammals from Annual Layers in Teeth and Bones*. Jerusalem: Israel Program for Scientific Translations.

Lü, Z., W. E. Johnson, M. Menotti-Raymond, N. Yuhki, J. S. Martenson, S. Mainka, S. Q. Huang, Z. H. Zheng, G. H. Li, W. S. Pan, X. R. Mao, and S. J. O'Brien. 2001. "Patterns of Genetic Diversity in Remaining Giant Panda Populations." *Conservation Biology* 15:1596–1607.

Peters, G. 1982. "A Note on the Vocal Behavior of the Giant Panda, *Ailuropoda melanoleuca* (David 1869)." *Mammalian Biology* 47:236–246.

———. 1985. "A Comparative Survey of Vocalization in the Giant Panda, *Ailuropoda melanoleuca* (David 1869)." *Bongo* (*Berlin*) 10:197–208.

Schaller, G. B., J. C. Hu, W. S. Pan, and J. Zhu. 1985. *The Giant Pandas of Wolong*. Chicago: Chicago University Press.

Wei, F. W., G. Z. Xu, J. C. Hu, and B. Li. 1988. 野生大熊猫的年龄鉴定 [The age determination for giant panda]. *Acta Theriologica Sinica* 8:161–165.

Xie, Z., and J. Gipps. 1997. 大熊猫谱系 [The giant panda studbook]. Beijing: China Association of Zoological Gardens.

Xu, G. Z., and F. W. Wei. 1988. 利用大熊猫牙齿进行年龄鉴定的制片技术 [Aging giant pandas from teeth sections]. *Journal of China West Normal University: Natural Science Edition* 9:114–117.

Zeng, G. Q., G. T. Jiang, W. X. Liu, Z. Xie, and N. L. Liu. 1994. 大熊猫全年尿中孕酮和 17β-雌二醇水平的变化 [Concentration of progesterone and 17β-estradiol in urine of giant pandas throughout the year]. *Acta Zoologica Sinica* 40:333–336.

Zeng, G. Q., G. T. Jiang, K. Q. Yang, W. X. Liu, Z. Xie, and N. L. Liu. 1990. 大熊猫生殖生物学研究: I. 大熊猫发情期血清和尿液中促黄体素, 孕酮和 17β-雌二醇含量的变化 [Studies on the reproductive biology of giant panda: I. Concentrations of LH, 17β-estradiol and progesterone in plasma and urine of giant pandas during estrous cycle]. *Acta Zoologica Sinica* 36:63–69.

Zeng, G. Q., X. Zhang, G. T. Jiang, W. X. Liu, Z. Xie, and N. L. Liu. 1992. 大熊猫生殖生物学研究: II. 大熊猫妊娠期尿中孕酮和绒毛膜促性腺激素样物质含量的变化 [Studies on the reproductive biology of giant panda: II. Changes of progesterone and chorionic gonadotropin-like substance levels in urine of giant pandas during pregnancy]. *Acta Zoologica Sinica* 38:429–434.

Zhang, H. M., K. W. Zhang, R. P. Wei, and M. Chen. 1994. 卧龙大熊猫的繁殖与人工巢 [Reproduction and artificial dens for giant pandas in Wolong]. In 成都国际大熊猫保护学术研讨会论文集 [International Workshop on Giant Panda Conservation and Research, Chengdu], pp. 221–225. Chengdu, China: Sichuan Publishing House of Science and Technology.

Chapter 9

Reproduction and Early Development of Cubs

Summary: From 1989 to 1998, we documented 11 parturition events from five radio-collared giant pandas in our Qinling study area. Prior to parturition, females moved to middle- to low-mountain areas, restricted their movements, and reduced their proportion of time active to about half its normal value (to 27% active). Parturition ($n = 10$) occurred between the dates of August 15 and August 25. Gestation length averaged 145–146 days (ranging from 129 to 157 days, $n = 10$). The interbirth interval averaged 2.17 years ($n = 6$), and the sex ratio at birth was 1:1 ($n = 6$). Three cubs increased mass at daily rates of 64, 60, and 46 g/day at ages of 24 to 102 days. Cubs opened their eyes for the first time at ages of 40–49 days and were first able to use their sense of sight at ages of 88–90 days. Dental eruption was first observed at ages of 75–88 days. We documented changes in body growth and pelage characteristics of cubs, as well as behaviors of females and cubs at den sites. Females stayed with a newborn at den sites for the first 5 days. When cubs were 6–14 days old, females left den sites occasionally to defecate, but they did not forage at all during this time period. In the case of the cub Xiaosan, we first documented its mother leaving the den site to forage on day 15. In this chapter we analyze behavioral characteristics of nursing females during the first 125 days after parturition. Cubs remained at den sites until 94–125 days past birth. After leaving their dens, cubs remained restricted to resting areas until the age of 5 or 6 months, at which time they were able to climb trees, where they typically rested while their mothers were foraging, and the two were apart for relatively long time periods. We distinguished five stages of developmental behavior of cubs within their first 8 months of life: (1) birth until day 9 or 10, a period of high mortality risk, (2) day 9–10 through day 40–60, a period of gradual growth, (3) day 40–60 to day 90–130, when cubs opened their eyes and increased in strength and activity, (4) day 90–130 until the age of about 5 months old, when cubs stayed at the daybed areas, and (5) after the age of 5 months, when cubs start their arboreal life. Litter size among Qinling pandas was

1.4, suggesting an annual birth rate of 0.654, and survival of cubs to their first birthday was 59.5%. The Qinling panda population is self-sustaining on the basis of their reproduction and cub survival rates.

Introduction

Modern zoos have been involved in captive breeding of pandas ever since the first panda, a 2-year-old named Suren, was shipped from China to New York in 1936. However, it was not until 27 years later, in 1963, that breeding success was finally achieved when a cub was born and raised at the Beijing Zoo. As of September 1997, a total of 185 cubs had been born in captivity worldwide. As the technology of captive breeding has developed and workers have gained more experience, the mortality rate among captive-born cubs has gradually declined. However, details regarding birth, nursing, and early development of cubs in the wild have a remained mystery.

In this chapter we provide information on the gestation length for female pandas in the wild and document parturition, female's cub-rearing behavior, and early development of cubs. We also provide important parameters relevant to population dynamics, as well as document observations on female behavior during cub rearing that will contribute to our scientific understanding of captive management.

Pregnancy, Parturition, and Cub Rearing

Each spring during our study, we endeavored to understand the condition of all adult females and judge whether or not they were pregnant. Most females showing signs of estrus in spring gave birth the following autumn. We made careful observation of our radio-marked pandas, documenting, summarizing, and analyzing their behavioral characteristics during gestation and parturition.

CONTINUOUS MONITORING

Determining the location of a mother-cub dyad allowed us to use radio telemetry to locate their den. We then set up our monitoring equipment (battery-powered video recorder and microphone in a waterproof container plus a remote monitor for viewing) near the den site. We generally placed the monitoring equipment where it was out of the way of the normal movements of the animals yet allowed us easy access to them. We were often able to use rock seams within the den itself to stabilize the equipment; our observations suggested that this did not interfere with normal behavior of the mother. Initially, the mother would occasionally lift her head and sniff at the monitoring equipment but would then continue her normal resting behavior. The 50-m length of our electronic cables allowed us to observe the pandas from a sufficient distance to avoid influencing them.

Our specific methods of observation were as follows: We listened to the radio-collared female's telemetry signal every 5 minutes. If the signal indicated that she was active, we then began the monitoring equipment to document mother-cub activity until they both became completely inactive. When telemetry suggested long periods of inactivity, we activated the monitoring equipment at 15 minute intervals to document the posture of both mother and cub.

Lack of electricity at our research station required us to travel the 14 km to Huayang town to recharge our equipment, and even then, recharging required at least 8 hours. This limited our ability to monitor animals continuously over long-time durations.

Activity and Movements during Pregnancy and Parturition

BEHAVIORAL CHARACTERISTICS OF PREGNANT FEMALES

Between 1989 and 1996, we monitored the diurnal activity patterns of 17 individual radio-marked pandas (Table 9.1). We calculated the percentage of time active within each individual monitoring bout (Schaller et al., 1985). Each time we monitored a den site, we first found locations where we could receive clear signals and then pitched our tents for a temporary camp at that spot. Because of topographic constraints and movements of the animals, we were often unable to obtain the desired number of daily signals. We considered daily data suitable for analysis if we obtained no fewer than 80 individual signals or were able to monitor the animals for at least 20 hours.

We calculated the percentage of time active from these series of continuous monitoring bouts on a monthly basis. The percentage of activity for all individuals held fairly steady, varying from 51.9% to 63.9% during November through July. However, activity dipped to 29.6%, 26.0%, and 20.1% during the months of August, September, and October, respectively. Our field investigations revealed that female pandas rarely left their den sites and were not very active prior to and following parturition; thus, their percentage of time active was quite low. Excluding lactating females, males and nonlactating females maintained their activity levels during August, September, and October (58.7%, 42.3%, and 51.2%, respectively). During August, September, and October, lactating females were active during 14.6%, 17.4%, and 36.2% of observation bouts, respectively, which was noticeably lower than the nonlactating females during the same months. During November, activity among females nursing cubs rose to 41.3%, equivalent to nonnursing females. In January, when cubs were 5 months old, the percentage of activity among nursing females was 8.9% lower than nonnursing females. By the time cubs were 6 to 7 months old (i.e., during February and March), females raising young were 2.0% to 4.7% less active

Table 9.1.

SAMPLE SIZES, BY DAYS OF MONITORING AND NUMBER OF DISCRETE BEHAVIOR RECORDS, OF EACH RADIO-COLLARED PANDA BY YEAR. A DASH (—) INDICATES NO DATA WAS AVAILABLE FOR THAT YEAR.

Name	Days of monitoring (Number of discrete behavior records)								Total
	1989	1990	1991	1992	1993	1994	1995	1996	
Females									
Yanghe	31.8 (3,056)	4.4 (427)	—	—	—	—	—	—	36.3 (3,483)
Jiaojiao	16.6 (1,592)	20.2 (1,939)	6.5 (620)	0.8 (72)	1.2 (113)	7.9 (759)	1.0 (97)	14.4 (1,387)	68.5 (6,579)
Baoma	—	3.8 (362)	—	—	—	—	—	—	3.8 (362)
Ruixue	—	—	—	—	1.5 (145)	10.4 (1,003)	0.9 (90)	—	12.9 (1,238)
Xiwang	—	—	—	0.8 (77)	—	6.1 (587)	1.1 (107)	5.4 (515)	13.4 (1,286)
Nüxia	—	—	—	—	1.0 (97)	5.6 (541)	—	13.1 (1,253)	19.7 (1,891)
Momo	—	—	—	—	0.6 (54)	6.5 (622)	—	1.0 (96)	8.0 (772)
Xiaosi	—	—	—	—	—	—	—	1.4 (130)	1.4 (130)
Males									
Huayang	17.3 (1,656)	—	—	—	—	—	—	—	17.3 (1,656)
Huzi	—	—	3.8 (363)	—	2.7 (259)	4.7 (451)	—	—	13.2 (1,265)
Daxiong	—	8.0 (764)	4.1 (391)	—	—	—	—	—	14.1 (1,358)
Xinxing	—	—	1.3 (126)	—	—	—	—	—	1.3 (126)
Xiaohuo	—	—	—	—	—	0.3 (25)	—	0.4 (36)	0.7 (61)
Dabai	—	—	—	—	—	2.5 (242)	1.0 (94)	1.8 (168)	5.3 (504)
Jiaoshou	—	—	—	—	—	3.9 (377)	1.0 (97)	—	4.9 (474)
0940	—	—	—	—	—	—	0.5 (47)	—	0.5 (47)
Xiaosan	—	—	—	—	—	3.4 (328)	1.0 (95)	4.9 (469)	9.3 (892)
Total	65.7 (6,304)	38.5 (3,695)	15.6 (1,500)	1.6 (149)	7.0 (668)	51.4 (4,935)	8.5 (819)	42.2 (4,054)	230.5 (22,124)

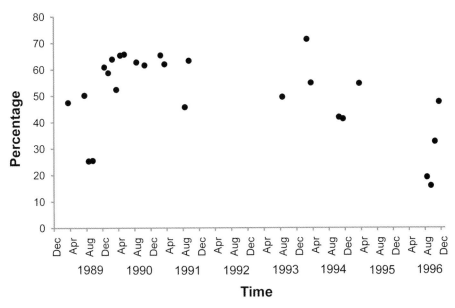

FIGURE 9.1. Monthly mean percentage of time active for adult female Jiaojiao, March1989 to November 1996 (some months are missing).

than females without cubs or with yearlings. Thus, our comparisons suggest that females with young cubs up to the age of 3 months exhibit reduced activity levels relative to females without young and that after that time, activity levels are similar.

We gained further insight into activity patterns by comparing individually monitored females during different portions of their annual cycles. We collected more data on Jiaojiao than any other radio-marked female and followed her for the longest time period. Jiaojiao produced cubs in August of the years 1989, 1992, 1994, 1996, and 1998. Unfortunately, we were only able to follow her continuously during the autumns of 1989 and 1996 (Lü, 1991). Jiaojiao was much less active during months in which she gave birth than during other months, and her activity bouts were of shorter duration (Figure 9.1).

PARTURIENT FEMALES' MOVEMENT IN AUTUMN

From the earliest period of our panda research in the Qinling, we noted that pandas in this area exhibited marked seasonal movements, migrating seasonally along an elevation gradient. The Qinling is characterized by two main types of bamboo that constitute the main diet for pandas: in midmontane areas below elevations of 2,000 m the principal bamboo species is *Bashania*, whereas *Fargesia* dominates above 2,000 m. During April–May, pandas generally migrate to higher elevations to forage on *Fargesia* shoots, gradually descending the following September to focus their attention on

Bashania until the subsequent spring (Pan et al., 1988; Lü, 1991). We thus viewed 2,000 m as a useful dividing line in the Qinling: when radio signals indicated that an animal's activity had moved above or below that contour and remained there for some period of time, we would conclude that the animal had transitioned in its yearly migration.

Our long-term observations in the Qinling demonstrated that pregnant females exhibited quite different movement patterns from other pandas as parturition approached. We observed movement patterns during 1989–1996 among adult and subadult pandas in autumn as they migrated from higher- to lower-elevation habitats (Table 9.2).

We categorized pandas as being either pregnant/parturient females, females not producing cubs in that year, or males (and we also summarized all pandas together). We calculated the mean date at which each class of panda made their downward migration in autumn, as well as the earliest and latest dates. Pregnant females began their downward migration, on average, in early August, earlier than other categories of panda (Figure 9.2). However, individual animals provided exceptions to the general patterns. In particular, we point out that in 1993, pregnant female Momo exhibited no downward migration, but rather stayed at elevations above 2,000 m, where she gave birth to her cub Guiye. Momo did not descend to the midmontane region below 2,000 m until after October 1. If we exclude this observation, the mean date of descent among pregnant females was July 29, and the latest descent advanced from October 1 to August 15 (lines with diamonds and squares, Figure 9.2). That being said, the female Jiaojiao descended rather earlier than other pandas regardless of reproductive condition (Table 9.2). We estimated the dates of descent for Jiaojiao by years in which she was pregnant (range July 20 to August 12) versus years in which she did not bear cubs (range July 26 to August 17), and a comparison indicates she descended earlier in years when she was pregnant than in other years.

In addition to moving down to lower elevations earlier than other pandas, pregnant females tended to find valleys or canyons to which they restricted their movements roughly 2 weeks prior to parturition. At that time, their movements became quite limited, often to within a 50-m radius of a central location. The daily locations of any given pregnant female during that time showed considerable proximity, demonstrating their sedentary behavior. This behavior contrasts with other adult pandas, which generally did not remain in one place for more than a week.

These reduced movements are not a function of habitat because when adult female Momo produced her cub at the high-elevation area in 1993, she also restricted her movements to near the birth site during the 2- to 3-week period afterward. As mentioned earlier, Momo did not descend to her winter range at lower elevations until October 1 of that year.

Table 9.2.

THE TIME EACH PANDA MIGRATED FROM ABOVE TO BELOW 2,000 M IN ELEVATION, 1989–1996. A YEAR WHEN A FEMALE DELIVERED A CUB IS INDICATED WITH AN ASTERISK.

Individual	Estimated age	Migrating date
	Females	
Yanghe	20	September 24, 1989
Jiaojiao	5*	August 1, 1989
	6	August 4, 1990
	7	July 27, 1991
	8*	July 20, 1992
	9	August 2, 1993
	10*	July 21, 1994
	11	August 16, 1995
	12*	August 11, 1996
Baoma	6	October 1, 1990
	7	October 1, 1991
Ruixue	7	September 1, 1992
	8*	July 14, 1993
	9	August 26, 1994
Nüxia	5	September 1, 1993
	6	August 26, 1994
Momo	5*	October 1, 1993
	6*	August 15, 1994
Xiwang	5*	August 4, 1997
	Males	
Huayang	21	September 1, 1989
	22	September 1, 1990
	23	September 1, 1991
Huzi	2	August 1, 1991
	3	August 11, 1992
	4	August 21, 1993
	5	September 20, 1994
Daxiong	16	September 1, 1992
Xinxing	7	September 19, 1991
	8	September 11, 1992
Dahuo	15	September 16, 1992
	16	August 26, 1993
	17	September 1, 1994
Xiaohuo	8	September 1, 1992
	9	September 1, 1993
Dabai	9	September 1, 1992
	10	September 1, 1993
	11	September 1, 1994
Jiaoshou	6	September 23, 1993
	7	September 23, 1994
0940	8	September 23, 1993
	9	September 23, 1994

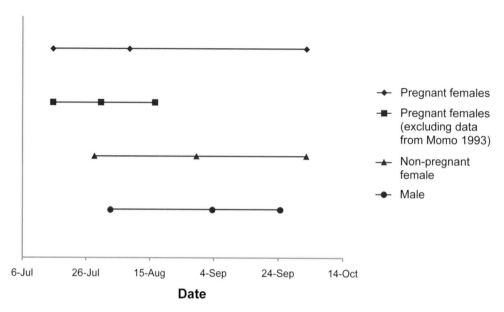

Legend:
- ◆ Pregnant females
- ■ Pregnant females (excluding data from Momo 1993)
- ▲ Non-pregnant female
- ● Male

X-axis (Date): 6-Jul, 26-Jul, 15-Aug, 4-Sep, 24-Sep, 14-Oct

Date

FIGURE 9.2. Earliest, mean, and latest dates of the adult giant pandas descending from summer to winter habitats, Qinling Mountains.

FEMALE ACTIVITY NEAR PARTURITION

In August 1997, Jiaojiao's first female offspring Xiwang was pregnant, and we were able to observe her behavior prior to parturition. Xiwang had been born in August 1992 in Shuidonggou and thus was 5 years old in August 1997. She descended from the summer range earlier than most other pandas and spent about 3 days in the vicinity of our research base. During that time, we observed her digging up an old animal bone, which she ate, following which she sat immobile in a grove of bamboo, and we noted no evidence that she foraged on bamboo while she remained in the grove (nor did we find any scats).

During continuous monitoring of Xiwang on August 16–18, we noted that her proportion of time active during the 48-hour period was only 20.4%. This percentage was even lower than that of her mother, Jiaojiao, during the 7-day period prior to her parturition of August 13, 1996 (27.1% active). In both cases, time active was roughly half that of nonpregnant and nonparturient females.

We also observed Jiaojiao during her estrus of March 11–13, 1989. On August 1 of that year, she descended to lower elevations earlier than other pandas; during August 6–14, her entire area of activity in Shuidonggou did not exceed 1 ha. On August 14–17, we monitored Jiaojiao continuously from 1000 to 1700 hours and noted a marked decline in her percentage of activity through time: from 48.0% on August 14 to 24% on August 15, 9.6% on August 16, and 12.9% on August 17.

Jiaojiao showed almost no movements from August 16 to 21 on the basis of the radiolocations. We inferred from these data that she gave birth on either August 16 or 17. This was Jiaojiao's first cub, the male we subsequently named Huzi. Because of rainstorms that prevented us from trudging up the mountain, we were unable to continue monitoring Jiaojiao and Huzi over the subsequent few days.

On August 21 the rain ceased, allowing us to again search for Jiaojiao. We finally encountered her in a small tributary of Shuidonggou, at 1,950 m, sitting under a 3-m-high stone wall, cradling her cub. Beneath her was a tangle of broken bamboo stems. We estimated the cub at this stage to be 15 cm in length with a 7-cm-long tail and his weight to be 200–250 g; his skin was pinkish, and his upper parts were covered with sparse, white fur.

Because she had evidently moved very little since August 16, we inferred that the rock wall where we found her was indeed the birth site. This wall, which faced southward and was approximately 150 m from the nearest water source, leaned outward at roughly a 30° angle, sheltering an area of perhaps 2 m² from the elements. A rough footpath lay about 80–100 m from the site, along which cane cutters frequently traveled.

Gestation Lengths and Interbirth Intervals of Qinling Female Pandas

PARTURITION OF RADIO-MARKED FEMALES

From 1988 through 1996, we radio marked a total of 12 female pandas; in order, they were Qingqing, Yanghe, Jiaojiao, Xiaof, Baoma, Shuilan, Ruixue, Keke, Xiwang, Boshi, Nüxia, and Momo. However, we have definite knowledge of parturition for only four: Jiaojiao, Ruixue, Nüxia, and Momo (Table 9.3). Ruixue and Baoma were captured when raising cubs; thus, we could only infer these cubs' birth dates, and they were not counted as cubs born to collared females.

After Jiaojiao gave birth to Huzi in 1989, we began our efforts to approach them closely. Initially, Jiaojiao was very alert to our presence. If we got too close to Huzi, Jiaojiao would rush at us with warning vocalizations and then lead the cub away from us. It was only 3 years later, in the autumn of 1992, after Xiwang was born, that Jiaojiao gradually allowed us to approach close enough to obtain detailed observations. Jiaojiao and Xiwang allowed us to observe them while in the den, at distances as close as 2 m. If we made any noise, Jiaojiao would lift her head and look at us, seemingly ill at ease. But unlike earlier, she did not leave the den with Xiwang and would resume her previous behavior. After Xiwang had begun leaving the den

Table 9.3.

CUBS PRODUCED BY COLLARED FEMALE GIANT PANDAS IN THE QINLING STUDY AREA, 1989–1998. U = UNKNOWN. A DASH (—) INDICATES NO OBSERVATIONS WERE MADE.

Mother	Partition date	Name	Gender	Status	Age at last observation
Jiaojiao	August 15, 1989	Huzi	M	Alive	7 years
	August 15, 1992	Xiwang	F	Alive	5 years
	August 17, 1994	Xiaosan	M	Alive	2.5 years
	August 20–21, 1996	Xiaosi	F	Unknown	3.5 months
	August 1998	Xiaowu	U	Alive	8 months
Momo	August 25, 1993	Guiye	F	Dead	3 months
	August 1993	None	M	Dead	~3 days
	August 17, 1994	None	M	Alive	1.5 years
Ruixue	August 21, 1993	Shigen	U	Unknown	2 days
	August 1995	None	U	Unknown	—
Nüxia	August 1996	None	U	Unknown	—

with her mother to forage in bamboo, any noises from us no longer elicited threats or aggressive behavior on the part of Jiaojiao.

Later, when Jiaojiao gave birth to Xiaosan and Xiaosi, she allowed us to approach closely for further observation, facilitating our systematic documentation of suckling behavior in 1994 and 1996. Momo also allowed us to approach but was less tolerant of us than was Jiaojiao. On one occasion, she threw her cub to the ground when we approached too closely. We thus reduced the frequency of our close observations of her.

The two other adult females were even more timid and never became accustomed to our close presence. In the case of adult female Ruixue, we were only able to conduct a single close observation of her cub Shigen. Thereafter, Ruixue concealed Shigen to the extent that we had no way to find it. We deduced, from her activity and movement, that Nüxia had produced a cub in 1996, but we were never able to find her and her cub during that time.

Among all 10 cubs, we were only able to determine the gender of the seven produced by Jiaojiao and Momo (three females, four males). In all cases but one, we determined gender by observation of external genitalia. The exception occurred in 1993, when Momo produced twins; we discovered the carcass of the second cub when she led the surviving cub, Guiye, from the den. Molecular analyses confirmed that the dead cub had been a male.

GESTATION LENGTH AND TIMING OF PARTURITION

The gestation period for mammals is defined as the time from fertilization to parturition. For free-ranging animals, we generally have no way to determine the exact date of fertilization, so we calculated gestation length as the time from observations of mating until parturition (Alt, 1989). The

Table 9.4.

CHRONOLOGY OF MATING, GESTATION, AND PARTURITION FOR FEMALES JIAOJIAO, MOMO, RUIXUE, AND NÜXIA, QINLING STUDY AREA, 1989–1998. A DASH (—) INDICATES DATE OR GESTATION PERIOD WAS NOT DETERMINED.

Mother	Last mating event observed	Parturition date	Gestation period (days)
Jiaojiao	March 11, 1989	August 15, 1989	157
	March 9–13, 1992	August 15, 1992	155–159
	April 10, 1994	August 17, 1994	129
	After April 8, 1996	August 20–21, 1996	At least 135–136
	March 1998	August 1998	—
Momo	March 31 to April 1, 1993	August 25, 1993	146–147
	March 25,1994	August 17, 1994	145
	March 7, 1996	—	—
Ruixue	March 25–27, 1993	August 21, 1993	147–149
	March 19–20, 1995	—	—
Nüxia	March 23–24, 1996	August 20–25, 1996	148–155

gestation lengths of female pandas in the Qinling varied from 129 to 159 days, with most falling between 145 and 159 days and a mean of 145–146 days (Table 9.4). Among cubs that we could monitor, birth dates all occurred between August 15 and 25.

INTERBIRTH INTERVALS

We were able to monitor the interbirth intervals for three females: Jiaojiao (3, 2, 2, and 2 years), Momo (1 and 2 years), and Ruixue (2 years; Table 9.4). Because one of Momo's cubs, Guiye, did not survive long, the interval between successful births for Momo in 1993 was 3 years, rather than 2 years. The effective interbirth interval (i.e., interval between births of cubs that survived their early period) was 2.33 years for all verified births and 2.17 years if we remove the cub Guiye from the analysis.

Behavior of Nursing Females at Den Sites

BEHAVIORAL CATEGORIES OF NURSING FEMALES (ETHOGRAM)

To quantify behavior of females and cubs, we developed an ethogram with 32 behaviors. Here we list and define some behavior categories of females and cubs while in dens that are relevant to maternal behavior (for a full list, see the Chinese edition):

1. Bridging behavior. While in dens, females adopted a "bridging behavior." The cub was lying on the ground, and the mother was lying atop it, with her belly protecting the cub. This type of position generally occurred when cubs were 30–70 days of age, before females exited the den. When the cub was positioned on top of the mother, we classified its posture

Reproduction and Early Development of Cubs   211

as either prone, lying on its side, or supine. When the mother was lying on its side, the cub occasionally was positioned such that it was wedged up next to the mother's buttocks in what seemed to be a sitting position. In fact, we considered this posture to be a type of lying on its side. When cubs were alone in the den, they typically adopted a prone position. Upon leaving the den, mothers typically placed cubs on their sides or in the supine position, but cubs often turned themselves over to adopt the prone position.

2. Nursing. Mothers bring the cub to a nipple and allow it to suckle.

3. Perianal licking. Mothers used their tongues to lick their perineum, which occurred when cubs were up to about 20 days of age and which we assume was to keep the area clean (Liu et al., 1993).

4. Lip smacking. Females used upward and downward movement of their lips to produce a smacking sound. This action probably reflected some type of nervousness on the part of the female. We noted in 1993 that Jiaojiao exhibited this type of behavior three to four times per minute, but in 1994 and 1996, she did this only once or twice per hour.

5. Huffing. Females exhibited a sound we interpreted as a mild type of threat (Schaller et al., 1985).

6. Moving the cub. The mother would use her mouth to change the position of the cub. While in the den, this usually behavior occurred after the mother returned from being outside or after she had dropped the cub on the ground. Additionally, females used this action to adjust the position of the cub prior to their ability to move about on their own.

7. Cub anal licking. Using her tongue, the mother would lick the anal region of the cub. We observed this behavior in the case of the cub Xiaosi until the age of about 7 weeks.

8. Cub grooming. Using her teeth, the mother would groom and pick at the fur of the cub. In 1996, we noted that after grooming herself once or twice, Jiaojiao would sometimes groom her cub Xiaosi. She would cease this activity upon the cub's squealing vocalization.

9. Slapping the cub. When holding the cub or sometimes when the cub was positioned atop the mother, she would use her front feet in an upward and downward motion, lightly slapping the cub. At times, we noted this behavior on the part of the mother even when the cub could not be physically reached by her.

While inside the den with her cub, we never documented any vocalizations on the part of the mother. We did document vocalizations by the cub toward the mother.

Behavior of Mothers during the Early Stage of Den Life

We wanted to monitor the activity of each mother-cub pair at every time interval. In practice, logistical constraints (e.g., battery life) allowed us to record behavior during only some of these intervals, and our continuous monitoring was characterized by gaps. From August 19 to October 18, 1994, monitoring bouts were generally about 24 hours in length and encompassed at least one complete cycle of entrance and exit of the female from the den. After this time, as females' time away from the den lengthened, we lengthened our recording bouts accordingly, so that we were able to record complete cycles of her den emergence and reentry. If the female carried the cub out of the den, we terminated that recording session. In 1996, we used a second set of criteria to set the length of recording bouts. We list the characteristics of the continuous recording bouts we conducted during 1994 and 1996 in Table 9.5.

Here we describe the life of the female during the very early development of her cub. Xiaosi was born on August 21, 1996. We considered this "day 0" of her life. During the 24-hour monitoring period of 1900 on day

Table 9.5.

AGE, INITIAL DATE, AND DURATION OF CONTINUOUS MONITORING BOUTS FOR CUBS XIAOSAN (1994) AND XIAOSI (1996). BOUTS MARKED WITH AN ASTERISK ENDED PREMATURELY WHEN THE MOTHER GATHERED THE CUB AWAY FROM THE DEN AND WE WERE UNABLE TO FOLLOW THEM.

Bout	Cub age (days)	Initiation of monitoring	Duration (days)
		Xiaosan	
1	1	August 18, 1994	72
2	13	August 30, 1994	59
3	23	September 9, 1994	24.4
4	32	September 18, 1994	24.5
5	41	September 27, 1994	29.5
6	48	October 4, 1994	24.9
7	56	October 11, 1994	24
8*	62	October 18, 1994	16.6
9	75	October 31, 1994	71.6
10	85	November 10, 1994	70
		Xiaosi	
1	2	August 22, 1996	24
2*	6	August 26, 1996	17
3	25	September 14, 1996	26.25
4	33	September 22, 1996	29
5	45	October 4, 1996	30
6	51	October 10, 1996	24
7	59	October 18, 1996	27
8*	76	November 4, 1996	17
9	81	November 9, 1996	44
10	91	November 19, 1996	63

1 until 1900 on day 2, Jiaojiao cradled the cub beneath her chin, neither eating nor drinking. On day 5 (August 26), Jiaojiao emerged from the den for 4 minutes. Because of heavy rain at the time, we were unable to detail her behavior outside the den, but we did observe footprints she had evidently produced. On day 6, we observed that water had evidently leaked into the den, and Jiaojiao had moved to a different den site.

Xiaosan was born on August 17, 1994, which we considered day 0 for her. We made a total of 46.75 hours of continuous observations of her during the day 1 to day 4 period, during which time Jiaojiao neither ate nor drank. On day 9 (August 26), we noted claw marks and white hair on the den floor as well as on rocks near the den entrance, suggesting that Jiaojiao had exited the den during some point. However, we documented no direct evidence that she had foraged or defecated anywhere near the den. On day 10 (August 27), we noted footprints made by Jiaojiao when she had entered the den. Additionally, we encountered a panda scat about 4–5 m from the den entrance, approximately 50–70 mm^3 in volume, that we considered produced by Jiaojiao. The outermost 7–10 mm layer of the aromatic scat was a grasslike green, whereas the inner portion of the scat was yellowish in color.

From 1455 hours on day 13 to 0200 hours on day 16, we recorded some 59 hours of remote recordings, noting that Jiaojiao exited the den seven times, and we provide summaries of her den emergence in Figure 9.3.

Early on the morning of September 1, we noted evidence that Jiaojiao had foraged on bamboo just outside the den entrance. Because we recorded that she had spent 19 minutes outside the den beginning late on the night of August 31 (at 2322 hours), whereas prior to this time she had not spent more

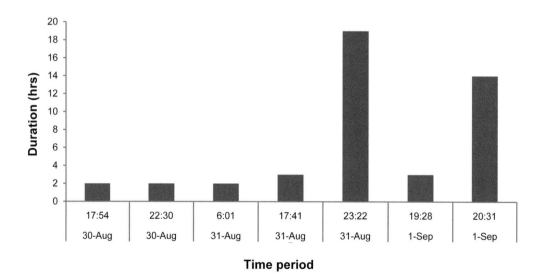

FIGURE 9.3. Duration (in hours) of times when adult female Jiaojiao left the den, spending time outside it, August 30 to September 1, 1994. Parturition in 1994 occurred on August 17.

214   Chapter 9

than 3 minutes outside the den, we inferred that she had indeed spent this time foraging on bamboo, the first such episode since the birth of Xiaosan 14–15 days prior.

It seemed clear to us that the changes in Jiaojiao's behavior shortly after the birth of Xiaosi was similar to what had occurred after she had given birth to Xiaosan: For the initial 4 days postpartum, she sat within the den holding her newborn cub, neither eating nor drinking. After the fifth day, she left the den occasionally to defecate. It was not until day 14 or 15 that she began to forage on bamboo outside the den. Between days 5 and 14, we were confident that she did not eat, but we cannot be certain that she took no liquids.

During this time of fasting, she was willing to leave the den itself for various reasons and was not completely restricted in her movements. In 1994, she left the den for the first time on August 25, i.e., day 9 or 10 since the birth of cub Xiaosan.

We categorized Jiaojiao's activities upon exiting the den as one of three types:

1. Excretion. We documented the first feces from Jiaojiao in 1994 on August 27. These feces were wet and sticky and exuded a strong odor. Upon examination, we noted that the outermost layer of some 7–10 mm were a grasslike green color, whereas the inner portion was yellow. Total volume was about 50 mm^3. The latest we documented this type of feces was on September 5, 1994, i.e., day 18, and this was also the first day we began to see the more typical scat consisting of bamboo residue.

2. Obtaining bedding for the den. On August 30, 1994, the den entrance had become wet with the rain that had been falling all day. Around dusk, Jiaojiao exited the den, bit off some bamboo branches, and dragged them back into the den to form a bed. We saw no evidence that she had consumed the bamboo while outside the den.

3. Moving the cub. Movements outside the den during the fasting period were recorded for Jiaojiao in 1994 and 1996 and Momo and Ruixue in 1993 for purposes of moving sites. Of these, Jiaojiao's movements were clearly caused by the weather. In both 1994 (with Xiaosan) and 1996 (with Xiaosi), rainfall had caused a relatively deep pool of water to develop near the den entrance, forcing her to move the cub to an alternate location at roughly day 6. We thus speculate that the reasons for Ruixue and Momo leaving their dens were similar.

TIMES OF ENTERING DENS AFTER THE MOTHER HAD BEGUN FORAGING

After mothers had begun exiting the den to forage, we categorized her activities as within or outside of the den. We defined a "den activity cycle" as the time between two successive exits of the den, including time when she remained in close proximity to the den.

The time spent by the mother inside and outside the den gradually changed as cubs matured. During Xiaosan's first 45 days of life, Jiaojiao always left the den between the times of 1100 and 1500 hours and returned between the times of 1500 and 1900 hours, with her departure tending to become earlier with time and her return gradually becoming later in the day. During the first 50 days raising Xiaosi in 1996, Jiaojiao's return to the den showed no such obvious pattern, but her departure did tend to become earlier with the maturation of the cub.

When Xiaosan reached the age of 48 days, we first documented Jiaojiao returning to the den after midnight. After this time, Jiaojiao's pattern was one of leaving the den during daylight hours and returning at night. In the case of Xiaosi, Jiaojiao first began returning to the den postmidnight on day 52. After this time, her pattern of emergence and return was similar to the obvious pattern she had displayed in 1994.

Life Outside the Den for Cubs

On the basis of behavioral differences, we categorized the activity of cubs after leaving the protection of the den until the age of 8 months into arboreal and terrestrial activity.

TERRESTRIAL ACTIVITY

After cubs had left the den, their mother led them to bamboo groves. At this time, cubs were not yet able to move about on their own and depended on their mother picking them up by their necks to move them. While suckling, mothers typically sat among dense bamboo, and it was very difficult for us to determine activity during these nursing bouts.

When females foraged for themselves, they would generally place the cub where they had been sitting together: we term these areas "daybeds." Our observations suggested that these areas were not provided with any specific bedding material; however, because they were generally characterized by leaf litter and various understory vegetation, the cub did not touch the ground directly. Also, we noted that the scats produced by the female were often strewn about in the area in the immediate vicinity of the daybed; cubs sometimes rested among these scattered feces.

While the cub lay by itself in the daybed, Jiaojiao usually foraged no more than 20 m away. At the merest alarm squeak of the cub, she would immediately return to the cub's side (Zhu, 1996). We did make an observation on January 15, 1995, where Xiaosan was located up against a cliff face on the mountain ridge while Jiaojiao was on the opposite side of the slope, a distance of about 50 m, but this was the farthest distance we documented Jiaojiao from the daybed.

After feeding, Jiaojiao would emit a bleating sound when approximately 5–10 m from the daybed. This behavior was in contrast to her silence upon returning from forays to reunite with either Xiaosan or Xiaosi in the dens.

ARBOREAL ACTIVITY

In the Qinling, when cubs had gained the ability to walk and climb trees, we considered that their terrestrial life had ceased and they had begun their period of arboreal activity. On January 15, 1995, when Xiaosan was 5 months of age, we discovered him sitting in a daybed under a nearby cliff. However, 15 days later, on February 1 Xiaosan had gained the ability to quickly climb trees; we observed him sitting about 4–5 m off the ground on the branch of a tree of approximately 30 cm diameter at breast height.

On January 17, 1992, when Xiwang was 5 months old, she was still unable to walk. The next opportunity we had to observe her was on March 5 (6.5 months of age), at which time she was sleeping on the branch of a pine tree about 3 m above the ground.

Protective Behavior of Females

While ensuring her own safety, Jiaojiao was fiercely protective of her cubs as well.

SHIELDING BEHAVIOR

In 1989, when Jiaojiao was nursing her first cub, Huzi, we noted that upon finishing a foraging bout and returning to the den she would linger at the den entrance, making a slight sideways detour, before quickly entering. We also noted little evidence that she foraged in the immediate vicinity of the den entrance. Additionally, we observed that if she sensed our presence, she would move in the opposite direction from the den, making loud vocalizations while moving.

In 1993 when Ruixue had given birth to Shigen and then again in 1996 after Momo had given birth to her cub, we spent considerable efforts trying to locate the dens. Initially, we used the information from telemetry to clarify that these females were located in the general vicinity of particular mountain valleys. However, when we approached the females, in both cases, they began to move higher up on the mountain ridges at a brisk pace. We believed these movements were designed to lead us away from their dens. These females adopted similar behaviors when their cubs had been left alone to sleep in a tree. We noticed that this type of behavior diminished only when the cubs had finished nursing.

AGGRESSIVE BEHAVIOR

Were an unwelcome intruder to come too close to a cub, the response of a female panda would invariably be an attack: either chasing and attacking the intruder or using howling, jaw snapping, sneezing, or biting at the trunk of a nearby tree. On February 4, 1990, as we were observing Jiaojiao leading her 6-month-old cub Huzi down the slope at Shuidonggou, they were approached by the subadult female Shuilan, who came to within about 20 m of the pair. Upon realizing how close she was, Jiaojiao immediately gave chase and swiped at Shuilan's face, producing a wound. A few hours later, Shuilan's radio signals indicated that she had fled the area, whereas Jiaojiao and her cub continued to use the original area. Later, in June 1990, we documented a case in the same location in which Jiaojiao chased a black bear away from the area. Similarly, in the spring of 1990, during our first attempts to approach Huzi for observation, Jiaojiao was located a few dozen meters away foraging. When she perceived us, she immediately ran back to where her cub was, howling at us. She charged toward us and then backed off a few steps, repeating this behavior several times. We can attest to how other animals might have reacted to such displays, as we ourselves felt fearful and threatened.

Natal Dens of Giant Pandas

DEN CHARACTERISTICS

Female pandas use dens during the time period following parturition to raise their cubs during their early development. We observed Jiaojiao's selection of dens as representative of other female pandas in the Qinling; thus, we provide an overview of the characteristics of her dens here.

During the cub-rearing period, we were generally unable to obtain continuous observation on Jiaojiao for periods of more than 3 days. Our observations were limited to situations in which Jiaojiao and her cub remained in the same den for at least 3 days. We documented the locations of each of Jiaojiao's den sites beginning in 1992 (Figure 9.4) and documented den characteristics (Table 9.6). All dens were located at elevations of 1,900–2,000 m.

The amount and type of bedding material within the den varied with cub age. Early on, when Jiaojiao rarely left the den, very little bedding material was provided, and what little there was consisted of bamboo stalks and small pieces of wood. By the time Jiaojiao began to forage outside the den frequently, the quantity of bedding material had increased and consisted primarily of tree branches and bamboo.

When we had the opportunity to inspect the entrances to dens B and C in detail, during 1992, 1994, and 1996, we found primarily thick wooden stems, mostly from pine and oak trees. On October 24, 1994, we documented tree branches of 85, 120, and 150 cm lengths, with corresponding diameters of 11, 8, and 5 cm. Jiaojiao had evidently bitten off pine branches from nearby trees and placed them in the den. In 1994 and 1996, after Jiaojiao had transferred her cubs from den C to B, the main bedding material in den B was fresh bamboo, and tree branches had also been spread about the entrance.

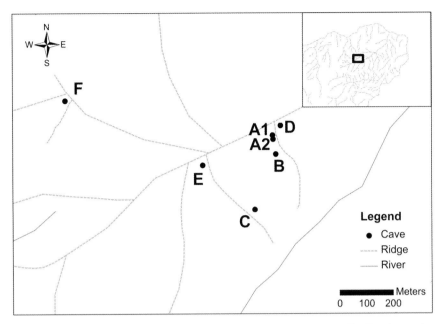

FIGURE 9.4. Sketch map showing geographic distribution of dens used by Jiaojiao in rearing cubs.

From August 27 to 31, 1994, we documented high and low temperatures in both the inner and outer sections of den B. The mean high temperature in the outer section of the den was 20.8°C, and the mean low was 13.9°C. In the inner portion of the den, the mean high and low temperatures were 18.3°C and 16.2°C.

MOVING FROM ONE DEN TO ANOTHER

Dens used by Jiaojiao when she was raising cubs Xiwang, Xiaosan, and Xiaosi, as well as durations spent at each, are listed in Table 9.7. Jiaojiao reused dens on multiple occasions within a single year. When raising Huzi, Jiaojiao used both dens C and F. In 1992, she used den A2 and B twice each; in 1994, she used den B twice, and in 1996, she used den D twice and den B three times. She also reused dens from previous years, as dens A2, B, and C were used for rearing Xiwang, Xiaosan, and Xiaosi, in turn.

Den B was used more frequently than any other den; den C was used for the longest accumulated time. We speculate that such reuse was related to the structural characteristics of these dens (Table 9.6). Den B had a small entrance and a spacious interior, making it particularly suitable for raising a cub. This den was also quite well concealed; although it was oriented to the southeast, sunlight barely entered inside, leaving the interior dark. Den C was similar in its structure and was similarly well hidden but, in contrast to den B, was much more exposed to the sun's rays. At midday, sunlight reached all the way into this den, resulting in it being relatively dry inside.

Table 9.6.

CHARACTERISTICS OF DENS USED BY JIAOJIAO FOR REARING XIWANG, XIAOSAN, AND XIAOSI.

Den	Elevation (m)	Aspect	Brightness	Structure and construction	Size, L × W × H (m)	Slope
A1	1,960	E	Relatively bright	Under an inclined boulder; entrance opens wide	Relatively small	Flat
A2	1,960	SW	Relatively bright	Under an inclined boulder; entrance opens wide	Relatively small	Flat
B	1,950	SE	Dark	Between two adjacent rocks; entrance small and triangular; inside spacious	0.9 × 1.3 × 0.7	30° inclining downward
C	1,920	SE	Bright	Between three large boulders; entrance small and triangular, perched 3 m above ground level	1.8 × 1.2 × 0.8	Inclined 15°; entrance 80 cm higher than interior
D		SE	Bright	Under a huge inclined boulder; entrance opens wide.	Large	Flat
E	2,000	SE	Bright	Rock den, interior and exterior sections; entrance spacious	1.8 × 1.6 × 1.0	Flat
F	1,900–2,100	SW	Relatively bright	Between two adjacent boulders; entrance spacious	3.0 × 1.0 × 0.7	Flat

Our observations suggested that moisture entering the den was among the causes for Jiaojiao to leave one den for another. Her three moves from den B (twice in 1994, once in 1996) all occurred when rain water seeped into the den, and each time we were able to confirm, after she had left, that water had pooled within the den. Additionally, our investigation of den C after she had left it on October 19, 1996, revealed that the ground was soaking wet.

Jiaojiao's behavior during movements from one den to another differed with the stage of cub development. During the cub's earliest development, even as the mother moved the cub from one den to another, the two were inseparable. Jiaojiao would move the cub by holding it physically in her mouth, and when she stopped moving, would cradle it in her arms. In contrast, during later stages of cub development, she often allowed the cub to be by itself while it foraged, staying close to it while actually moving as well as when cradling it.

During a rainstorm on September 2, 1994, at the age of only 16 days, Xiaosan was moved by Jiaojiao from den B. As Jiaojiao held Xiaosan in her mouth, we could hear the cub's sharp cries, at which point Jiaojiao would

Table 9.7.

SEQUENCE AND DURATION (DAYS) OF USE OF JIAOJIAO'S DENS WHILE REARING EACH CUB.

Cub	Birth den	Second den	Third den	Fourth den	Fifth den	Sixth den	Seventh den	Total time in dens (days)
Xiwang								
Den	A1	A2	B	C	B	E	F	
Duration	<3	<15	<2	71	<6	<5	2	125
Xiaosan								
Den	B	C	B	E				
Duration	16	50–53	13	2–5				94
Xiaosi								
Den	B	D	C	B	D	B	A2	
Duration	6	6–9	40–43	Unknown	Unknown	9	Unknown	>76

stop and cradle her cub, which resulted in a diminishing and ultimately cessation of the cubs' crying.

On October 25, when Xiaosan was 50 days old, Jiaojiao led the cub away from den C but did not arrive at den B until on October 31. Jiaojiao and Xiaosan spent the intervening 6 days in a mountain ravine located between the two dens. On October 26, we observed Jiaojiao cradling Xiaosan within a bamboo grove, and we inferred that they were resting. We then observed Jiaojiao to change positions, lying on her side, whereupon Xiaosan climbed upon her abdomen. We frequently observed Jiaojiao adopt this type of posture with her young cubs when inside dens.

On October 28, we discovered Xiaosan resting by himself in a daybed located in front of a stony cliff. When we first saw Jiaojiao she was only 5 m from us, but upon sensing us, she left the immediate area. On the subsequent day, when we again observed Xiaosan at the same daybed, Jiaojiao was foraging about 20 m from the cub.

In 1992, as Jiaojiao was in the process of transferring Xiwang from den E to den F, they similarly spent 5 to 6 days within bamboo groves, at which time Jiaojiao would forage somewhat distantly from her cub.

At a certain point in the development of the cub, dens were no longer used for the safety and security of the cub. Although cubs were occasionally stashed by their mother in caves or rock crevices, she no longer stayed with them as she did during the denning period. At this point, we considered the denning period to have ceased.

Calculated from December 18, 1992, when we last observed Jiaojiao leaving den F with Xiwang in her mouth, the denning period for Xiwang lasted for 125 days. In 1994, Jiaojiao left her den life with Xiaosan on November 19;

thus, Xiaosan's time in the den lasted for 94 days. In 1996, the last observation we made of Xiaosi was on December 4, and at this time the cub had already left den A2. However, in this case we were unable to determine if Jiaojiao and Xiaosi subsequently used other dens; thus, we estimated her den life as having lasted no fewer than 105 days.

Cub Maturation and Growth

At times when the mothers voluntarily left the den we took the opportunity to weigh cubs, to take various body size measurements, and to examine dental characteristics, pelage color, and condition, as well as the cub's reaction to our own research activities. Because of logistic constraints, we were unable to use identical equipment to weigh cubs each year; indeed, the cubs' growth by itself required us to use slightly different equipment to weigh them as they grew. From 1992 to 1996, we used spring scales with capacities of 5 and 20 kg and steel springs with capacities of 25 and 30 kg. Although the spring scales had the advantages of being small and easy to carry, they were not as precise as the steel scales. To maintain consistency, we adjusted all scales to each other. When weighing cubs, we placed them in appropriately sized cloth bags or backpacks and later subtracted those weights from the total weight obtained. We measured cubs using steel tapes scored at 1 mm as well as leather tapes scored at 1 cm.

CHANGES IN BODY MASS

We documented the conditions of growth and maturing of Qinling panda cubs primarily on the basis of observations and measurements of Xiwang, Xiaosan, and Xiaosi. We also used the opportunity to compare wild and captive cubs by measuring the captive cub Jingxin (Table 9.8).

We recorded the weights of all wild and captive cubs to the age of approximately 150 days (Figure 9.5). We were able to weigh the cub Guiye once: at the age of about 80 days. Guiye's weight was only 40%–60% the weight of the other panda cubs. Increases with time were similar among the other panda cubs, with Xiaosi being the lightest among those we weighed.

Table 9.8.

		Growth rate (g/day)					
MEAN GROWTH RATE OF GIANT PANDA CUBS AT VARIOUS STAGES OF DEVELOPMENT (DAYS SINCE BIRTH). A DASH (—) INDICATES NO WEIGHTS WERE OBTAINED.							
Cub	Number of weights	Days 0–30	Days 31–60	Days 61–90	Days 90–120	Days 120–150	Mean
Jingxin	32	51	74	79	80	97	76
Xiwang	9	—	43	67	—	—	64
Xiaosan	14	—	63	54	68	—	60
Xiaosi	22	50	54	35	23	—	46

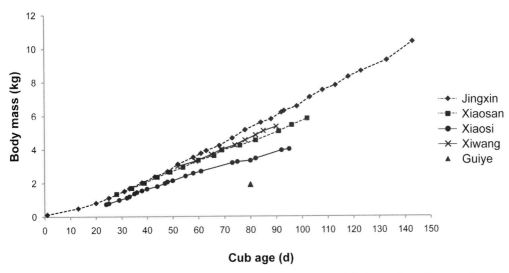

FIGURE 9.5. Body weights of wild cubs Xiaosan, Xiwang, Xiaosi, and Guiye and captive cub Jingxin at different ages.

Examining growth in cub weight by 30-day periods (Table 9.8), we see that the captive-bred cub Jingxin displayed an increasing trend in daily weight gain with time. Its daily growth rate exceeded that of free-ranging cubs during all time periods examined. In contrast, Xiaosi's pattern of growth displayed a pattern of decreasing daily weight gain with time, and by the 90–120 day period it was only about a third of that of Xiwang. This decrease was probably related to the low number times (only three) that Xiaosi was weighed. Regardless, Xiaosi had the slowest weight gain of all our measured cubs during this period of time.

We suspect that Xiaosi's slow growth rate was related to her infestation with external parasites, mainly ticks of the family Acarina. We first documented ticks on Xiaosi on September 15, 1996, when she was 25 days old. We found six more on October 18, and the largest was 7–8 mm long, engorged with blood. Further physical examinations continued to yield evidence of a larger number of ticks. We note, however, that prior to this, we had not documented external parasite infections on panda cubs.

DEVELOPMENT OF CUBS FROM BIRTH UNTIL 8 MONTHS

We defined five periods of growth within the first 8 months that describe cub development (Table 9.9). We found that our free-ranging cubs were similar to what is known from captivity (Beijing Zoo, 1974; Zhang et al., 1996) with respect to changes in skin color, growth of fur, and development of eyes and ears. We thus inferred that the behaviors of captive-bred mothers and cubs were also similar to free-ranging animals of similar ages and developmental stages in the Qinling. On the basis of these data, we can then

Reproduction and Early Development of Cubs 223

Table 9.9.

DEVELOPMENT STAGES OF CUBS AGED 0 TO 8 MONTHS AND THE CHARACTERISTICS OF BOTH THE DAM AND THE CUB AT EACH STAGE.

Stage	Time range	Distinguishing characteristics	Cub's physical characteristics	Cub's behavioral characteristics	Dam's behavioral characteristics
1	Birth to 9–10 days	Black pigmentation of cub's skin	Black pigmentation of the skin	Held by the dam; most of the time not visible	Holds the cub; does not leave the den; does not feed or drink
2	9–10 days to 40–50 days	Cub's eyes opening	Hair growth	Can be left alone in the den	Goes out feeding; gradually increases the time staying outside
3	40–50 days to 90–130 days	Stop living in caves and start using lairs	Starting teething and has eyesight at the end of this stage	More capable of movements	Time outside (longest 48 hours) dens exceeds that of inside; puts objects at the entrance of the den
4	100–140 days to 150–180 days	Cubs climbing trees	Physical growth	Stays near den; new sounds occur (Zhu, 1996); walks with no difficulty	Mouth taking cub when moving around; will not leave the cub over 50–100 m away; bleating when returning to cub
5	after 150–180 days		Physical growth	Stays in trees when alone; follows the mother to move around	May leave the cub a long distance away for a long time (Lü, 1991); bleating when returning to cub

evaluate the relationships between behavior under captivity and reproductive success, which would provide assistance to future captive rearing efforts.

Evaluation of Female Panda Breeding Status

ASSESSMENT OF REPRODUCTION IN THE WILD AND IN CAPTIVITY

Our understanding of the age of first reproduction in captive pandas (Xie and Gipps, 1997) is that they begin breeding at the age of 5–6, which is basically in accord with our data from free-ranging animals in the Qinling. In contrast, because of differences in environmental conditions and the fact that humans exert some control over the situation in captivity, males in the wild begin breeding at a somewhat later age.

The existing literature on captive breeding tends to produce the impression that mother pandas are inept at caring for their young and that they often roll over on their cubs and smother them. However, our detailed behavioral observations of Jiaojiao under free-ranging conditions in the Qinling illustrated that panda mothers can use various successful postures with which to cradle young cubs while in their dens without crushing them. At the slightest vocalization from the cub indicating discomfort, we noted that the mother would immediately respond in an adaptive way (Lü, 1993; Liu, 1996; Zhu, 1996). Furthermore, mother pandas adopted aggressive and effective protective behaviors to deter possible predators from their cubs (Lü, 1991). These actions all suggest that panda females are indeed fully capable of caring for their cubs.

We have summarized some of the existing literature on reproductive parameters of brown (*Ursus arctos*), American black (*U. americanus*), polar (*U. maritimus*), and sloth (*Melursus ursinus*) bears (Table 9.10). We present these data for purposes of comparison with pandas and to examine whether pandas exhibit a systematic disadvantage. In the Qinling, free-ranging females become reproductively mature at 4.5 years and produce their first offspring at 5 years, which is a bit older than some populations of American black bears but younger than some others as well as all brown bears. Although the mean litter size of the panda (1.45) is the lowest among those data available to us, their interlitter interval is shorter than four of the American black bear populations and longer than the other one. Birth rate is a function of age at first reproduction, litter size, and interlitter interval, all of which are interrelated (Bunnell and Tait, 1980, 1981). Taken together, reproductive rates of giant pandas are intermediate in their values with respect to those of brown, polar, and American black bears.

For wild American black bears (*Ursus americanus*), the reproductive rate of individuals suggests considerable variation. In east central Ontario, Canada, 25% of females accounted for 66% of total litters. However, we believe this level of individual heterogeneity does not characterize free-ranging pandas. Our observations of free-ranging individuals in the Qinling did not suggest such levels of individual heterogeneity. From radio tracking employed since 1989, only Yanghe (for whom we had limited tracking time) and Xiaosi (not monitored as an adult) failed to reproduce as adults. From continuous monitoring of Jiaojiao, Ruixue, and Momo, we documented interlitter intervals of 2 or 3 years. Thus, we have found little individual heterogeneity in reproduction among free-ranging adult females.

Annual survival rates of panda cubs were lower than seen among any of the brown bear populations examined (range 0.625–0.89; Aune et al., 1994), although higher than the single American black bear population (0.48; LeCount, 1987). One possible reason for this is that this particular black

Table 9.10.

REPRODUCTIVE PARAMETERS OF SELECTED BROWN, BLACK, POLAR, AND SLOTH BEAR POPULATIONS. NCDE = NORTHERN CONTINENTAL DIVIDE ECOSYSTEM. A DASH (—) INDICATES DATA ARE NOT AVAILABLE.

Species	Population	Litter size	Interbirth interval	Natality (cub/year)	Age of first production	Source
Brown bear ($Ursus\ arctos$)	Kodiak Island	2.23	3	0.74	5.0	Stringham (1990a)
	Lower Alaska Peninsula	2.30	3	0.77	4.4	Stringham (1990a)
	McNeil River	2.10	3.6	0.58	5.9	Stringham (1990a)
	Eastern Brooks Range	1.77	≥4	0.44	10.1	Stringham (1990a)
	Western Brooks Range	1.98	≥4.1	0.48	7.4	Stringham (1990a)
	Southern Yukon	1.70	≥4	0.42	≥7.5	Stringham (1990a)
	Northern Yukon	2.00	≥4	0.57	7.0	Stringham (1990a)
	Northwest Territories	1.83	3.8	0.48	≥8.0	Stringham (1990a)
	Glacier National Park, Canada	2.00	—	—	≥5[a]	Stringham (1990a)
	Yellowstone, 1959–1970	2.18	3.2	0.68	5.7	Stringham (1990a)
	Tuktoyaktuk Peninsula	2.30	≥3.3	≤0.70	6.4	Stringham (1990a)
	South central Alaska	2.8	3	1.07[a]	5.0	Stringham (1990a)
	NCDE Montana	2.14	2.69	—	5.7 (4–7)	Aune et al. (1994)
	Yellowstone	—	—	—	6.15	Knight and Eberhardt (1985)
Black bear ($U.\ americanus$)	Western Montana	1.7	≥3.0	0.57	6.6	Stringham (1990b)
	Northeastern Montana	2.38	2.28	1.04	6.3	Stringham (1990b)
	Lowell, Idaho	1.65	3.57	0.46	5.0	Stringham (1990b)
	Council, Idaho	1.9	3.23	0.59	4.8	Stringham (1990b)
	North Carolina	2.45	2.16	1.13	4.0	Stringham (1990b)
	Kenai Peninsula	1.9	2.56	0.74	4.4	Stringham (1990b)
	Pennsylvania	2.9	2.00	1.45	3.6	Kordek and Lindzey (1980)
	Northeastern Pennsylvania	3.0	1.82	1.64	3.18	Alt (1989)
	Maine	—	2–3	—	4–6	McLaughlin et al. (1994)
	East central Ontario	1.9 (winter), 2.5 (summer)	2	0.6 (age 5–7), 1.2 (age 8–18)	6 (5–8)	Kolenosky (1990)
Polar bear ($U.\ maritimus$)	Alaska	1.63	3.6	0.45	5.4	Lentfer et al. (1980)
	Wrangel Island	1.85 (1969), 1.68 (1970)	—	—	—	Uspenski and Kistchinski (1972)
Sloth bear ($Melursus\ ursinus$)	Chitwan, Nepal	1.6	—	—	—	Laurie and Seidensticker (1977)

[a] Small sample size may influence estimate.

226　Chapter 9

bear population faced unusual ecological conditions. In that Arizona study, reproduction was found to be density dependent; 50% of cub mortality was attributed to infanticide, with only 25% coming from predation (LeCount, 1987). However, in the Qinling, except for the one twin cub that was abandoned by its mother, we documented only one additional cub loss. Regardless of cause, the magnitude of cub mortality was far lower than recorded among American black bear populations. Data constraints preclude us from making a thorough comparison between survival rates of giant pandas and other bears, but from the available material, it certainly does not appear that Qinling panda survival is particularly low compared with other bear species.

THE STRATEGY OF PANDA REPRODUCTIVE BEHAVIOR

Female pandas reproduce, on average, only every 2.33 years. If a female enters her reproductive years at age 5 and continues through age 20, she can have at most six to seven litters during her lifetime. Thus, only by ensuring the success of these reproductive events can she maximize her adaptive capability. At the same time, reproducing requires energy, and the low nutritive quality of the panda's bamboo diet forms an energetic constraint (Schaller et al., 1985). Our research in the Qinling has made clear that even in years in which reproduction does not take place, forage resources are important for both sexes, and securing these resources is key to reproductive success.

The gestation period for giant pandas is among the longest of bears (Kleiman, 1983), but the facts that the period of implantation is quite brief and that the newborn cubs are very small decrease the resource drain on the mother. Lactation requires a large amount of resources: lactating females can lose up to 40% of their weight during their fasting period (Oftedal, 1993). Thus, it is incumbent on the mother panda to complete her period of fasting as early as possible in order to reduce consumption of her own resources. To maintain condition, she must allocate some time for her own foraging even while caring for her young and fragile cub. A female raising multiple cubs within a litter would face difficulties in providing sufficient nutrition and resources for the survival of all. By raising only a single cub, she stands a much better chance of providing sufficiently for it. We see then that giant pandas adopt a strategy through both mating and cub rearing that is designed to provide the highest probability of reproductive success.

Comparing the interlitter interval of giant pandas in the Qinling to those of some bear populations from the temperate and cold zones of North America reveals that they are 2–3 years shorter, suggesting that reproductive events occur twice as fast. This interbirth interval is what causes the overall reproductive rates of pandas to occupy an intermediate level when compared with other bear species. We see also that juvenile survival of panda cubs is lower than that documented for North American brown bears but higher than for American black bears.

Reproduction and Early Development of Cubs   227

Giant panda females also produce milk that is higher in protein content (22% by energy content) than that of American black bears (14% by energy content). We speculate that this difference arises from differential evolutionary pressures faced by the two species: panda females begin foraging again only 10 days postpartum, whereas American black bear females fast during the entire period of lactation.

Pandas are indeed a specialized bear that subsists on forage (bamboo) of low nutritional value. However, although they have the longest gestation length and produce the relatively smallest and least developed offspring of any carnivore (Kleiman, 1983; Oftedal and Gittleman, 1989), none of these characteristics have prevented giant pandas in the Qinling from successfully reproducing and prospering. Thus, we have rationale for asserting that pandas are successful specialists. Whether we are considering them from an evolutionary viewpoint or from the perspectives of physiological and ecological pressures, they have shown their ability to persist and their potential to increase.

References

Alt, G. L. 1989. "Reproductive Biology of Female Black Bears and Early Growth and Development of Cubs in Northeastern Pennsylvania." PhD thesis, West Virginia University.

Aune, K. E., R. D. Mace, and D. W. Carney. 1994. "The Reproductive Biology of Female Grizzly Bears in the Northern Continental Divide Ecosystem with Supplemental Data from the Yellowstone Ecosystem." *Bears: Their Biology and Management* 9:451–458.

Beijing Zoo. 1974. 大熊猫的繁殖及幼兽生殖发育的观察 [Observation on reproduction and cub development of giant panda]. *Acta Zoologica Sinica* 20:139–147.

Bunnell, F. L., and D. E. N. Tait. 1980. "Bears in Models and in Reality: Implications to Management." *Bears: Their Biology and Management* 4:15–23.

———. 1981. "Population Dynamics of Bears—Implications." In *Dynamics of Large Mammal Populations*, edited by C. W. Fowler, and T. D. Smith, pp. 75–98. New York: John Wiley.

Kleiman, D. G. 1983. "Ethology and Reproduction of Captive Giant Pandas (*Ailuropoda melanoleuca*)." *Zeitschrift für Tierpsychologie* 62:1–46.

Knight, R. R., and L. L. Eberhardt. 1985. "Population Dynamics of Yellowstone Grizzly Bears." *Ecology* 66:323–334.

Kolenosky, G. B. 1990. "Reproductive Biology of Black Bears in East-Central Ontario." *Bears: Their Biology and Management* 8:385–392.

Kordek, W. S., and J. S. Lindzey. 1980. "Preliminary Analysis of Female Reproductive Tracts from Pennsylvania Black Bears." *Bears: Their Biology and Management* 4:159–161.

Laurie, A., and J. Seidensticker. 1977. "Behavioural Ecology of the Sloth Bear (*Melursus ursinus*)." *Journal of Zoology* 182:187–204.

LeCount, A. L. 1987. "Causes of Black Bear Cub Mortality." *Bears: Their Biology and Management* 7:75–82.

Lentfer, J. W., R. J. Hensel, J. R. Gilbert, and F. E. Sorensen. 1980. "Population Characteristics of Alaskan Polar Bears." *Bears: Their Biology and Management* 4:109–115.

Liu, D. Z. 1996. 圈养大熊猫 (*Ailuropoda melanoleuca*) 行为生态学研究 [Studies on behavioral ecology of captive giant pandas]. PhD thesis, Beijing Normal University.

Liu, W. X., Z. Xie, N. L. Liu, and X. Wang. 1993. 大熊猫分娩活动的观察 [Observation on the parturient activities of the giant panda]. *Acta Theriologica Sinica* 13:241–244.

Lü, Z. 1991. 秦岭大熊猫的种群动态，活动方式和社会组织 [Population status, activity pattern, and social structure of giant pandas in Qinling]. PhD thesis, Peking University.

———. 1993. "Newborn Panda in the Wild." *National Geographic* 183:60–65.

McLaughlin, C. R., G. J. Matula Jr., and R. J. O'Connor. 1994. "Synchronous Reproduction by Maine Black Bears." *Bears: Their Biology and Management* 9:471–479.

Oftedal, O. T. 1993. "The Adaptation of Milk Secretion to the Constraints of Fasting in Bears, Seals, and Baleen Whales." *Journal of Dairy Science* 76:3234–3246.

Oftedal, O. T., and J. L. Gittleman. 1989. "Patterns of Energy Output during Reproduction in Carnivores." In *Carnivore Behavior, Ecology, and Evolution*, edited by J. L. Gittleman, pp. 355–378. Ithaca, NY: Cornell University Press.

Pan, W. S., Z. S. Gao, Z. Lü, Z. K. Xia, M. D. Zhang, L. L. Ma, G. L. Meng, X. Y. She, X. Z. Liu, H. T. Cui, and F. X. Chen. 1988. 秦岭大熊猫的自然庇护所 [The giant panda's natural refuge in the Qinling Mountains]. Beijing: Peking University Press.

Schaller, G. B., J. C. Hu, W. S. Pan, and J. Zhu. 1985. *The Giant Pandas of Wolong*. Chicago: Chicago University Press.

Stringham, S. F. 1990a. "Grizzly Bear Reproductive Rate Relative to Body Size." *Bears: Their Biology and Management* 8:433–443.

———. 1990b. "Black Bear Reproductive Rate Relative to Body Weight in Hunted Populations." *Bears: Their Biology and Management* 8:425–432.

Uspenski, S. M., and A. A. Kistchinski. 1972. "New Data on the Winter Ecology of the Polar Bear (*Ursus maritimus* Phipps) on Wrangel Island." *Bears: Their Biology and Management* 2:181–197.

Xie, Z., and J. Gipps. 1997. 大熊猫谱系 [The giant panda studbook]. Beijing: China Association of Zoological Gardens.

Zhang, G. Q., H. M. Zhang, M. Chen, T. M. He, R. P. Wei, and S. A. Mainka. 1996. "Growth and Development of Infant Giant Pandas (*Ailuropoda melanoleuca*) at the Wolong Reserve, China." *Zoo Biology* 15:13–19.

Zhu, X. J. 1996. 野生大熊猫 (*Ailuropoda melanoleuca*) 幼仔声音的发育和声谱分析 [Vocality development and sound spectrum analysis of giant panda cubs (*Ailuropoda melanoleuca*)]. Master's thesis, Peking University.

Chapter 10

Qinling Bamboo and the Feeding Strategy of Giant Pandas

Summary: In this chapter, we discuss the feeding habits of giant pandas, analyze feeding behavior, and estimate panda daily food consumption. The major foods of the Qinling giant panda are bamboo of the genera *Bashania* and *Fargesia*. Pandas select bamboo species for foraging based on season, bamboo culm diameter, and bamboo age. We introduce a novel method of analyzing behavior. Giant pandas adopt specific behavioral patterns while foraging. Movements and activities of individual pandas foraging on the same bamboo species are quite similar to one another. However, feeding behavior varies as a function of age, sex, and other factors. By comparing the same individual during various physiological stages, we found feeding behavior tends to stabilize, and the animal becomes more skillful and rapid in its feeding as it gains experience and becomes reproductively mature. However, past a certain age, tooth wear and declines in other physiological functions destabilize and reduce the pace of feeding behavior. We categorized panda feeding time into three stages: preparation time, time spent pausing among feeding bouts, and time actually consuming forage. Pandas can potentially adopt three strategies to increase daily consumption without increasing the total time spent foraging: increase the proficiency with which they process bamboo, reduce preparation time, or reduce time spent pausing between feeding bouts. We assessed various approaches to estimating daily food consumption and developed and used a new method. Subadults feeding on *Bashania* leaves and culms had daily food consumption values of 18.0 ± 4.3 kg/day (fresh weight), whereas older individuals on the same forage had values of 13.7 ± 2.1 kg/day. Adults foraging on new *Bashania* shoots had daily food consumption values of 41.7 ± 5.6 kg/day. Within the 346 km^2 of the Qinling in which pandas are densely distributed, we estimated total yearly biomass of first- and second-year growth of *Bashania* leaves, culms, and shoots (i.e., those parts eaten by pandas) as 134,534 metric tons (t). We estimated the edible portions of *Fargesia* (old shoots from the current year and new culms from the previous year) as 85,515 t. The 79 pandas living within the area of dense distribution consume an estimated 1,139 t of *Bashania* yearly (including new shoots, new leaves, and new culms), or about 0.8% of available *Bashania*, and 331 t of *Fargesia*, or about 0.4% of that available. Thus, Qinling pandas consume an insignificant portion of the available bamboo.

Giant Panda Diet

PRINCIPAL FOODS OF THE GIANT PANDA

Bamboos all belongs to Poaceae, the grass family, and are distinguished from other plants in this family by their unique woody culms. Bamboos also produce complex branches at the bamboo nodes, have extensive networks of underground rhizomes, and have a flowering cycle several decades long (Chapman and Peat, 1992). Bamboos are widely distributed within tropical and subtropical regions but only a few animal species depend solely on bamboo for survival. Aside from pandas, some other mammals, a few birds, and some insects are known to feed on bamboo shoots, culms, and leaves. Mammals in China that use bamboo include takins (*Budorcas taxicolor*), serows (*Capricornis sumatraensis*), tufted deer (*Elaphodus cephalophus*), dwarf musk deer (*Moschus berezovskii*), wild boars (*Sus scrofa*), porcupines (*Hystrix hodgsoni*), Asiatic black bears (*Ursus thibetanus*), bamboo rats (*Rhizomys* spp.), and red pandas (*Ailuris fulgens*; Musser, 1972; Petter and Peyrieras, 1975). Among birds, golden pheasants (*Chrysolophus pictus*), koklasses (*Pucrasia macrolopha*), blood pheasants (*Ithaginis cruentus*), and Temminck's tragopans (*Tragopan temminckii*) occasionally feed on bamboo. However, among these animals, only the bamboo rat survives solely on bamboo; all others use only specific bamboo parts and only seasonally, and their total bamboo consumption is low. The wild giant panda, on the other hand, is a significant consumer of bamboo (Sheldon, 1937; Wang and Lu, 1973).

The wild giant panda's selection of bamboo as its major food source carries with it important evolutionary consequences. When we regard the giant panda's physical characteristics, although it retains sharp claws and sharp canines, its stocky body and relative lack of agility hinder its ability to capture animal prey. In the remote mid- and high-mountain regions where pandas persist, although the fauna is diverse, the abundance of potential animal carcasses is low, making the prospect of scavenging a difficult one for pandas. A reliance primarily on plant material for nutrition would depend on access to an abundant, reliable plant species. Bamboos are characterized by their wide geographic distribution, relatively high annual yield, and comparatively long life cycle. Together, these characteristics have led pandas to choose bamboo as their primary forage.

Researchers have discovered regional variation in bamboo species that constitute the primary forage by pandas: Wolong features six species commonly eaten by pandas, including *Bashania faberi*, *Fargesia spathacea*, and *F. nitida*. Pandas eat five species in the Liangshan, although only one species in some areas of the Min Shan (such as Wanglang; Qin, 1985). We have observed pandas to feed on culms, leaves, shoots, and branches of bamboo but never to excavate rhizomes for forage. The southern slopes of the mid-Qinling support nine species (representing five genera) of bamboo, including monoaxial, biaxial, and multiaxial species (Pan et al., 1988). Species of the genera *Bashania* and *Fargesia* are the major wild bamboos in this region and also constitute the major foods for pandas. Incidentally, our field observations have twice (once in April 1989 and again in September 1996) documented free-ranging pandas foraging on golden bamboo (*Phyllostachys nigra*), a species artificially introduced to the area.

OTHER FOOD ITEMS CONSUMED BY PANDAS

Pandas also eat other species of plants and fungi; they do scavenge on carcasses of dead animals and have also been known to ingest certain types of soil. In the Min Shan region, studies have reported that pandas include the culms, leaves, and bark of a number of species in their diet, of which the most numerous are a horsetail (*Equisetum hiemale*), as well as pondweed (*Potamogeton* spp.) and other aquatic plants; Chinese gooseberry (*Actinidia* spp.) and other vines; Chinese necklace poplar (*Populus lasiocarpa*), Sichuan bramble (*Rubus setchuenensis*), and other shrubs; forbs, including Chinese lovage (*Lingusticum* spp.), cow parsnip (*Heracleum candicans*), Tatarian aster (*Aster tataricus*), snow lotus (*Saussurea* spp.), qiang huo (*Notopterygonium incisum*), and chameleon plant (*Houttuynia cordata*); and trees, including juniper (*Sabina squamata*), dragon spruce (*Picea asperata*), mountain loquat (*Ilex franchetiana*), *Litsea moupinensis* var. *szechuanica*, *Helwingia japonica*, and others (Zhu and Long, 1983). In Wolong, researchers have documented giant pandas feeding on wild Chinese angelica (*Angelica* spp.), *Polyporus frondosus*, and the bark of coniferous trees, including Faber's fir (*Abies fabri*), Chinese hemlock (*Tsuga chinensis*), and Armand's pine (*Pinus armandi*). In addition, a captive-reared panda at the Chengdu Zoo consumed three species of grass. Captive pandas at Yingxionggou in Wolong have been recorded as eating the leaves of a grass (*Deyeuxia scabrescens*) and shrub (*Buddleia davidii*) species. A preliminary estimate is that no fewer than 25 wild plant species have been documented as having been consumed by giant pandas. However, the proportion of diets made up by these other species is quite small, generally <1% of total food consumption. In fact, in all our years of field observation in the Qinling, we have yet to observe pandas feeding on material other than bamboo. At times, when feeding on bamboo, pandas that we observed consumed forbs growing nearby, but this appeared to be incidental and not deliberate selection by the panda. The discrepancy between our observations and those of other researchers may be due to true variation between these small panda populations or may have arisen solely from sampling errors.

In 1974 the researchers at Wolong heard local residents recall that they had once discovered "the remains of a small rodent" in the stomach of a giant panda and that a panda had been seen capturing a bamboo rat. The Wolong team also documented hair from a golden monkey (*Rhinopithecus roxellanae*) in giant panda feces, and at the Wanglang Nature Reserve, researchers encountered panda feces containing hair, bones, and hooves of musk deer (*Moschus* spp.). At the Yingxionggou captive enclosure in Wolong, when caretakers offered mutton for the pandas to eat, four out of seven individuals immediately consumed the meat (Schaller et al., 1985). We encountered similar situations during our observations of free-ranging pandas in the Qinling (Table 10.1).

These data show that free-ranging giant pandas are not completely herbivorous; when they encounter edible decomposed meat or bones, they welcome the opportunity to augment their diets. We consider such situations common. Additionally, our observations indicate that females are more apt to supplement their diet from scavenging during gestation or lactation. For example, the two occasions when Jiaojiao ate bones or wild boar skin (in 1990 and again in 1994) both occurred while she was pregnant. Additionally, we noted Xiwang eating pork bones that had been discarded as human trash at Shanshuping in July 1997, a time her behavior suggested was just prior to parturition. These observations suggest that pandas feed on decomposed meat or bones as a result of physiological needs, perhaps in order to supplement nutrients such as vitamins or essential amino acids. Our speculation

Table 10.1.

OBSERVATIONS OF FREE-RANGING PANDAS IN THE QINLING FORAGING ON ITEMS OTHER THAN BAMBOO, 1984–1997.

Date	Individual	Sex	Location	Item
March 1985	Dandan	F	Foping	Chicken and skin of pig
March 1987	Unknown	Unknown	Unknown	Hair and hoof of *Moschus berezovskii*[a]
February 1989	Unknown	Unknown	Tudigou	Carcass of *Budorcas taxicolor*
December 1990	Jiaojiao	F	Shuidonggou	Bone and meat
July 1991	Shuilan	F	Shuidonggou	Chicken
October 1991	Unknown	Unknown	Liaojiagou	Meat[a]
1992–1993	Huzi	M	Shuidonggou	Chicken
February 1993	Nüxia	F	Shuidonggou	Bone
March 1994	074	M	Shuidonggou	Bone of *Elaphodus cephalophus*
October 1994	Jiaojiao	F	Shuidonggou	Skin, rib, and thighbone of *Sus scrofa*
July 1997	Xiwang	F	Shanshuping	Discarded pork bones

[a] Located in feces.

should be confirmed with further research because we lack evidence of a direct link between these behaviors and female reproductive conditions. Our results could also stem from the fact that because we focused our research on the animals during certain critical portions of their reproductive cycle, we had more opportunity to observe these instances during those life history events. Similarly, we cannot rule out the possibility that the frequency with which pandas eat carrion reflects nothing more than the chance occurrence that would attend opportunistic foraging.

The Wolong researchers also documented gray colored, clay-like soil in droppings of a panda named Zhenzhen. A captive panda named Jiajia also consumed similar soil when outside Wolong's exclosure at Yingxionggou (Schaller et al., 1985). During our own research in the Qinling, we documented Huzi ingesting silt while crossing a river in May 1995. Whether these phenomena arise from a requirement to supplement their diets with a particular micronutrient of importance or to aid in digestion awaits further research.

Giant Panda Selection of Bamboo

SEASONAL FORAGE SELECTION: THE ECOLOGICAL SEASON OF THE GIANT PANDA

In considering panda food habits, investigators in Wolong categorized the year into three ecological seasons: spring (April–June), when pandas fed mainly on new shoots of *Fargesia spathacea*; summer (July–October), when pandas fed mainly on the leaves of *Bashania faberi*, and winter (November–March), when pandas fed mainly on old shoots, culms, and leaves of *B. faberi* (Schaller et al., 1985).

Our understanding of panda food habits in the Qinling comes from the accumulation of firsthand data beginning in 1984, including radio telemetry data and direct observations. Panda diets in the Qinling also showed a clear seasonality. As mentioned above, pandas primarily ate two kinds of bamboo: *Bashania*, which grows at 900–1,900 m, and *Fargesia*, which grows at 1,800–3,000 m. These two bamboos are also the dominant types in the Qinling region.

At its lowest limits (~800 m), *Bashania* begins sending up shoots in early April, and shoots appear progressively later with elevation. By late April, *Bashania* shoots in most Qinling panda habitats have grown to the point (~25–50 cm) where they are suitable for panda foraging. At this point, most pandas feed primarily on *Bashania* shoots. *Fargesia*, on the other hand, does not begin sending up shoots until early May even at its lower distributional limit at 2,000 m, and its shoots also appear progressively later with elevation. *Fargesia* shoots only become suitable for pandas (>50 cm height) toward the end of May. As a result, pandas successively migrate upward, moving into *Fargesia* groves to forage on the current year's shoots as well as older shoots from the previous year. This continues until September, when the previous years' *Fargesia* shoots have grown to heights of about 2 m and become lignified, at which point panda gradually begin descending back toward *Bashania* groves, where they feed on bamboo leaves. They continue subsisting on little but *Bashania* leaves through the following February. Beginning in March, pandas begin adding new culms of *Bashania* to their staple diet of *Bashania* leaves; the frequency of occurrence for culms reaches 19%–27%. We assume that the addition of culms is related to seasonal changes in the nutritional content and palatability of *Bashania* leaves.

Because *Bashania* and *Fargesia* grow at disparate elevations, we used the elevation of radio-collared pandas to infer the species they were feeding on. Individual pandas exhibited considerable heterogeneity in the specifics of their seasonal elevational movements. The general pattern was for pandas to migrate from the low-elevation *Bashania* to *Fargesia* groves in late May or early June, but a minority of individuals delayed their upward movements until late June or early July. This individual heterogeneity reflected different resource use strategies. Pandas generally remained at high elevations feeding on *Fargesia* forest for 2–3 months, migrating back to low-elevation *Bashania* in late August or September. However, a few individuals remained in *Fargesia* until late September or early October. Furthermore, signs of feeding as well as droppings have been found in high-elevation *Fargesia* groves as late as December. This suggests that individual pandas may live year-round at these higher elevations, further evidence of differing resource utilization strategies among individual pandas. For the most part, pregnant females descended earlier than the other pandas (i.e., late July or early August) in order to search for a den and give birth. Table 10.A.1 provides detailed data on elevational movements of radio-collared individuals.

On the basis of field observations, we estimated the frequency of foraging on various bamboo species and parts and the percentage of time foraging on these items by month, which we used to delineate ecological seasons of panda foraging (Figure 10.1).

Pandas fed almost exclusively on *Bashania* leaves during January and February. In March and April, they began adding *Bashania* culms to their diets. Beginning in April, new *Bashania* shoots began appearing in diets. The frequency of new *Bashania* shoots in diets peaked in May and declined in June as the frequency of new *Fargesia* shoots suddenly increased. During July and August, pandas fed solely on *Fargesia* shoots, both the fresh ones of that year and the previous years' shoots. From September through February, pandas fed almost exclusively on *Bashania* leaves, with only a few individuals occasionally feeding on old *Bashania* culms or old *Fargesia* shoots (displaying the individual heterogeneity mentioned earlier; Figure 10.1).

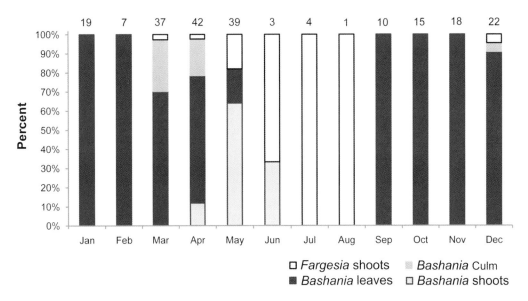

FIGURE 10.1. Percent frequency of observations of pandas feeding on various bamboo species and parts by month, with the total number of observations at the top of each bar, Qinling study area, 1985–1996.

From September through February, pandas feed primarily *Bashania* leaves, indicating that during this time period, *Bashania* leaves are fresh, tender, and moist and their nutritional value is comparatively high (Figure 10.1). By March or April, after a winter of growth, these leaves' cell walls have gradually thickened, and lignin and cellulose content have increased, with a corresponding decrease in cell content. In addition, because of the arid spring climate with little precipitation, bamboo leaves have generally become desiccated. At this time, pandas begin adding *Bashania* culms to their diet. Although culms are not the best choice in terms of nutrition or palatability, they would still be able to compensate for some nutritional deficiencies caused by the aging and drying of bamboo leaves. By April, following the difficulty of this lean period, new shoots of *Bashania* gradually emerge. At this time, pandas cannot wait to begin feeding on these newly emerging shoots, which serve to make up for the nutritional deficiencies of winter and spring. In May, *Bashania* shoots reach their growth peak and become pandas' main nutrition source. In May and June, with the emergence and growth of new *Fargesia* shoots, these gradually replace *Bashania* as pandas' primary forage. During this period, pandas will also feed on previous years' *Fargesia* shoots. This remains the case until late August, at which time *Fargesia* shoots gradually mature and become woody, and pandas begin abandoning them. However, this is just the time at which leaves of the

Feeding Strategy of Giant Pandas 237

low-elevation *Bashania* are again palatable and abundant, so they become the new object of feeding for the pandas.

SELECTION OF BAMBOO PARTS, AGE, AND THICKNESS

Selection of Bamboo Culms and Leaves

Both direct observations and scat analyses provided evidence that *Bashania* leaves constitute the main forage for pandas, with culms becoming important only in March and April (at which time, panda scats consisted of approximately 85.3% leaves and 14.7% culms). When a panda feeds on *Bashania* culms, it usually eats only the central section, selecting only the leaves from the outer portion and discarding the rest. However, occasionally pandas will consume the entire bamboo culm, allowing the leaves to fall off without eating them. We speculate that this is highly dependent on bamboo freshness. By contrast, when feeding on old shoots (that is, new culms) of *Fargesia*, pandas generally consume very few leaves, feeding almost entirely on new culm after peeling away the outer tegument. As when feeding on *Bashania*, the central portion is eaten, and the remainder discarded.

Selection of Bamboo Thickness and Age

Pandas are very selective regarding bamboo age and culm thickness. When foraging on *Bashania,* they primarily select relatively thick first- and second-year growth. Our observations suggested that pandas preferred first-year *Bashania*. When foraging on the more developed and larger second-year growth, we noted that pandas often ceased foraging after consuming only one or two handfuls. We made fewer observations of pandas feeding on *Fargesia*; our preliminary impressions were that pandas also preferred first- or second-year growth of new shoots or culms, but we do not yet know if the pandas also select for *Fargesia* culm diameter. Observations of free-ranging pandas indicate that they exhibit selectivity when feeding on bamboo shoots. For example, in May 1993, our observations of Huzi eating *Bashania* shoots indicated that he invariably consumed all nearby shoots >1 cm in diameter. These shoots were generally >25 cm high.

These feeding behaviors display one type of adaptation to their habitat: given limited time for foraging, pandas invariably ingested the most nutritious, easily digested, and easily obtained food items. Examples included rotten meat, bones, and the relatively tender portions of bamboo. Similar criteria were displayed in selection of bamboo shoots. The fleshy part of a bamboo shoot makes up a relatively larger portion (compared with the sheath) in a thicker than a thinner shoot. Thus, within any unit of time, pandas would obtain more nutritive value and caloric content by eating the thicker shoots.

Giant Panda Feeding Behavior

WITH ZHANG YINGYI

During the 7-year period from 1991 to 1997, we recorded video footage of five free-ranging pandas at various developmental stages as they foraged on various dietary items. We then selected the clearest portion of this footage for computerized analysis, which allowed us to conduct detailed behavioral analyses. We obtained 158 minutes of video-recording bouts and present details by individual panda and time period of monitoring (Table 10.2). We conducted analyses of feeding behavior on the basis of individual differences, as well as longitudinal variation within each individual as they progressed through various developmental stages.

METHODS

We digitized recorded images using a microVIDEO DC30 video card at a speed of 30 frames/s (NTSC format) or 25 frames/s (PAL format). To analyze these video bouts, we first conducted a preliminary observation of each video section during which we categorized each behavior present (specific categorizations were needed to standardize observations). Formal analyses proceeded by documenting the exact time of each behavior, the type of growth stage of the bamboo, and the mode employed by the panda to consume it. Unusual behaviors were also noted with their times and details of the panda's behavior. We also recorded ancillary data on the age and sex of the panda and the time and location of the feeding bout.

We ordered each event sequentially and categorized all bamboo feeding into 12 behavior categories (see the Chinese edition for details of each behavior). We identified an additional four categories of events that occurred after foraging on one culm of bamboo and prior to selecting the subsequent culm. On the basis of the sequence of movements, we divided feeding bouts into nine stages. We defined three additional stages (items 10–12 below) of the entire feeding bout based on the type of behavior.

1. Time spent grasping bamboo: The time expended from when the panda raised its paw to grasp bamboo until it grasped the bamboo culm, which it eventually ate. In calculating this duration, we subtracted the time required for the panda to reach for the bamboo.
2. Time spent biting off bamboo: The time from when the panda grasped the bamboo to when the bamboo was twisted and bitten off. In calculating this, we subtracted the duration from starting to pick up bamboo until the bamboo was actually grasped.
3. Time spent gathering a handful of bamboo leaves: The time expended from commencing until completing gathering.

Feeding Strategy of Giant Pandas   239

Table 10.2.

FEEDING BEHAVIOR ON LEAVES OF *BASHANIA* AT VARIOUS DEVELOPMENT STAGES FOR FOUR INDIVIDUAL GIANT PANDAS, QINLING STUDY AREA, 1993–1997. THE MEAN IS PRESENTED FOR EACH CATEGORY, WITH THE VARIANCE IN PARENTHESES. THE SAMPLE SIZE IS GIVEN BELOW EACH ENTRY.

Individual	Time period	Age (years)	Handling time per handful(s)	Mouthfuls per handful	Bites per mouthful	Chews per mouthful	Time spent per bite(s)	Time spent per chew(s)	Time per mouthful(s)
Xiwang	May 1994	1.75	23.7 (+7.1) $n=12$	3.8 (+1.2) $n=17$	3.6 (+1.8) $n=61$	8.5 (+3.0) $n=43$	0.6 (+0.1) $n=63$	0.5 (+0.0) $n=63$	6.6 (+2.4) $n=43$
	January 1994	1.5			2.7 (+1.2) $n=6$	5.7 (+1.0) $n=7$	0.6 (+0.1) $n=6$	0.5 (+0.0) $n=7$	4.3 (+0.5) $n=5$
	December 1994	2.44	21.9 (+5.9) $n=7$	4.0 (+1.1) $n=9$	2.6 (+0.9) $n=28$	6.1 (+1.9) $n=29$			5.5 (+0.3) $n=7$
	March 1995	2.58			2.0 (+0.0) $n=7$	6.0 (+0.0) $n=6$	0.6 (+0.1) $n=7$	0.5 (+0.0) $n=6$	4.5 (+0.2) $n=6$
	March 1997	4.75	20.6 (+5.8) $n=30$	4.7 (+1.3) $n=30$	2.6 (+0.8) $n=141$	5.3 (+1.3) $n=141$			4.5 (+0.8) $n=30$
Jiaojiao	December 1994	9.33	15.5 (+2.2) $n=9$	4.2 (+0.4) $n=9$	2.4 (+0.7) $n=38$	4.9 (+0.9) $n=38$			3.6 (+0.3) $n=28$
	January 1995	9.4			2.8 (+0.5) $n=5$	4.5 (+0.6) $n=5$	0.5 (+0.0) $n=5$	0.5 (+0.0) $n=5$	3.6 (+0.3) $n=5$
Huzi	August 1991	2	23.6 (+9.7) $n=13$	5.8 (+1.6) $n=16$	2.6 (+1.1) $n=72$	5.3 (+1.2) $n=66$	0.6 (+0.1) $n=110$	0.6 (+0.1) $n=105$	4.5 (+1.2) $n=66$
	August 1991	2	24.1 (+9.4) $n=16$	5.7 (+1.5) $n=17$	2.5 (+0.8) $n=85$	5.0 (+1.1) $n=85$	0.6 (+0.1) $n=120$	0.6 (+0.1) $n=116$	4.2 (+1.0) $n=85$
			25.1 (+10.1) $n=25$	5.3 (+2.0) $n=29$	2.5 (+1.0) $n=110$	5.1 (+1.4) $n=110$			4.8 (+1.3) $n=79$
	October 1994	5.17			2.0 (+0.0) $n=6$	4.9 (+0.7) $n=11$	0.5 (+0.1) $n=11$	0.5 (+0.1) $n=11$	3.8 (+0.5) $n=11$
	May 1995	5.75			2.0 (+0.0) $n=3$	4.7 (+1.2) $n=3$	0.5 (+0.1) $n=3$	0.5 (+0.0) $n=3$	3.5 (+0.6) $n=3$
Da Shun	October 1993	16		4.1 (+0.7) $n=56$	4.1 (+1.8) $n=227$	8.9 (+2.4) $n=227$			

240   Chapter 10

4. Time spent eating a handful of leaves: The time from the end of gathering to the end of eating the handful of leaves.
5. Time spent eating a mouthful of leaves: The time from the start of the first bite on this mouthful of leaves to the end of the last chewing motion.
6. Time spent biting off leaves: The time from the start of the first bite to the end of the last bite.
7. Time spent chewing a mouthful of leaves: The time from the end of the last bite to the end of the last chewing motion.
8. Time spent ripping tegument: The time from the start of ripping tegument to the end of ripping tegument.
9. Time spent feeding on a culm: The time spent actually biting off and chewing on a culm. In calculating this, we subtracted the time spent ripping tegument from the time of starting to feed on the culm to the end of feeding on the culm.
10. Total preparation time: The sum of the time spent picking up, biting, gathering leaves, and ripping tegument within any single, continuous observation bout.
11. Total feeding time: The sum of time spent feeding on leaves and culms within any single continuous observation bout.
12. Total time spent pausing: The sum of pauses within any single continuous observation bout.

BEHAVIOR WHILE GATHERING AND FEEDING ON *BASHANIA* LEAVES

The majority of video recordings we made were of pandas feeding on *Bashania* leaves, and these filmed activities of the various individual pandas contained many commonalities. We have thus selected some representative sections for detailed description.

Most *Bashania* selected by pandas for feeding were first-year growth, possessing primary branches only, generally three to four branchlets to each joint and three to four leaves on each branchlet. When feeding in bamboo groves on mountain slopes, the panda would generally lean its back against the slope while reaching for a nearby bamboo culm to gather and eat the leaves. When on relatively flat ground, the panda would grasp bamboo while sitting, stretching upward to gather the leaves, and then return to the sitting position to begin eating. Pandas appeared very languid when feeding on bamboo leaves; only when reaching for leaves did they extend themselves with any evident effort or turn over to stand up. At most, pandas would advance a single step or so before returning to their original sitting position while drawing the bamboo toward the mouth.

Specific motions and activities while grasping bamboo varied, as this was part of the process of forage selection. Forage selection can be considered to have begun before eating as the panda finished the previous bamboo culm. Toward the end of chewing the last handful of bamboo leaves, the panda began looking around to find the next desirable bamboo culm. In almost all cases, the panda needed only reach once to find acceptable bamboo. However, exceptions occurred when the panda changed its mind about the acceptability of some bamboo. Examples included the bamboo being too far away, the bamboo being too difficult to draw in, the bamboo springing back after the panda accidentally lost control of it, and the panda succeeding in pulling the bamboo inward only to decide, upon close inspection, that it was unacceptable. Occasionally, if the target bamboo culm was a bit far from the animal, it would become tangled with other, less desirable culms, requiring the panda to use one paw to separate out the target bamboo while using the other paw to grab it.

After grasping bamboo, some pandas immediately began biting off leaves, whereas others first twisted the culm before gathering leaves. We noted some diversity among pandas in methods used to select and grab bamboo. For example, during an observation bout of Xiwang in March 1997, bamboo that had not been bitten into fragments comprised only 38.5% of the total ($n = 26$). In contrast, an observation bout of Huzi in August 1991 showed that bamboo that had not been bitten off comprised 74.4% of the total ($n = 39$). Such differences may have been functions of age, season, or food conditions.

We summarize the movements made by pandas in biting bamboo as follows: As the panda reaches out a front paw to draw a bamboo culm, it grasps the bent-over bamboo culm with the other front paw. After grasping the culm with both paws, it cocks its head to one side, biting the culm with its premolars, and breaks off a piece of bamboo through the combined pressure exerted by holding and biting. The length of the bamboo piece thus broken evidently has no relationship to bamboo age; rather, it is a function of the panda's posture when feeding as well as its distance from the bamboo.

Following one of Xiwang's feeding bouts in March 1997, we measured 15 of the bamboo pieces she had broken in this way. Lengths of bamboo pieces she had broken off while in a standing position ($n = 3$) approximated her own height; when she fed in a sitting position ($n = 8$), bamboo pieces were consistent with the height of her mouth while sitting; when feeding while reclining, bamboo pieces were similar to her head height while in that position ($n = 4$). Moreover, bamboo pieces were longer when she had broken them from a farther distance.

After biting off a bamboo culm, the panda would use its paws to draw it close and gather the leaves toward one side of its mouth. Pandas exhibited considerable dexterity in handling the culms and leaves while gathering leaves together for eating. Typically, the panda would use one paw to grasp the bamboo culm while the other paw pulled a branchlet with leaves toward one side of its mouth. At times, the panda would bite directly on a

242   Chapter 10

branchlet, running the branch through its mouth from one side to the other, stripping off the leaves as it did so. This action would be repeated many times, until such time as the leaves had accumulated to a number worth eating. During a March 1997 feeding bout, we documented each branchlet Xiwang drew toward herself: at 3.5 leaves per branchlet, she accumulated approximately 44.5 ± 7.5 leaves per bunch (n = 13). After stripping off and gathering the leaves, the panda would then reach with the opposite paw to tightly grip the bamboo leaves, put them into its mouth, and bite down several times. Prior to each bite, the panda would grasp the leaves with its paws, turning them slightly so that they were bundled tightly. It would then place the bundle into one side of its mouth and bite the leaves off with its premolars. At the same time, it would gradually relax the tension of the paw holding the bamboo, moving it outward slightly in order to grasp another bundle of leaves for the next bite. During most feeding bouts, the panda would alternate between using left and right premolars. After a few bites, the panda would chew several times before swallowing. We calculated the sequence, beginning with the first bite to swallowing, as one mouthful. The panda would take several mouthfuls to complete eating a single handful of bamboo leaves.

We noted heterogeneity among individual pandas in the number of mouthfuls used to eat one handful of bamboo leaves, the number of bites per mouthful, and the number of chews per mouthful, although these were relatively consistent within individuals. In our March 1997 recording bout of Xiwang during which she fed on 30 bamboo handfuls, 63% of handfuls required 5–6 mouthfuls, and 30% required 3–4 mouthfuls. She generally began the bout with 3–4 bites per mouthful, continuing with 2 bites per mouthful, and again finishing the mouthful with 3–4 bites. In all cases, she chewed 5 times before swallowing. When we observed Huzi's eating behavior as a subadult and later as an adult, he exhibited similar patterns, except that he did not begin a mouthful with 3–4 bites, but rather started immediately with the 2 bites/mouthful pattern. Observations of Jiaojiao's eating style in December 1994 (n = 10) were similar to those of Huzi. Our observations also revealed idiosyncratic feeding behaviors that characterized individual pandas when feeding on bamboo leaves. For example, Xiwang invariably held the leaves in the left corner of her mouth, using her right paw to grasp them, whereas Jiaojiao and Huzi held the leaves in the right corner of their mouths, using their left paws to grasp them.

Behavior while Gathering and Feeding on *Bashania* Culms

Because the outer teguments of *Bashania* culms were quite lignified and had little nutritional value, pandas generally ripped them off in order to feed on the inner portions. However, in a very small number of situations, we noted pandas eating the entire culm, but only when it was very soft, tender, and not yet lignified. During our March 1997 recording of Xiwang, only 1 of 12 culms was eaten in its entirety, evidently because few culms were still tender. After biting off a culm, the panda generally gripped it quickly using either both paws or only a single paw, after which it used its premolars to bite down on the culm's base, ripping the tegument off with its jaws. This behavior would then be repeated on the opposite side of the panda's mouth. Next, the panda would insert the bamboo culm into its mouth at an angle and begin chewing. The culm length inserted was generally similar to the width of the panda's lower neck. After finishing one mouthful, the panda would tear off the tegument and eat another mouthful. The tearing movements continued without interruption, similar to the way in which people peel bark from sugarcane. If these tearing motions did not succeed in separating the tegument from the culm, the panda would resort to using its incisors. If the panda happened to encounter a particularly stubborn node, it would use its molars to bite it off, discarding the branchlets and leaves attached to that node. Pandas also sometimes used their molars to bite off branchlets, in which case they would also discard them. A panda usually permitted the unwanted tegument, nodes, and leaves to fall on its belly or on the ground nearby; if they remained in its mouth, it would chew unproductively several times before allowing them to fall.

During Xiwang's March 1997 feeding bout mentioned in the previous section, she took an average of 2.6 bites (standard deviation [SD] = 0.7) and 3.8 (SD = 1.5) chews per mouthful. She usually consumed only the middle portion of the bamboo culm. Of the 12 culms, she consumed at most 4 sections (equating to roughly half the culm). She averaged 6.1 (SD = 3.3) mouthfuls/bamboo culm (n =11). In contrast, during an earlier March 1994 bout when Xiwang was still a cub, she averaged 3.4 (SD = 1.7) bites (n = 107) and 8.5 (SD = 3.4) chews per mouthful (n = 99) and spent 6.5 s (SD = 2.2 s) eating time/mouthful (n = 98).

On March 20, 1996, we made video recordings of Xiaosan feeding on *Phyllostachys* at the mouth of Shuidonggou, the first detailed observation of pandas feeding on this type of bamboo. *Bashania* was present in the immediate surroundings, but Xiaosan appeared interested only in the *Phyllostachys*. This introduced bamboo has more secondary and tertiary branches and a greater number of leaves than *Bashania*. On the other hand, *Phyllostachys* leaves are smaller, about one-fourth to one-third the size of *Bashania* leaves. When feeding on *Phyllostachys*, Xiaosan exhibited similar patterns to those when feeding on *Bashania*: first biting off the culm and then tearing off the tegument before consuming it. One difference was that when he encountered a node, Xiaosan used his teeth to bite it apart and used his paw to catch the side branches that fell. After he finished eating part of the culm, he grasped the branches and leaves together, biting and chewing them. This difference was probably because *Phyllostachys* leaves are small and numerous, so that gathering them separately would require spending time and energy.

Because *Phyllostachys* nodes are relatively hard, he placed culms deeply into his mouth when biting, so we suspect he bit them off using his carnassials. In *Phyllostachys*, side branches have smaller branchlets, making them difficult to bite off. Both bites ($\bar{x}$ = 6.0, SD = 2.4) and chews ($\bar{x}$ = 8.1, SD = 3.6) per mouthful were higher than we documented for pandas eating *Bashania* branchlets. In addition, we observed that because the side branches were difficult to bite, Xiaosan occasionally used his paw to pull them outward to aid in biting them off. However, when eating *Phyllostachys* culms, we noted little difference from eating *Bashania* culms: bites ($\bar{x}$ = 2.7, SD = 0.9) and chews ($\bar{x}$ = 4.0, SD = 2.0) per mouthful (n = 20) were both similar to our data for Xiwang eating *Bashania* culms in March 1997.

BEHAVIOR WHEN GATHERING AND FEEDING ON NEW SHOOTS OF *BASHANIA*

New *Bashania* shoots emerge quickly in April of each year. These new shoots are moist, crisp, highly palatable, and easily consumed, so naturally, they are immediately favored by pandas. Our understanding of feeding behavior of pandas eating fresh bamboo shoots is based on analyses we conducted of three videotaped feeding bouts, as well as numerous direct observations we made during radio tracking.

Because new shoots are easily broken off at the base, pandas need only to pull vigorously to get them. A panda eating new shoots generally adopted a standing posture, used its front paw to pull the shoot toward itself, and then lowered its head to bite the lower section of the shoot. It would then shake its head energetically, pulling out the bamboo shoot, and, using its paw to control the shoot, begin feeding. When foraging on particularly short and slender shoots, a panda would generally lower its head and bite the shoot tip while shaking its head and pulling, consuming the shoot while standing. This behavior, which we assume saves time and energy, contrasts with the sitting posture adopted when dealing with coarser, taller shoots. The sheath of the shoot, which is primarily composed of cellulose, has no particular nutritional value and is difficult to eat, so pandas generally removed it, eating only the fleshy part. We observed three ways pandas used to remove the sheath: In most cases, the panda would insert the shoot base into its mouth at an angle, using its first premolar to bite it. It would then pull upward with its jaw while grasping the shoot with its paws and pulling in the opposite direction, which would pull off the sheath. In a small number of cases, the panda would use its canines to wedge the shoot, using its head and paw together to pull the shoot out from its sheath. In a very small number of cases, the panda would use its incisors to remove the sheath, holding it horizontally and biting on one side to pull the sheath away. After the sheath was stripped away it was generally allowed to fall, although sometimes the panda would spit out a sheath that had lodged in its mouth while it shook its head or grab the sheath with its paw and drop it.

Table 10.3.

DATA FOR THE FEEDING TIME OF THREE INDIVIDUALS FEEDING ON *BASHANIA* SHOOTS, SPRING 1995. WE PRESENT THE MEAN PLUS VARIANCE, WITH SAMPLE SIZE INDICATED UNDER EACH VALUE.

Individual	Age (years)	Bites per mouthful	Chews per mouthful	Time per bite(s)	Chewing time(s)	Time per mouthful(s)
Huzi	5.75	1.2 (+0.6)	2.8 (+1.4)	0.6 (+0.4)	0.5 (+0.1)	1.8 (+0.5)
		23	21	16	21	13
Jiaojiao	9.75	1.0 (+0.0)	5.5 (+5.6)	0.7 (+0.3)	0.4 (+0.1)	2.5 (+0.2)
		5	7	3	6	2
Xiwang	2.75	1.4 (+0.6)	3.3 (+1.5)	0.5 (+0.1)	0.6 (+0.2)	2.8 (+1.9)
		5	4	5	4	4

Panda feeding behavior while consuming the fleshy parts of bamboo shoots was similar to those described when feeding on culms and leaves. In all cases, pandas first bit, then chewed, alternating left and right sides. We recorded the behavior of three individual pandas feeding on bamboo shoots in spring 1995, documenting the number of bites and chews per mouthful while consuming the fleshy part of the shoot, as well as the time spent for each action (Table 10.3). All values were smaller than those for pandas feeding on culms and leaves, differences we attribute to the higher palatability of shoots.

On May 11, 1993, while tracking and observing Huzi feeding on bamboo shoots, we noted that at each location where he stopped to sit, he would consume every bamboo shoot >1 cm in diameter. His searching time varied, but was never >10 s. When he was unable to find fresh shoots within about 10 s, he would shift his position slightly, moving perhaps 0.5 or 1 m. Just before finishing consuming all shoots within reach, he would begin scanning his surroundings. When moving a slightly greater distance (>3 m), he would use his mouth to pull up any odd bamboo shoots encountered en route and continue to an area with denser shoots before consuming them. Occasionally, when he encountered relatively small shoots, we noted him standing on three legs, using his free front paw to obtain and eat them. We also noted that when two shoots were in very close proximity to one another, he occasionally used both paws simultaneously, grabbing a shoot with each.

We also observed Huzi eating shoots on May 16, 1993, and again on the exact same date 2 years later in 1995, both times recording each moment that he broke off a shoot. These observations allowed us to quantify his time spent on each activity, including reaching for the shoot, removing the sheath, eating the fleshy inside part, and pausing between bouts. During the 1993 observations, Huzi moved among different ecological patches three times. We defined an ecological patch as a small area used by a panda in the process of foraging. His mean feeding time per shoot in each of these patches varied from 31.0 to 52.7 s. In addition, we also conducted a comparative experiment in which we provided some fairly large shoots directly to Huzi and recorded his time spent eating them under conditions in which he had no need to search for them himself, and his feeding time was comparable to the natural foraging (mean time per shoot 44.8 s).

With ecological patches considered a random factor in a one-way analysis of variance of these May 16, 1993, observations, we found no significant differences among these three groups. This result indicates that time spent feeding on shoots was unaffected by ecological patch, at least within the relatively short time periods of our observations. Although we were unable to determine the panda's criteria for shifting among ecological patches, these results suggest that some criteria must exist. Although providing bamboo shoots directly to Huzi saved him the time otherwise needed to search for shoots, those we gave him were generally quite large, requiring him to spend more time removing sheaths and extract the fleshy inner portion. This may explain why, when comparing the combined data on Huzi's time spent foraging freely within the three ecological patches in 1993 to his time spent with shoots we provided him in a two-sample t test, we found no significant difference. That is, the time Huzi saved in searching for shoots was spent on eating the inner fleshy part of the relatively large shoots we provided him. However, a two-sample t test comparing combined consumption times in the 1993 observations with those in the 1995 observation was significant. We speculate that this difference was caused by variation in the canopy density and culm thickness of the shoots between the two years.

DEVELOPMENT OF INDIVIDUAL FEEDING BEHAVIOR

The process of behavioral development among individual animals is one of continuous improvement. Through learning, cubs begin to master basic feeding skills and, as they gain experience later in life, continue to perfect them. From birth to roughly 2 years of age, milk from their mother generally forms the young panda's main source of nourishment. At 1 year of age, cubs begin learning how to obtain and eat bamboo, but at first this activity is only playful, and they gain very little nutritional value from it. As they grow older, cubs become more proficient at feeding, their consumption of bamboo increases, and bamboo gradually replaces milk as their main food. After the age of 2, young pandas leave their mother and begin living independently, thus entering the subadult stage. At this time, their feeding behavior patterns have begun to take shape but are still flexible. Throughout the subadult stage, feeding behavior continues to develop and improve. At about 4.5 years, pandas are sexually mature, participate in mating, and are considered adults, at which time their modes of feeding behavior have become fixed.

Feeding Strategy of Giant Pandas   247

We made numerous observations of free-ranging cubs as their feeding behavior evolved. In April 1995, we observed Xiaosan, then aged 7–8 months, engaged in gnawing on bamboo while up in a tree. Although he bit at the bamboo, he did not actually bite off or chew bamboo as subadults or adult do, and thus we interpreted his behavior as playing and learning only. On April 16, 1993, when we gave a culm of bamboo to Xiwang, then a 9-month old cub old, she put it into her mouth and began biting, eventually breaking it off. She then gathered some bamboo leaves together and placed them the corner of her mouth, using her paw to stroke them. Then, on her own, she grasped a nearby culm with a paw, biting it off at the base and, half reclining on the slope, bit it energetically. Her demeanor and posture during this activity seemed no different from an adult panda, except that she focused only on biting and did not actually chew the bamboo or swallow. The previous winter, when Xiwang was 4 or 5 months old, we had seen her biting a bamboo culm, but this may have been no more than a conditioned reflex. With increasing age, this behavior gradually became an essential part of her behavioral repertoire.

We first observed Xiwang feeding on bamboo shoots on May 5, 1993, when she was only 8.5 months old. After climbing down from a tree, she arrived at a small slope and adopted a half-reclining, half-sitting position. When we arrived, we offered her a tender bamboo shoot about 40 cm long, which she accepted without hesitation. At first, she seemed to not know what to do, sniffing it here and there, until she began sniffing the shoot's tip. At that point, she turned the bamboo shoot over, seemingly inadvertently putting the thick end into her mouth, and began eating the shoot's fleshy interior. After eating the fleshy portion that was already exposed and available, she began peeling off the sheath. She first bit the main portion of the shoot, then forcefully drew her head back and, using both claws, grasped both sides of the shoot and pushed outward. At this point, the sheath fell to her abdomen, but her attention was completely focused on eating the shoot. When she was almost finished eating the first shoot, we provided her with a second one. With what appeared to us as some puzzlement, Xiwang grabbed the remains of the first shoot, extended her head to smell the new shoot, and opened her mouth without hesitation to take it. She seemed to take no further interest in the first shoot.

On one occasion, we observed her holding a large sheath with her left paw and a bamboo shoot in her right paw. She took a bite from the center of the shoot but seemed not to like it much, at which point she bit into the sheath in her left paw, which elicited a similar response. At this point she let the sheath drop.

Xiwang generally ate slowly, but our impression was that she was no stranger to foraging procedures. Once, when we placed a bamboo shoot in front of her, she immediately bit its central portion and then spat it out. We inferred from this that her foraging behavior was still in development and that her learning process was still ongoing. This process stimulated the development of her digestive system. It also decreased the burden of lactation on her mother, Jiaojiao, allowing her to recover and gain strength in preparation for the subsequent reproductive cycle.

Our video recordings of Xiwang and Huzi allowed us to compare feeding behaviors of individuals at different ages and development stages. Huzi's feeding behavior as a 2-year-old was quite different from his later, adult behavior and appeared very unskilled. At the age of 2, he had just entered the subadult stage, having recently left his mother to live independently. At this time, the number of mouthfuls per handful of leaves was significantly greater than during his adult stage, although the number of bites and chews per mouthful were similar during both stages. Thus, he ingested less food per mouthful as a subadult than as an adult, and he performed a greater number of bites and chews per unit amount of food. In addition, the number of bites and chews per handful of bamboo leaves displayed high variance when Huzi was a subadult, whereas both activities had low variance later, when he was an adult. This suggests that as a 2-year-old, Huzi had not yet adopted the consistent feeding pattern he ultimately did as an adult. Also, he was less skilled as a 2-year-old in gathering and grabbing bamboo and exhibited some movements that later disappeared. For example, as a subadult we observed him to switch the sides of his mouth while biting off bamboo, whereas he generally used only one side as an adult. As a subadult, we occasionally observed that he would gather very few leaves in a handful and eat only one mouthful. We never observed this behavior when he was an adult.

Comparing Xiwang's behaviors while feeding on *Bashania* leaves at 19 months to those at the age of 4.5 years also revealed significant variation. As a juvenile, her feeding movements, as with Huzi, appeared unskilled. Although the number of mouthfuls/handful she used as a juvenile did not differ significantly as an adult, both bites and chews per mouthful were significantly greater. Thus, as with Huzi, she used a greater number of bites and chews per unit of food as a juvenile than as an adult. Also, like Huzi, Xiwang exhibited considerable bout-to-bout variation in her feeding behavior as a juvenile and significantly less as an adult, which further indicated behavioral instability as a juvenile. As a juvenile, Xiwang appeared careless in gathering bamboo leaves, often biting off leaves and allowing many to fall off the culms without gathering them. However, as an adult, she appeared to be very organized in her gathering behavior. She would pull toward herself specific branchlets she desired. As a juvenile, we twice observed her to gather a handful of bamboo leaves from two different bamboo culms and twice observed her to lose control of bamboo she was gathering, causing it to spring backward. We also observed her to reach for a culm of bamboo whose leaves she had already gathered, to attempt to bite off a culm without success (afterward abandoning it), and to gather a handful with so few leaves that she ate only a single mouthful. None of these behaviors reappeared when we observed her as an adult.

Similarly, we observed variation in Xiwang's behavior when feeding on *Bashania* culms during the two developmental stages. At 19 months, she would first bite the bamboo culm before moving her paw outward for some distance along the culm before grasping it again and only then begin biting movements. Each outward paw movement took about 1 s. In contrast, as an adult, she generally moved her paw outward in preparation for the next bite at the same time she bit the culm. This difference was further evidence that she gained skill in feeding with age. As a juvenile, Xiwang's bites and chews per mouthful were higher than as an adult, as was their variability, further suggesting behavioral instability at this young age.

The increasing stability, agility, and speed of movements involved in feeding behavior that we observed as pandas progressed through juvenile, subadult, adult, and old-age stages accompanied the developmental process of the teeth and other organs of the digestive system, increases in strength, improvements in physiological function, and the accumulation of feeding experience. By the time they had reached adulthood, pandas had developed a skilled and consistent feeding pattern. However, with further increases in age, tooth wear and declines in other physiological functions contributed toward a deterioration of that skill and consistency.

In January 1993, we captured Da Shun, who was old and in poor health. From tooth wear and other physiological characteristics, we estimated his age at approximately 19 years. We were able to videotape his behavior feeding on *Bashania* leaves during his time in captivity and documented that his ratios of bites and chews per mouthful were significantly greater than those of Xiwang in adulthood and had higher variance. His feeding pattern was unstable, similar to what we documented for Xiwang when she was an infant.

SENSORY FUNCTION DURING FEEDING

Our observations indicated that the senses of smell and sight played rather important roles during feeding, with smell seemingly of greater importance than sight. During the process of moving among ecological patches or after selecting a feeding site, a panda would invariably lower its head and sniff at the base of the bamboo before deciding which to select. Further, a panda would often pool a bunch of leaves together in front of its nose to sniff prior to actually gathering them in to feed. Similarly, when foraging on shoots, pandas occasionally brought the fleshy part toward its nose and sniffed it after tearing away the outer sheath but before commencing eating.

However, when sitting at a feeding site and selecting from among the surrounding culms or shoots, sight may play a more important role. We noted that as it ate the last one or two bites, a panda would search carefully around the vicinity for food items of the appropriate size before reaching for them. The above observations all occurred during daytime, and it is possible smell may play a decisive role in searching for and selecting bamboo when foraging during nighttime.

Taste also plays a role in bamboo selection. We once observed Xiaosan bring a *Phyllostachys* branch to his nose, sniff at it, and chew it briefly only to spit it out without eating it. We also observed Xiwang to occasionally reject a culm after consuming some of the leaves but with many others remaining uneaten. We speculate that the remaining leaves were insufficiently palatable to her. Huzi also occasionally dropped a handful of leaves before he had completely finished eating it, turning to feeding on another batch of leaves.

Allocation of Feeding Time

On March 21, 1997, at Shanshuping, while documenting Xiwang feeding on *Bashania,* we noted that she used three different feeding sites in the space of a little over half an hour. She first fed on a slope, then moved a few dozen meters to a relatively flat site, and finally moved to a site about 5–6 m away beside a rock. We selected continuous feeding segments from each site for analysis as follows: Using each bamboo culm as a basic unit, we classified every movement recorded during the bout as earlier discussed, as well as the exact moment when each started and its duration. We grouped all behaviors into three primary types, preparing, feeding, and pausing, as well as summarized each for the three different feeding sites she used (Table 10.4). Finally,

Table 10.4.

ALLOCATION OF FEEDING TIME BY XIWANG WHILE FEEDING ON *BASHANIA* AT THREE DIFFERENT FORAGE SITES, MARCH 21, 1997. WE PRESENT THE PERCENTAGE OF TIME FOR PREPARATION, FEEDING, AND PAUSING IN PARENTHESES. GRASPING, BITING, GATHERING, AND TEARING ARE PART OF PREPARATION TIME. OBSERVATION TIMES WERE 513, 399, AND 824 S FOR THE SLOPE, FLAT AREA, AND AREA NEAR THE BOULDER, RESPECTIVELY. A DASH (—) INDICATES NO OBSERVATIONS.

	Allocation of time (s)		
Behavior	Leaves, on slope	Stalks and leaves, flat area	Stalks and leaves, near boulder
Grasping	7.54	12.05	43.09
Biting	10.16	16.20	32.01
Gathering	232.58	46.57	184.37
Tearing tegument	0	56.51	126.64
Preparation	242.29 (0.47)	112.16 (0.28)	321.27 (0.38)
Feeding on leaves	236.54	89.06	286.19
Feeding on stalks	—	174.69	194.77
Total time feeding	236.54 (0.46)	263.75 (0.66)	480.96 (0.58)
Pausing	34.50 (0.07)	23.66 (0.06)	22.57 (0.03)

we calculated Xiwang's allocation of feeding time at the three different feeding sites.

Xiwang's food items varied among the three sites. At the first site (the slope), she ate only leaves, but at the subsequent two sites she ate both leaves and culms. The proportion of the total bout spent eating was lower at the first site than at the two other sites. Averaging times spent at each of the three different sites provided an indication of her feeding behavior on a daylong basis. Overall, time spent pausing (as we defined it) constituted about 5% of her day, whereas feeding constituted about 57%. The remaining 38% of her time was spent obtaining bamboo, gathering leaves, ripping off tegument, and other preparatory activities. Because she spent very little time grasping and biting and because the former was done simultaneously with chewing the previous mouthful, time spent on these movements can safely be ignored. By contrast, the largest part of preparation time was spent on gathering leaves and ripping off tegument.

ESTIMATING DAILY FOOD CONSUMPTION

Researchers have adopted various methods to estimate daily food consumption of various species. Panda researchers have generally adopted the following methods:

1. Direct measurement. In a captive setting, it is possible to weigh the bamboo consumed each day directly (Wang, 1989), and this method has often been used in zoos and generally has been considered to be the most precise method available. For pandas in captivity, one can measure the total weight of the bamboo before the panda feeds and then measure the weight of discarded branches afterward. Subtracting the latter from the former yields the fresh weight of the panda's consumption. We adopted this method for pandas of the Qinling region during the mid-1980s. However, this method inevitably has some shortcomings. First, most wild-born pandas in captivity were brought in ("rescued") by people because they were sick or old or had other abnormalities. Thus, we must consider the degree to which these animals are representative. Further, even if these sampled individuals are normal and healthy pandas, they have lower energetic requirements than free-ranging animals because they are spared much of the physical exertion of searching for food. Thus, estimates will doubtless be smaller than those for free-ranging pandas, although it is difficult to estimate the extent of this bias. Finally, captive pandas are usually provided other food items to supplement bamboo, which no doubt affects estimates of daily food consumption.

2. Inversely estimating from feces. This method proceeds from data on the ratio of the feces mass to food mass and then from extrapolations of the number of scats excreted daily (from field observations). This method was used in the Wolong research project during the 1980s (Schaller et al., 1985).

An advantage of this method is that it is simple in that estimating the mass of feces voided by captive pandas is easy. However, it is quite another matter to estimate the volume of feces voided daily under free-ranging conditions because we cannot easily follow the exact route of a panda. Also, the ratio between feces and food must still be developed using data from captive animals. Further, the ratio of fecal to food mass will vary with season, region, diet, and water content.

3. Time-speed method. This method proceeds by taking the product of mean daily feeding time and food biomass ingested per unit time. However, because both the time spent and rate of ingestion per time vary across age classes and physiological conditions, it requires a large volume of behavioral data from free-ranging animals.

We already know that daily food consumption is likely to vary by age, sex, and physiological condition. Also, pandas adopt different feeding strategies at different times of the year in satisfying their nutritional needs. However, our empirical data were far from exhaustive, and we had no way to obtain information from every conceivable condition that might affect a pandas' daily food consumption. Thus, we have made reasonable extrapolations based on our observations and ancillary data.

Daily Food Consumption of Free-Ranging Pandas

Our primary method of estimating daily food consumption for free-ranging pandas in the Qinling was the time-speed method, which we were able to employ on the basis of our many years of observation.

DAILY FOOD CONSUMPTION OF GIANT PANDAS FEEDING ON *BASHANIA* LEAVES

We used our tracking and behavioral observations of radio-collared individuals to estimate the feeding (hours/day) and consumption (kg/hour). From the previous discussion of seasonal food selection, recall that free-ranging pandas spend September to April feeding primarily on *Bashania* leaves. Thus, an analysis of the daily food consumption when feeding on *Bashania* leaves explains a good portion of the nutritional supply for the entire year.

On April 6, 1996, we sampled 226 first-year growth bamboo leaves, which weighed a total of 118.2 g; thus, mean biomass/leaf was 0.52 g. The proportion of feeding time spent feeding on leaves varied depending on the food source. When Xiwang fed only on bamboo leaves, she spent 46.1% of her time feeding. In two of the three video segments, however, she fed on both leaves and culms. Her mean time spent feeding including all three bouts was 34.3%. Thus, we have two alternative values for v, the feeding rate. When Xiwang fed only on leaves,

$$v_1 = (3,600 \times 0.461 \times 42.3 \times 0.52)/20.6 = 1.77 \text{ kg/hour}, \tag{1}$$

where (seconds per hour × proportion of time feeding × leaves per handful × biomass per leaf) is divided by the handling time per handful (Table 10.2). When she fed on both leaves and culms within a single day, her feeding rate was

$$v_2 = (3,600 \times 0.343 \times 42.3 \times 0.52)/20.6 = 1.32 \text{ kg/hour.} \tag{2}$$

Thus, we obtain Xiwang's daily food consumption E during different periods of time when feeding only on *Bashania* leaves (Table 10.5):

$$E_1 = 24 \, (A - 0.108 - 0.01) \, v_1, \tag{3}$$

where the value in parentheses is the proportion of time active A minus nonfeeding activities.

To obtain daily food consumption, we would apply this equation to the proportion of the diet consisting of leaves only, which can be estimated from fecal analysis or from Figure 10.1. Our data from fecal analysis

Table 10.5.

SUBADULT PANDA XIWANG'S PERCENTAGE OF ACTIVITY AND WEIGHT OF FOOD PER DAY.						
Date of observation	Age	Mean total time of observation (hours)	Percentage of daily activity	Time feeding per day (hours)	Weight of food per day (kg)	Percentage of bamboo leaves in food
November 2–28, 1992	3 months	17.75	7.0[a]	–1.15[a]	—	—
February 5–7, 1994	1 year, 6 months	21.5	51.2	9.46	16.74	100
March 7–8, 1994	1 year, 7 months	50.5	66.3	13.08	23.15	72
November 10–13, 1994	2 years, 3 months	67.25	52.8	9.84	17.42	100
March 12–13, 1995	2 years, 7 months	21.75	64.4	12.62	22.34	72
August 22–23, 1996	4 years	7.75	25.8[b]	3.36[b]	—	—
September 9, 1996	4 years, 1 month	19.25	35.1	5.59	9.89	100
September 14–15, 1996	4 years, 1 month	25	44.0	7.72	13.66	100
October 4–5, 1996	4 years, 2 months	26.25	48.6	8.83	15.63	100
October 10–11, 1996	4 years, 2 months	19	57.9	11.06	19.58	100
October 18–19, 1996	4 years, 2 months	13.25	62.3	12.12	21.45	100
November 9–10, 1996	4 years, 3 months	13.5	55.6	10.51	18.60	100

[a] Obtained when Xiwang was 3 months old. At that time, Xiwang's daily activity was low, her percent of walking per day was almost zero, and she fed on milk, so it is inappropriate to calculate her food intake at that time by this method.

[b] Obtained in August 1996. At that time, Xiwang fed on the shoots of *Fargesia* at high altitudes. We do not know the panda's speed of intake of *Fargesia*, so we cannot calculate her daily food intake.

indicated that *Bashania* leaves occupied 85.3% and culms occupied 14.7% by weight. On the basis of estimates of her percent time spent active from 1992 to 1996 radio tracking, we estimate Xiwang's daily food consumption was 18.01 ± 4.33 kg/day (n = 10).

DA SHUN

We estimated from tooth wear that Da Shun was close to 18 or 19 when we first discovered him in October 1993. He appeared to be suffering from some disease because we noted that he was relatively inactive and lethargic. We thus take him to represent older, sicker pandas. Our observations on him indicated that, when feeding on *Bashania* leaves, he obtained 38.6 leaves per handful (± 6.03; n = 15), consumed 4.05 mouthfuls per handful (± 0.70; n = 56), had 4.10 bites per mouthful (± 1.80; n= 227), and chewed 8.90 times per mouthful (± 2.39; n = 227), and he required 56.1 s in total per handful of bamboo leaves (± 8.29; n = 10). From this, as well as previously presented data on bamboo, we estimate Da Shun's feeding rate while consuming *Bashania* leaves as 1.29 kg/hour.

Using this rate to represent older pandas and applying the percent time active from continuous radio monitoring of Yanghe and Huayang, we estimated daily food consumption for these older pandas (Table 10.6). The mean percent time active for these older pandas was 56.0% (i.e., 13.44

Table 10.6.

PERCENTAGE OF ACTIVITY OF OLD GIANT PANDAS HUAYANG (A MALE OF >20 YEARS) AND YANGHE (A FEMALE OF >19 YEARS) AND THEIR WEIGHT OF FOOD PER DAY.

Individual	Date of observation	Mean total time of observation (hours)	Percentage of daily activity	Time feeding per day (hours)	Weight of food per day (kg)
Huayang	January 19–20, 1989	25.0	44.0	8.0	9.97
	January 25–26, 1989	24.0	49.0	9.2	11.52
	February 3–4, 1989	24.0	63.5	12.6	16.01
	February 8–9, 1989	24.0	59.4	11.7	14.74
	February 14–15, 1989	24.0	56.3	10.9	13.78
	March 13–15, 1989	70.5	57.4	11.2	14.12
	November 15–20, 1989	118.3	57.5	11.2	14.15
Yanghe	February 17–18, 1989	26.3	48.6	9.1	11.39
	February 20–23, 1989	85.0	53.8	10.3	13.00
	February 24–25, 1989	25.3	49.5	9.3	11.67
	March 13–17, 1989	117.8	66.7	13.4	17.00
	April 14–18, 1989	120.0	58.5	11.4	14.46
	May 13–18, 1989	120.3	55.1	10.6	13.41
	October 14–18, 1989	103.5	50.7	9.6	12.04
	December 2–5, 1989	60.5	59.5	11.7	14.77
	January 14–17, 1990	78.8	67.6	13.6	17.28

hours). Subtracting the time spent walking and performing other nonfeeding activities gives the mean time spent feeding. Multiplying by the feeding rate v provides an estimate of daily food consumption of older pandas when feeding on *Bashania* leaves:

$$E = 13.71 \pm 2.06 \text{ kg/d } (n = 16).$$

ESTIMATING ADULT DAILY FOOD CONSUMPTION OF *BASHANIA*

We used Xiwang's feeding rate to represent other adults and applied the percent time active of other healthy adults (Jiaoshou, Huzi, Nüxia, and Jiaojiao) to estimate individual-specific daily food consumptions (Table 10.7). We noted some variability among individuals: as an adult, Huzi had higher daily food consumption than the other pandas, whereas Jiaojiao had lower consumption. However, because no interindividual differences were statistically significant, we take their mean daily food consumption as a standard.

BASHANIA SHOOTS

Most data in this section came from observation we made of Huzi during 1995 when he was a healthy adult of 6 years old. Observations we made of him feeding on *Bashania* shoots are from May 16, 1995 (Table 10.A.2). We made recordings for 20 minutes, during which Huzi ate 37 bamboo shoots.

We analyzed these data by considering two categories of bamboo shoot diameter: shoots of diameters of 1–2 cm ($n = 21$) and diameters >2 cm ($n = 16$). We calculated shoot volume consumed (diameter × length) and also feeding

Table 10.7.

GIANT PANDAS' WEIGHT OF FOOD PER DAY IN ADULTHOOD (FRESH WEIGHT).

Individual	Age and sex	Date	Total time of observation (hours)	Percentage of daily activity	Time feeding per day (hours)	Weight of food per day (kg)
Jiaoshou	Adult male	February 5–8, 1994	73.3	56	10.85	19.20
		March 7–8, 1994	21	50	9.41	16.65
		March 12–13, 1995	24.3	53.6	10.27	18.18
Huzi	Adult male	February 7–8, 1994	27.3	68.8	13.92	24.64
		March 7–8, 1994	21.8	58.6	11.47	20.31
		October 31 to November 2, 1994	46.5	58.1	11.35	20.09
Nüxia	Adult female[a]	September 7–8, 1993	24.3	42.3	7.56	13.38
		February 5–8, 1994	71.5	55.9	10.82	19.16
		March 7–8, 1994	21.8	66.7	13.42	23.75
Jiaojiao	Adult female[a]	March 13–15, 1989	64.3	47.5	8.81	15.59
Average				55.8	10.79	19.09

[a] No cub.

256 Chapter 10

Table 10.8.

DIFFERENT GIANT PANDAS' WEIGHT OF FOOD PER DAY WHEN FEEDING ON *BASHANIA* SHOOT.

Individual and statistic	Sex	Date of observation	Total time of observation (hours)	Percentage of daily activity	Duration of activity (hours)	Gross weight of food per day (kg)	Net dry weight of food per day (kg)
Huzi	M	May 24–25, 1995	48	65.1	12.8	99.1	43.6
Yanghe	M	May 13–18, 1989	120.3	55	10.4	80.5	35.4
Jiaojiao	F	May 19–20, 1990	24.8	68	13.5	104.5	46.0
Average fresh weight						94.7	41.7

rate. We also documented the condition of 48 bamboo shoots at panda feeding locations. Our calculations were that 42 shoots with a diameter of 1–2 cm weighed 2.01 g/cm³ (including the sheath), whereas 6 shoots with a diameter >2 cm weighed 1.86 g/cm³. On the basis of these weights, we estimated the biomass (including sheaths) of bamboo shoots eaten by Huzi as 5.28 kg/hour of shoots with a diameter of 1–2 cm and 10.97 kg/hour of shoots with a diameter >2 cm. Given the ratio of diameter sizes in Huzi's diet, we thus estimated that he consumed 7.74 kg/hour of *Bashania* shoots (including sheaths). Using Huzi's feeding rate as a standard, we then estimated daily food consumption of shoots (gross weight with sheaths) for radio-marked individuals on the basis of their individual activity data. Finally, we estimated dry matter consumption by applying a ratio of 5:1 (fresh:dry), which factored in sheath weight and water content (Table 10.8). Mean daily food consumption while feeding on *Bashania* shoots was 41.7 kg/day (±5.6 kg/day, $n = 3$).

There may be biases in calculations conducted this way arising from variability in biomass/unit length of shoots sampled from differing times and places. In addition, the ratio of large (diameter >2 cm) to medium-sized (diameter of 1–2 cm) bamboo shoots consumed by pandas was not a constant, but rather was a function of the condition of the shoots at the specific feeding site. Despite this limitation, we take this calculation as roughly accurate in depicting panda daily food consumption while feeding on *Bashania* shoots. In May 1983, we conducted 24-hour tracking of an adult male in the Foping Nature Reserve and recorded his daily consumption of new shoots of *Bashania* over 3 days as 43.6, 45.2, and 57.0 kg/day (including sheaths).

FARGESIA SHOOTS

In June 1985, the former Qinling Forestry Bureau successfully rescued and treated a female giant panda and named her Qingqing. We estimated that she was close to 4 years old at the time and, after some recovery time, weighed 56 kg. During the 15 days before she was released back into the wild, she consumed an average of 38 kg (fresh weight) of *Fargesia* shoots daily. Because she was a subadult, her diet was supplemented by adding a multivitamin glucose solution to her drinking water. Thus, we estimate that the daily food consumption of a free-ranging adult female would have been some one-third greater, i.e., 50.7 kg/day.

Estimate of Daily Food Consumption of Parturient Females

We must consider parturient females, such as Jiaojiao, Ruixue, and Nüxia, separately. We will use data from 1996 when Jiaojiao gave birth to Xiaosi as an example to assess the magnitude of error in our estimate of her daily food consumption. Our observational data indicated that activity declines for female pandas in the time period prior to parturition. At 7 days prior to parturition, Jiaojiao's percent time active had already declined to about half her normal rate. For several days after giving birth, mothers generally fast, resting in their dens, neither eating nor drinking. As the cub grows, they gradually begin leaving their dens to feed, and daily rate of activity gradually increases, returning to a normal rate approximately 2 months post-parturition. Until this time, however, mothers' daily lives differ considerably from their normal patterns, particularly in that their movements are greatly restricted. Thus, it would be inappropriate to apply the above method to estimate daily food consumption and nutritional conditions. For example, our calculations would suggest that Jiaojiao consumed 7.3 kg/day during August 13–15, 1996, the 5–7 days before she gave birth, a time during which she was actually fasting. Similarly, our estimation would suggest that Jiaojiao's daily food consumption increased during the period August 22–23 to October 18–19 (i.e., when her cub was 2 to 60 days old), but we do not view these results as reflecting the true situation. As mentioned in chapter 1, understanding the critical nutritional condition of females shortly prior to parturition and during lactation is a difficult problem in panda research.

Distribution of Bamboo in the Qinling

METHODS

We estimated the area occupied by bamboo using visual interpretation of satellite imagery to produce a weighted estimate of the area of each forest type and bamboo community. Using field sampling of standardized sample plots, we estimated the percentage of bamboo forest area and the aboveground biomass of bamboo in selected portions of the intensive study area (i.e., the 380 km² region centered on the Xinglongling, with the upper reaches of the Xushui River to the north and the line between Huayang and Zhichang in the south). We quantified only the primary types of bamboo used by pandas in this area, i.e., *Fargesia* and *Bashania*.

Table 10.9.

DISTRIBUTION AND AREA OF *BASHANIA* AND *FARGESIA* BAMBOO IN A 380-KM2 STUDY AREA WITHIN THE XINGLONGLING MOUNTAINS.

Plant community type	Plant type area (km^2)	Percentage area		Area of bamboo (km^2)	
		Fargesia spp.	*Bashania fargesii*	*Fargesia* spp.	*Bashania fargesii*
Alpine scrub and meadows	1.71	60	0	1.0	0
Upper montane larch forest	5.43	30	0	1.6	0
Upper montane fir forest	94.04	50	0	47.0	0
Upper montane birch forest	48.67	50	0	24.3	0
Montane mixed coniferous (Armand's pine, Chinese hemlock) and broadleaf forest	98.22	5	20	4.9	19.6
Montane mixed broadleaf forest	45.26	0	30	0	13.6
Lower montane mixed oak-pine forest	51.73	0	60	0	31.0
Lower montane oak and Chinese pine forest	9.28	0	0	0	0
Lower montane shrub	10.87	0	0	0	0
Mixed shrub and logged residue	12.25	5	25	0.6	3.1
Clear-cut area	0.78	0	0	0	0
Cultivated fields	1.77	0	0	0	0
Total	**380**			**79.5**	**67.3**

RESULTS

We estimated the areas occupied by each vegetation and bamboo community type (Table 10.9). *Fargesia* groves occupied approximately 79.5 km², comprising 20.9% of the total area of 380 km², whereas *Bashania* groves occupied 67.3 km², 17.7% of the area. Because pandas were densely concentrated within some 346.2 km², we would thus estimate that *Fargesia* occupied some 72.5 km² of this panda core zone, whereas *Bashania* occupied 61.4 km². These values are slightly below that presented in Table 10.9 because of our use of a smaller area. The estimate here for *Fargesia* may understate reality somewhat because in our calculations we excluded areas between 1,900 and 2,400 m that flowered during the early 1980s, which have since recovered and were used by pandas for foraging during summer.

Annual Biomass of *Bashania*

NEW *BASHANIA* SHOOTS

A survey was conducted of *Bashania* shoot natality in the Foping and Yangxian areas of the Qinling region during the 1980s (Pan et al., 1988; Tian, 1990). We present those results in Table 10.10 and results from our own

Feeding Strategy of Giant Pandas   259

Table 10.10.

BASHANIA SHOOT NATALITY IN FOUR EARLIER SURVEYS. AVERAGE SHOOT NATALITY FOR THESE STUDIES IS 1.67 SHOOTS/M^2.

Time period	Investigators	Area	Elevation (m)	*Bashania* bamboo shoot natality (shoots/m^2)
1985	Tian Xingqun and Liu Qijian	Foping County	1,100–1,700	1.19
1985	Tian Xingqun and Liu Qijian	Yang County	1,400–1,800	0.59
May 1985	Present study	Foping County	1,300	1.3
July 1986	Present study	Huayang hollow	1,700	3.6

Table 10.11.

BASHANIA SHOOT NATALITY IN SPECIFIC AREAS. MEAN SHOOT DENSITY FOR THESE OBSERVATIONS IS 3.08 SHOOTS/M^2.

Time period	Area	Elevation (m)	Exposure	*Bashania* shoot natality (shoots/m^2)
July 5, 1986	Liaojiagou	1,600	West	5.38
July 7, 1986	Shuangchagou	1,600	West	0.90
July 8, 1986	Yizuouyequ	1,800	West	0.5
April 22, 1993	Guojiawan	1,400	South	1.50
April 22,1993	Daheiwan	1,340–1,390	North and south	3.30
April 22, 1993	Xiaoheiwan	1,340–1,360	North	2.00
April 25, 1993	Daheiwan	1,380–1,440	South	8.90
April 25, 1993	Xiaoheiwan	1,400–1,430	North	4.50
May 2, 1993	Shuidonggou-Chaijiawan	1,820	Ridge	6.00
May 2, 1993	Shuidonggou-Chaijiawan	2,080	Ridge	1.00
May 2, 1993	Shuidonggou-Chaijiawan	1,920	Ridge	1.00
May 5, 1994	Shuidonggou	1,600–1,750	South	4.00
May 11, 1995	Shanshuping	1,500–1,700	North and south	1.12

research team's surveys in Table 10.11. Combining these two tables (i.e., means of 1.67 and 3.08 shoots/m^2), we estimate the mean density of new *Bashania* shoots throughout the southern slopes of the Qinling was 2.38 shoots/m^2.

BIOMASS/NEW SHOOT

During our survey of *Bashania* shoots, we weighed a select number of new shoots in order to estimate that annual biomass of new *Bashania* shoots; we applied the mean biomass per shoot (0.12–0.27 kg/shoot) to the density of shoots (shoots/m^2; Table 10.11) and multiplied by the area occupied by *Bashania* (in m^2). Thus, we estimated the annual biomass of new *Bashania* shoots as 0.165 kg/shoot × 2.38 shoot/m^2 × 61,355,500 m^2 study area = 24,094 t.

260   Chapter 10

BASHANIA CULMS AND LEAVES

We estimated the biomass of *Bashania* leaves and culms from 12 sample plots on southerly and northerly exposures of river valleys during February 1986, March 1987, and April 1989. We take these to represent the biomass of *Bashania* leaves and culms available to pandas annually. The mean density of bamboo culms was 28.2 culms/m^2 (data for individual plots shown in Table 10.12). Of these, a mean of 5.56 culms/m^2 were first-year growth, 8.13/m^2 culms were second-year growth, and 15.27 culms/m^2 were >2 years old. At a mean biomass of 27.5 g/culm, we estimated mean the fresh biomass of first- and second-year bamboo leaves at 0.36 kg. At a mean biomass of 112 g/culm, the estimated mean fresh biomass of first- and second-year culms was 1.44 kg. At a mean biomass of 20 g/culm, the estimated mean fresh biomass of older leaves was 0.24 kg. Finally, at an estimated mean of 70 g/culm, the estimated biomass of older culms was 1.13 kg.

Thus, we estimated the total biomass of *Bashania* in the study area as follows: biomass of *Bashania* leaves = biomass/m^2 of leaves (expressed in kg/m^2) × total area of *Bashania* groves (m^2) = (0.36 kg/m + 0.24 kg/m) × 61,355,500 m^2 study area =36,813 t. Of this, first- and second-year leaves constituted 22,087 t. The biomass of *Bashania* culms was 157,683 t. Of this, the biomass of first- and second-year culms was 88,351 t. Therefore, the total biomass of first- and second-year growth leaves, culms, and shoots (i.e., annual biomass of *Bashania* palatable to pandas) was 134,534 t.

BIOMASS OF *FARGESIA*

In August 1987, we conducted a survey of *Fargesia* at elevations of 2,400–2,950 m in which we documented the biomass of current year bamboo shoots, new culms from the previous year, and older culms. For current year shoots, this survey resulted in values of 0.081 kg/shoot and 5.33 new shoots/m^2, i.e., biomass of new bamboo shoots in *Fargesia* groves of 0.43 kg/m^2. The density of previous year culms was 12 culms/m^2, with an average biomass of 0.063 kg/culm, i.e., biomass of previous year culms of 0.75 kg/m^2. The density of older bamboo culms was 108 culms/m^2, and mean biomass was 6.675 kg/m^2. Thus, we estimated the aboveground biomass of *Fargesia* within the area of densely concentrated pandas as current year shoots (31,162 t) + previous year bamboo culms (54,352 t) + older bamboo culms (483,737 t). The total biomass of current and first-year culms (i.e., palatable to pandas) was thus 85,514 t.

Bamboo Use by Giant Pandas

On the basis of our estimates of daily food ingested, we estimate bamboo use by species and plant parts. We then explore the ability of the bamboo in the region to provide food for the pandas, as well as the pandas' effects on bamboo.

Table 10.12.

BAMBOO STEMS AND LEAVES BY AGE IN 12 QUADRATS SAMPLED FOR BAMBOO BIOMASS ON SOUTHERLY AND NORTHERLY EXPOSURES OF RIVER VALLEYS DURING DECEMBER 1986, MARCH 1987, AND APRIL 1989. ALL SAMPLING AREAS ARE 1.0 M^2, EXCEPT FOR QUADRAT 12, WHICH IS 4.0 M^2. THE NUMBER OF SAMPLES IS GIVEN IN PARENTHESES; A DASH (—) INDICATES NO SAMPLES WERE RECORDED IN THAT AGE GROUP.

Quadrat	Sample date	Location	Elevation (m)	Bamboo age	Biomass (kg)		
					Stems	Leaves	Total
1	December 5, 1986	Valley	1,750	First year (19)	2.4	0.49	2.89
				Second year (22)	2.0	1.28	3.28
				Older (19)	1.45	0.16	1.61
2	December 5, 1986	North-facing slope	1,780	First year (5)	1.0	0.20	1.20
				Second year (14)	0.50	0.20	0.70
				Older (9)	1.35	0.25	1.60
3	December 5, 1986	South-facing slope	1,780	First year (12)	1.05	0.25	1.30
				Second year (20)	0.90	0.30	1.20
				Older (—)	0	0	0
4	December 5, 1986	North-facing slope	1,780	First year (4)	0.30	0.10	0.40
				Second year (33)	1.00	0.50	1.50
				Older (—)	0	0	0
5	December 9, 1986	Valley	1,750	First year (1)	0.02	0.005	0.025
				Second year (4)	0.44	0.062	0.502
				Older (22)	0.42	0.180	0.60
6	March 25, 1987	Riparian	1,680	First year (6)	0.96	0.04	1.00
				Second year (5)	1.20	0.23	1.43
				Older (21)	2.10	0.72	2.82
7	March 25, 1987	Riparian	1,680	First year (2)	0.08	0.01	0.09
				Second year (—)	0	0	0
				Older (28)	2.80	0.49	3.29
8	March 25, 1987	North-facing slope	1,730	First year (3)	0.24	0.02	0.26
				Second year (3)	0.45	0.06	0.51
				Older (14)	0.98	0.27	1.25
9	March 25, 1987	North-facing slope	1,730	First year (4)	0.75	0.05	0.80
				Second year (2)	0.24	0.03	0.27
				Older (23)	1.05	0.25	1.30
10	March 25, 1987	North-facing slope	1,730	First year (2)	0.30	0.02	0.32
				Second year (3)	0.22	0.03	0.25
				Older (19)	0.80	0.10	0.90
11	March 27, 1987	Riparian	1,460	First year (7)	1.94	0.16	2.10
				Second year (2)	0.48	0.12	0.60
				Older (8)	1.52	0.25	1.77
12	April 24, 1989	In gully		First year (7)	1.80	0.22	2.02
				Second year (14)	3.38	0.95	4.33
				Older (66)	—	—	—

NEW *BASHANIA* SHOOTS

The daily requirement we earlier estimated for *Bashania* shoots of 94.7 kg included the fleshy part of the shoot, as well as the discarded sheath and all other portions of the shoot. Thus, we consider this value to represent the daily use of *Bashania* shoots. On the basis of the time during which *Bashania* sends up shoots as well as our understanding of panda migration patterns, we further estimate that pandas spend approximately 30 days/year feeding on new *Bashania* shoots. Thus, the annual use of new *Bashania* shoots was 2.84 t.

BAMBOO LEAVES AND CULMS

Above we estimated that adult pandas eat 18.01 kg/day while feeding on *Bashania* leaves and culms, of which 15.63 kg are leaves and 2.38 kg culms. Very little is wasted when pandas feed on *Bashania* leaves, so the biomass of leaves eaten can be used as an estimate of leaf use. Such a simple equation of ingestion with use does not hold with culms, however, because pandas generally break off about 50%, most of which dies within a year. Thus, calculations of bamboo culm use should account for both biomass eaten directly and biomass damaged but not actually ingested. However, because all damaged bamboo was also removed from the available crop by the panda, we can simply estimate use as the amount damaged. Our radio telemetry data indicated that pandas spent approximately 7 months/year feeding on *Bashania* culms and leaves; tracking radio-collared pandas showed that pandas were located in *Bashania* groves for an average of 267.5 days/year. Thus, we calculated annual use of bamboo leaves as 3.7 t.

When feeding on *Bashania* leaves, pandas most often select first- and second-year leaves (which we earlier estimated weigh approximately 23.5 g/culm). To meet their requirements, pandas must thus feed on the leaves of approximately 665 new culms. We then assume that 50%, (n = 333) were damaged. Taking the biomass of each of these culms as 99.4 g, the annual use of *Bashania* culms is thus 7.86 t.

USE OF *FARGESIA*

Pandas' use of *Fargesia* consists mainly of old shoots and new culms. Daily ingestion of old *Fargesia* shoots was similar to daily use of new *Bashania* shoots, averaging 94.7 kg/day (Table 10.8). We also documented that pandas spend an average of 82.7 days/year feeding on *Fargesia* shoots (n = 42). Thus, annual use of *Fargesia* shoots was 4.19 t.

We used these values to estimate the proportion of the bamboo resources actually used by pandas for each bamboo species. For *Bashania*, we estimated annual production of shoots as 24,094 t and annual use per panda as 2.84 t, i.e., only 0.01%. We estimated annual biomass of first- and second-year leaves as 22,088 t and annual use per panda as 3.71 t, or 0.02%. We estimated the biomass of first- and second-year culms as 88,352 t and annual use per panda as 4.19 t, or 0.005%. At an assumed density of 0.228 individual/km^2 there would have been 79 pandas residing within the area of concentration, and together, they would have used only a small portion of bamboo production annually. Accounting for new shoots, new leaves, and new culms, use of *Bashania* amounted to 1138.8 t, some 0.8% of annual *Bashania* production. Use of *Fargesia* amounted to some 331.2 t, or 0.4% of annual production.

Thus, bamboo within our study area appeared to provide sufficient resources for pandas, and pandas had very little effect on bamboo. It is very unlikely that bamboo would suffer large-scale damage as a result of herbivory from pandas, nor is it likely that pandas foraging on new shoots would affect the age structure of the vegetation in a way that would hasten aging or senescence of bamboo groves.

Our observations beginning in the 1980s suggest that the Qinling's southern slopes are a functioning biotic community. None of our radio-marked pandas died; all made annual migrations among areas dominated by the two different bamboo species, feeding on them seasonally. Even in the years of mass bamboo flowering (see chapter 1), the system remained robust. We see no evidence of pandas being limited by the amount or quality of bamboo or the pandas in turn limiting the bamboo resource.

Note

Chapter originally appeared as Chapter 11 in Chinese edition published by Peking University Press, Ltd.

Appendix

Table 10.A.1 provides detailed data on elevational movements of radio-collared individual pandas. We only show the years when we have data during summer months. Table 10.A.2 presents observations we made of Huzi during 1995 when he was a healthy adult of 6 years old. We made observations of him feeding on *Bashania* shoots on May 16, 1995. We made recordings for 20 minutes, during which Huzi ate 37 bamboo shoots.

Table 10.A.1.

GIANT PANDAS' TIME IN SUMMER HABITAT (IN *FARGESIA* BAMBOO) AND WEIGHT OF BAMBOO SHOOTS USED FOR FOOD.

| Individual | Sex | Age | Year | Time in summer habitat | | | Average duration over all years (days) | Weight of bamboo shoots for food for all years (kg) |
				Start date	End date	Duration (days)		
Jiaojiao	F	5	1989–1990	June 22, 1989	July 31, 1989	39	40	3,815.1
		6	1990–1991	June 15, 1990	August 3, 1990	49		
		7	1991–1992	June 22, 1991	July 26, 1991	34		
		8	1992–1993	June 21, 1992	July 19, 1992	28		
		9	1993–1994	June 26, 1993	August 1, 1993	36		
		10	1994–1995	June 14, 1994	July 20, 1994	36		
		11	1995–1996	June 16, 1995	August 15, 1995	60		
Huzi	M	2	1991–1992	June 22, 1991	July 26, 1991	34	54	5,090.1
		3	1992–1993	June 16, 1992	August 15, 1992	60		
		4	1993–1994	May 26, 1993	July 31, 1993	66		
		5	1994–1995	June 16, 1994	August 10, 1994	55		
Xiwang	F	1	1993–1994	June 26, 1993	August 1, 1993	36	36	3,409.2
		2	1994–1995	June 25, 1994	July 30, 1994	35		
Dabai	M	9	1992–1993	July 1, 1992	August 31, 1992	61	78	7,355.0
		10	1993–1994	June 1, 1993	August 31, 1993	91		
		11	1994–1995	June 11, 1994	August 31, 1994	81		
Daxiong	M	~16	1990–1991	June 20, 1990	August 20, 1990	61	61	5,776.7

(continued)

Feeding Strategy of Giant Pandas   265

Table 10.A.1. (continued)

Individual	Sex	Age	Year	Time in summer habitat		Duration (days)	Average duration over all years (days)	Weight of bamboo shoots for food for all years (kg)
				Start date	End date			
		15	1991–1992	June 21, 1991	September 18, 1991	89		
		16	1992–1993	June 21, 1992	September 10, 1992	81		
		17	1993–1994	June 16, 1993	September 15, 1993	91		
Xiaohuo	M						66	6,202.9
		8	1992–1993	June 16, 1992	August 25, 1992	70		
		9	1993–1994	July 1, 1993	August 31, 1993	61		
Nüxia	F						78	7,386.6
		5	1993–1994	June 10, 1993	August 30, 1993	81		
		6	1994–1995	June 11, 1994	August 25, 1994	75		
Jiaoshou	M						100	9,422.7
		6	1993–1994	May 21, 1993	August 31, 1993	102		
		7	1994–1995	May 26, 1994	August 31, 1994	97		
Ruixue	F						81	7,639.1
		7	1992–1993	June 11, 1992	August 30, 1992	80		
		8	1993–1994	June 10, 1993	August 31, 1993	82		
		9	1994–1995	June 6, 1994	August 25, 1994	80		
Baoma	F						116	10,937.9
		6	1990–1991	June 6, 1990	September 30, 1990	116		
		7	1991–1992	June 7, 1991	September 30, 1991	115		
Huayang	M						72	6,818.4
		22	1989–1990	June 19, 1989	August 31, 1989	73		
		23	1990–	June 21, 1990	August 31, 1990	71		

Yangne	F	20	1989–1990	June 12, 1989	September 23, 1989	103	103	9,754.1
Momo	F	5	1993–1994	June 1, 1993	September 30, 1993	121	95	8,996.5
		6	1994–1995	June 6, 1994	August 14, 1994	69		
		7	1995–1996					
Xinxing	M	7	1991–1992	June 1, 1991	September 19, 1991	110	98	9,280.6
		8	1992–1993	June 5, 1992	August 30, 1992	86		
Weiming	M	9	1993–1994	June 1, 1993	September 22, 1993	113	113	10,701.1
		10	1994–1995	June 1, 1994	September 22, 1994	113		
129	M	17	1992	June 1, 1992	October 5, 1992	126	126	11,932.2
Shui an	F	3	1991	June 7, 1991	October 3, 1991	88	88	8,333.6

Table 10.A.2.

HUZI'S BEHAVIOR WHILE FEEDING ON *BASHANIA* SHOOTS ON MAY 16, 1995.

Time stamp for start and end	Utilizable time (s)	Length of bamboo shoot (cm)	Diameter of bamboo shoot (cm)
45″–1′30″	45	100	2
1′30″–2′20″	50	100	2
2′20″–2′55″	35	80	2
2′55″–3′35″	40	60	2
3′35″–4′00″	25	40	2
4′05″–4′35″	30	80	2
4′35″–5′15″	40	80	2
5′15″–5′50″	35	50	2.5
5′50″–6′25″	35	50	1.5
6′25″–6′55″	30	40	1.5
6′55″–7′25″	30	40	1.5
7′35″–7′55″	20	40	1.5
7′55″–8′25″	30	40	2
8′25″–9′00″	35	40	2
9′00″–9′55″	55	90	2
9′55″–10′30″	35	40	2
10′45″–11′15″	30	50	1.5
11′15″–11′45″	30	40	2
11′45″–12′35″	50	40	2
35″–60″	25	25	1
1′00″–1′25″	25	25	1
1′40″–1′55″	15	25	1
1′55″–2′10″	15	25	1
2′10″–2′20″	10	25	1
2′20″–2′40″	20	30	1
2′40″–3′00″	20	30	1
3′00″–3′30″	30	30	1
00″–10″	10	30	1
10″–30″	20	40	1.5
30″–53″	23	50	1.5
55″–1′30″	35	60	1.5
1′35″–2′10″	35	80	2
2′15″–3′00″	45	70	1.5
3′00″–4′20″	80	100	2
4′20″–4′50″	30	40	1.5
4′50″–5′10″	20	25	1.5
5′10″–5′20″	10	25	1.2

References

Chapman, G. P., and W. E. Peat. 1992. *An Introduction to the Grasses (Including Bamboos and Cereals)*. Wallingford, UK: CAB International.

Musser, G. G. 1972. "The Species of Hapalomys (Rodentia, Muridae)." *American Museum Novitates* 2503:1–27.

Pan, W. S., Z. S. Gao, Z. Lü, Z. K. Xia, M. D. Zhang, L. L. Ma, G. L. Meng, X. Y. She, X. Z. Liu, H. T. Cui, and F. X. Chen. 1988. 秦岭大熊猫的自然庇护所 [The giant panda's natural refuge in the Qinling Mountains]. Beijing: Peking University Press.

Petter, J. J., and A. Peyrieras. 1975. "Preliminary Notes on the Behavior and Ecology of *Hapalemur griseus*." In *Lemur Biology*, edited by I. Tattersall and R. W. Sussman, pp. 281–286. New York: Plenum Press.

Qin, Z. S. 1985. 四川大熊猫的生态环境及主食竹种更新 [Giant panda's bamboo food resources in Sichuan, China and the regeneration of the bamboo groves]. *Journal of Bamboo Research* 4:1–10.

Schaller, G. B., J. C. Hu, W. S. Pan, and J. Zhu. 1985. *The Giant Pandas of Wolong*. Chicago: Chicago University Press.

Sheldon, W. G. 1937. "Notes on the Giant Panda." *Journal of Mammalogy* 18:13–19.

Tian, X. 1990. 秦岭大熊猫食物基地的初步研究 [Studies on the food base of giant pandas in the Qinling Mountains]. *Acta Theriologica Sinica* 10:88–96.

Wang, S., and C. K. Lu. 1973. "Giant Panda in the Wild." *Natural History* 82:70–71.

Wang, X. Q. 1989. 圈养大熊猫全年食竹量的观察 [Observation on annual bamboo consumption of captive giant pandas]. *Sichuan Journal of Zoology* 8:28–28.

Zhu, J., and Z. Long. 1983. 大熊猫的兴衰 [The vicissitudes of the giant panda]. *Acta Zoologica Sinica* 1:93–104.

Chapter 11

Acquisition of Nutrients

Summary: Through analyses of the digestion process of giant panda forage (primarily *Fargesia* shoots), we determined that pandas digested and absorbed about 90% of the parenchyma cell content and a portion of the cell wall known as hemicellulose, and we also observed residue from cellulose. Through chemical analysis of panda feces, we concluded that giant pandas can digest 91% of the crude protein, 15.2%–18.1% of the hemicellulose, and 0.48%–10.95% of the cellulose found in bamboo. The giant panda must obtain a certain ratio of amino acids in order to meet its daily nutrient requirement. Through statistical analysis of the giant panda's daily intake of food, we calculated the daily energy requirement of a healthy adult panda to be 14,000–17,000 kJ. In comparison, our calculations suggested that daily energy acquired by normal adult was about 26,762 kJ. Daily intake of forage biomass (from which we inferred daily energy acquisition) was lower than this figure by 18.1%–25.5% for pandas that were ill or elderly. Parturient females' daily forage intake fluctuated before and after parturition. Pandas are not able to use bamboo with the same efficiency as the sympatric bamboo rat. Two additional bamboo species (*Phyllostachys sulphurea* and *Indocalamus longiauritus*) exist in the Qinling pandas' habitat, and these species may potentially be used as secondary food resources. We found no important differences between Qinling and Wolong in panda nutrition or energy supply.

In 1981, Ellen Dierenfeld, then at Cornell University, was the first to analyze the nutritional status of pandas when she studied two captive-raised giant pandas at the National Zoo in Washington, DC, as part of her master's degree (Dierenfeld, 1981). Her research initiated the new field of giant panda nutrition. In 1983, the first field studies to obtain information on panda nutrition began in Wolong when diets were documented and chemical analyses of feces to estimate absorption of nutrients were conducted. These researchers found that giant pandas utilized not only the majority of the cell content from bamboo but also hemicellulose from the cell walls (Schaller et al., 1985), a unique capability among carnivores. We began our studies of diets and nutrition of pandas in the Qinling in 1984 and documented a portion of our results in *The Giant Panda's Natural Refuge in the Qinling Mountains*, published in 1988 (Pan et al., 1988). In recent years, we have continued to make progress in this field. In this chapter, we discuss Qinling giant pandas' nutritional strategies and their adaptive evolutionary significance based on our direct observations.

Morphological Analysis of *Fargesia* Shoots

In studying the utilization of various nutritive elements in foods, researchers have generally turned first to chemical analyses to determine the content of numerous substances in both forage and feces. If the animal's daily forage consumption and defecation rates are known, the digestibility and absorption of various materials in food can be calculated. Although chemical approaches are fast and accurate, they are of little help in understanding the mechanics of digestion. In light of this, we supplemented chemical analyses with morphological studies to better understand panda use of bamboo. Morphological analysis, providing direct, visual information, conveys data not only on the digestibility of various materials but also on the mechanism and sequence of digestion.

During the Wolong research on forage and feces begun in 1983, researchers prepared samples in paraffin sections and analyzed them using optical microscopy (Li et al., 1984; Schaller et al., 1985). They documented substantial alterations in the morphology of food contents as they passed through the panda's gastrointestinal tract. We have continued this line of research in the Qinling since 1993, although more recently we have also introduced the use of ultrathin sectioning and transmission electron microscopy (TEM), which has expanded our understanding of digestion on a microscopic level.

Materials and Methods

MATERIALS

Beginning in 1993, materials used in our research came from the Changqing Nature Reserve, located on the southern slope of the Qinling. We obtained bamboo samples from species known to be preferred by pandas and from areas where the pandas were known to occur. Our main efforts were centered in panda summer habitat and focused on the main summer forage, *Fargesia*. We simultaneously collected both fresh bamboo and panda feces for purposes of comparison.

METHODS

We prepared samples using either standard paraffin sectioning, semithin sectioning, or ultrathin sectioning and observed them using an optical microscope or TEM. Standard paraffin sections were 7 μm thick, fixed using formaldehyde, and dyed with para-aminosalicylic acid. Semithin sections were 3 μm thick, fixed using glutaraldehyde-osmic acid, and dyed with toluidine blue. For both standard and semithin sections, we used a BH-2 Olympus optical microscope for observation and photography. Ultrathin sections were 500–600 nm thick, fixed using glutaraldehyde-osmic acid, and dyed with uranyl acetate–lead citrate; these sections were observed and photographed using a JEM-100CX TEM.

Results

PARAFFIN SECTIONS USING THE OPTICAL MICROSCOPE

Structure of Fresh Fargesia Samples (For Comparison)

Using an optical microscope, samples can be categorized into the following cell types (listed from outer to inner layers; Figure 11.1):

- Epidermal layer: These cells were located on the outside layer of the shoot and consisted of only a single layer. In cross section, they appeared rectangular, their walls were slightly thickened, and stomata were evident as part of their structure.
- Cortical layer: These cells were located interior to the epidermal cells. The layer was about three to four cells thick ($n = 4$, according to our experimental observations), and in cross section they were slightly larger than epidermal cells.
- Parenchyma cells: Widely distributed throughout the interior of the shoot, these cells were relatively large and had thin walls. Those nearest the cortical layer had slightly thicker walls and were slightly larger than those located elsewhere. Those in the layer of five to six cells closest to the pith tended to be slightly flattened in shape ($n = 4$), and in cross section they appeared somewhat more rounded than those located elsewhere. In longitudinal section, parenchyma cells appeared rectangular.
- Vascular bundle: Under cross-sectioning, vascular bundles were basically diamond shaped, and their size varied with their location. Layers varied from five to six cells ($n = 4$) according to their size. Diameters of vascular bundles gradually increased from outer to inner sections. We categorized vascular bundles as three types: (1) fibrous cells, (2) phloem, and (3) xylem. Under cross-sectioning, fibrous cells were almost round, with thick walls and relatively small cavities, and their interiors were cellular. These cells were generally located on the exterior of vascular bundles and are also known as fibrous caps. Phloem cells were generally located near epidermal cells. They included four to eight sieve tubes ($n = 8$) and were surrounded by small-sized parenchyma and accompanying cells. Within phloem cells were irregularly shaped ducts that gave rise to xylem. The xylem was composed of two

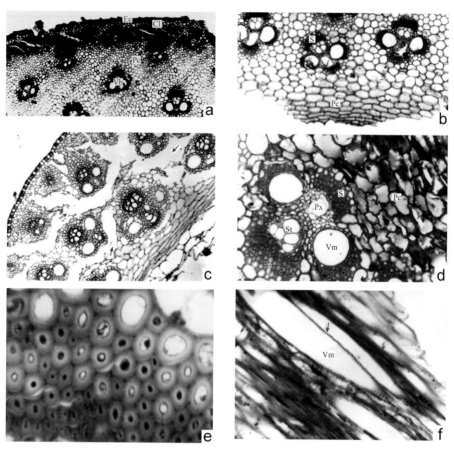

FIGURE 11.1. (a) and (b) *Fargesia* shoot, shown in its normal structure. Paraffin sections, transverse sections, with 33× magnification. Ec = epidermal cell; Cl = cortical layer; Pc = parenchyma cell; Vb = vascular bundle; S = sclerenchyma. (c) *Fargesia* shoot from panda feces. The parenchyma cell after digestion is shown. Paraffin sections, transverse sections, with 33× magnification. The small arrow in bottom center of the slide indicates the lacuna after digestion. (d) *Fargesia* shoot, showing the vascular bundle structure. Semithin section, transverse section, with a magnification degree of 3.3 × 20. Vm = metaxylem vessel; St = sieve tube; Px = protoxylem. (e) Sclerenchyma cells from a *Fargesia* shoot. Semithin section, transverse section, magnification degree of 3.3 × 100. (f) Tracheal elements from a *Fargesia* shoot, showing the aperture. Semithin section, transverse section, magnification degree of 3.3 × 100.

large, ladderlike ducts located between the phloem and the protoxylem. Small cells were neatly arranged within the ducts.

Fargesia *Shoots from Feces*

The contours of various cell layers could be identified under a microscope, but most of the cell content was not observable (some 90% of cell content was lost, *n* = 15). Cell wall chromatin in fecal samples was lighter than in fresh samples. In some areas, parenchyma cells had disappeared entirely, leaving only vascular bundles, which had thinner fibrous walls than in fresh samples (Figure 11.1).

SEMITHIN SECTIONS USING OPTICAL MICROSCOPE

Fargesia *Shoots*

As observed under the microscope, samples appeared blue, but otherwise, their general structure was similar to that observed in paraffin sections. Exceptions were that the color of the cell walls was a deeper color, and cytoplasm was lighter in color. Most cytoplasm was in good condition (Figure 11.1).

Under higher magnification of 40× and 100×, we were able to relatively clearly identify the primary and secondary walls of parenchyma and sclerenchyma cells (Figure 11.1), of which primary walls were characterized by a darker color. In cross section, a single pit on the metaxylem was clearly identifiable. In longitudinal section, the distribution of round pits within ducts was apparent. On a separate longitudinal section, pits were densely distributed on duct walls, separating ducts into distinct sections (Figure 11.1).

Semithin sections were clear with good contrast, allowing us to quantify elements on a very small scale. Measurements of Fargesia shoot elements in cross section were as follows: Epidermal cells had a radial direction of 16.7 μm and a tangential direction of 7.8 μm ($n = 3$). Cortex cells were ovoid in shape, with a diameter of 11.7 μm ($n = 3$). Parenchyma cells were ovoid in shape, with a diameter of 46.7 μm ($n = 20$). The metaxylem duct diameter was 15.8–95.7 μm ($n = 31$), and the primary xylem diameter was 34.0–56.8 μm ($n = 15$). The parenchyma cell wall was 1.6 μm thick ($n = 100$), and the sclerenchyma cell wall was 2.3 μm thick ($n = 10$).

Fargesia *Shoots in Feces*

As observed under the microscope, individual samples displayed cytoplasmic structure, but in most samples only the dyed blue cell walls were evident. Cell wall contours were faint and indistinct (Figure 11.2). Parenchyma cell walls tended to be thinner (Figure 11.2; see also Table 11.1). Cell walls of ducts also showed evidence of damage (Figure 11.2). Using the same sections and dying procedure, color in cells from feces was much lighter than that in fresh samples. Data analyses show significant differences in parenchyma cell wall thickness between samples from fresh ($n = 120$, $\bar{x}_1 = 1.615$, $s_1^2 = 0.1356$) and fecal samples ($n = 120$, $\bar{x}_1 = 1.000$, $s_2^2 = 0.0657$; $t = 13.71$, $P < 0.05$).

ULTRATHIN SECTIONS

Fargesia *Shoots*

Using TEM microscopy, the following cells types were observed in fresh *Fargesia* shoots:

- Parenchyma cells: Under cross-sectioning, parenchyma cells were oblate, and cell walls were thin. Protoplasm and the nucleus were found

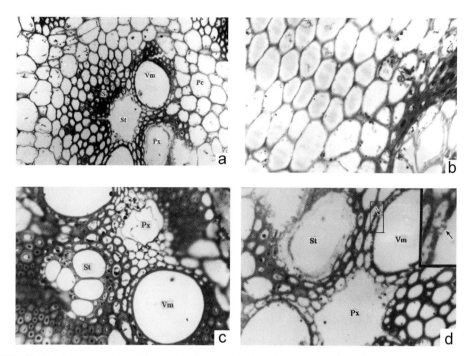

FIGURE 11.2. (a) *Fargesia* shoot in panda feces, showing the vascular bundle structure after digestion. Semithin section, transverse section, with a magnification degree of 3.3 × 20. Vm = metaxylem vessel; St = sieve tube; Px = protoxylem; S = sclerenchyma; Pc = parenchyma cell. (b) Parenchyma cells of *Fargesia* shoots in panda feces, showing the situation after digestion. Semithin section, transverse section, with a magnification degree of 3.3 × 40. (c) The vascular bundle of the *Fargesia* shoot, showing the natural trachea element semithin section, transverse section, with a magnification degree of 3.3 × 40. (d) The vascular bundle of the *Fargesia* shoot in panda feces, showing the trachea element after digestion. Semithin section, transverse section, with a magnification degree of 3.3× 40 (3.3 × 100 for the inset). The arrow indicates ruptures in trachea elements.

Table 11.1.

CELL WALL THICKNESS OF PARENCHYMA CELLS IN *FARGESIA* SHOOTS AND PANDA FECES (330× MAGNIFICATION). WE EXAMINED THREE SAMPLES OF EACH USING A TRANSVERSE CUT AND TWO SAMPLES USING A STRAIGHT CUT. FOR EACH SAMPLE WE MADE 20 MEASUREMENTS OF CELL WALL THICKNESS. SD = STANDARD DEVIATION.

| | Cell wall thickness (μm) | | | |
| | Fresh bamboo | | Panda feces | |
Section type	Sample	Mean (±SD)	Sample	Mean (±SD)
Transverse	1	1.70 (±0.34)	1	1.18 (±0.29)
	2	1.78 (±0.50)	2	1.03 (±0.20)
	3	1.58 (±0.34)	3	0.98 (±0.20)
Straight cut	1	1.48 (±0.34)	1	0.93 (±0.24)
	2	1.60 (±0.35)	2	0.85 (±0.29)

276 Chapter II

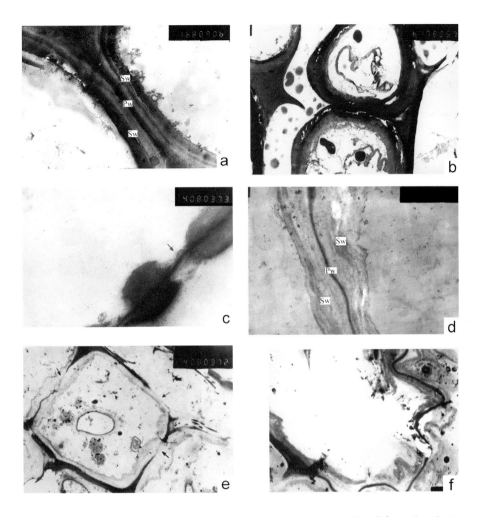

FIGURE 11.3. (a) The cell wall of a normal *Fargesia* shoot, showing the cell wall from the plant. Ultrathin section, transverse section, with a magnification of 16,000×. Pw = primary wall; Sw = secondary wall. (b) Sclerenchyma cells of a *Fargesia* shoot. Ultrathin section, transverse section, with a magnification of 6,100×. (c) An aperture of a tracheal element from a *Fargesia* shoot. Ultrathin section, transverse section, with a magnification of 40,000×. (d) The cell wall of a normal *Fargesia* shoot, showing the cell wall following digestion. Ultrathin section, transverse section, with a magnification of 16,000×. (e) *Fargesia* cell shoot, showing ruptures and loss of material. Ultrathin section, transverse section, with a magnification of 4,000×. (f) *Fargesia* cell shoot, showing loss of the secondary wall. Ultrathin section, transverse section, with a magnification of 5,300×. The arrows indicate loss of the secondary wall.

within the cell interior. On longitudinal sectioning, parenchyma cells were rectangular and had pits.

- Sclerenchyma cells: Cell layers were distinct, allowing discrimination of the relatively darker colored primary cell wall from the secondary cell wall. Within the secondary wall, three layers were evident: two relatively dark layers surrounding a thicker and more lightly colored inner layer (Figure 11.3).

- Fiber cells: These cells were quite thick with small cavities and appeared relatively dark (Figure 11.3).
- Vessels (or trachea elements): Morphology appeared similar to that observed under optical microscopy, although the structure of the pits on the cell walls was more easily identified. These vessels consisted of pit cavities and membranes. Pit membranes, which separated adjacent cells, were areas between cell layers and primary cell walls. Pit cavities were actually depressions within the secondary walls that opened up in the direction of the cell lumen. These kinds of pits on the secondary wall always appeared in pairs (Figure 11.3).

Fargesia *Shoots from Feces*

No clear differences were observed in sclerenchyma cells or vessels from those seen in fresh samples. Little cell content was evident in parenchyma cells, and cell walls were light in color, relatively sparsely distributed, and poorly discriminated. However, obvious changes had occurred in sclerenchyma cell walls through the process of digestion (Figure 11.3). There was evidence of breakage and some loss of the primary walls (Figure 11.3). Secondary walls also displayed compressions and distortions (Figure 11.3).

Discussion of Morphology Results

These results allow us some inference regarding digestion and absorption of nutrients by pandas. Substantial differences were evident in comparing fresh bamboo shoot cells with those in panda feces, of which the most notable was cytoplasm within parenchyma cells. Digestion appeared to result in the disappearance of most cytoplasm, whether viewed in standard paraffin, semithin, or ultrathin sectioning. This observation strongly suggests that pandas are able to digest and assimilate nutrients within bamboo cell cytoplasm. Parenchyma cells constituted a substantial portion of *Fargesia* shoots and had more cytoplasm than other kinds of cells as measured by volume. As is well known, cytoplasm contains the largest part of the cell's protein, as well as portions of its lipid and carbohydrate content. Digestion and assimilation of parenchyma cytoplasm may provide substantial nutritive benefits for giant pandas and form a major source of their overall nutrient intake. Thus, it is natural for us to assume that cytoplasm volume likely constitutes a criterion used by pandas in selecting forage. Using this type of economic calculus, pandas should select plant parts with relatively abundant cytoplasm, such as leaves and soft, delicate shoots; they should avoid plant parts with little cytoplasm, such as old culms. Indeed, our field observations support this hypothesis.

We also examined morphological changes of food items before and after digestion. Cell walls of *Fargesia* shoots seen in feces, notably parenchyma cells, appeared lighter than in fresh samples, and this difference was more obvious in secondary than primary walls under electron microscopy

(Figure 11.3). We consider the different hues to be due to the dying process as well as the constituents and structure of cell walls.

Our current knowledge of plant cell walls suggests that they are primarily composed of cellulose, hemicellulose, pectin polysaccharides, proteins, phenolic compounds, fatty acids, and minerals. Their specific composition varies depending on the type of cell as well as plant species, but generally speaking, cellulose comprises the largest and most important portion, from 20% to 50% by dry weight (Li and Wu, 1993). In bamboo, cellulose constitutes 27.8%–46% (Schaller et al., 1985) of the cell wall. The chemical structure of cellulose is a linear polymer of β-($1\rightarrow4$)-D-glucopyranose units (Shen et al., 1980). The length of each chain, as well as its residues, is a number ranging from several hundred to several thousand and differs for each species of plant as well as different parts of the plant. The function of cellulose is closely related to its structure, and it forms the basic framework of the cell wall.

Hemicellulose is made up of a mixture of polysaccharides and constitutes about 20% to 35.5% (Li and Zhang, 1984) of the cell wall by dry weight (about 23.1%–35.5% in bamboo; Schaller et al., 1985). Hemicellulose is composed of nine polysaccharides, including xylan, galacturonic acid, and mannose, which are also the main constituents of cell walls. These components, together with a type of pectin, form the polysaccharide matrix, which fills in the framework made by the cellulose molecules.

Lignin, a highly incorporated substance found in the cell walls of phenolic compounds, constitutes 15% to 35% of the cell stem (by dry weight; Li and Zhang, 1984). Researchers from Wolong found that lignin constituted 8.6% to 16.2% of bamboo (Schaller et al., 1985). Other studies have shown lignin to constitute 26% of bamboo culms by dry weight (Li and Wu, 1993). Lignin includes polymers of high molecular weight closely related in chemical structure, whose basic unit is a phenylpropionic alkyl, allowing for a stronger cell wall. Other components of cell walls are of minor importance, and we did not focus our studies on them.

In most cases, plant cell walls can be classified into three layers: the primary and secondary walls and the layer between them. This intermediate layer forms shortly after cell differentiation and consists mainly of pectin. Upon maturation, this layer becomes difficult to distinguish from the primary wall; the structure consisting of both is called the complex middle layer. The primary wall, which is located at the outer edge of the cell wall, is the first to emerge and is composed mainly of cellulose and pectin, along with small amounts of hemicelluloses and other polysaccharides. Lignin is also deposited after cell wall maturity. The secondary wall, located in the inner portion of the cell wall, is an additional wall formed on the primary wall after cell growth ceases. It is generally thicker than the primary wall and can be further classified into different layers arising from differing orientations of the cellulose microfibrils of which it is made. The secondary wall mainly consists of cellulose and hemicellulose, with a large amount of lignin and other substances. Secondary walls are characterized by a structure of pits, which function in the exchange of material among cells (Li and Zhang, 1984).

A complete description of the cell wall structure is a large undertaking. Since the 1970s, plant physiologists and biochemists have expended considerable efforts proposing various models that describe the cell wall. Despite considerable variation among these models, all share a common element in that they propose that cellulose forms the main structural polysaccharide of the cell wall framework, within which exist hemicellulose and pectin, within which other substances are widely distributed. Lignin deposits in the framework function to connect it with these other elements (Li and Wu, 1993).

Staining via the electron microscope functions by depositing a concentration of atoms with high molecular weights in specific areas, thus enhancing the visual contrast in those particular areas. The principle is that atoms with high molecular weights more strongly scatter electrons aimed at them than the atoms comprising most tissues (e.g., C, O, N, and H). These higher-weight atoms are enhanced with a coloring agent, generally uranium, lead, manganese, or other heavy metals (Zhu et al., 1983). The quantity of these substances in the samples, i.e., the degree of color differentiation, is directly proportional to the quantity bound in the sample. Each element in the coloring agent possesses a unique binding capacity to each substance in the sample. We used a combination of uranyl acetate and citric acid lead as the coloring agent to increase the contrast of proteins, nucleic acids, and carbohydrates and to aid in observation. Upon deposition of these elements in the tissue, these structures were identifiable under electron microscopy by their degree of staining.

We believe the large staining differences in the cell wall before and after digestion was clearly because of changes in the cell wall, primarily a decrease and, in some cases, a complete disappearance of the primary constituents of cell walls through the process of digestion. Bamboo cell walls are primarily composed of hemicellulose, cellulose, and lignin (Pan et al., 1988). Of these, lignin is known to be indigestible, but cellulose forms the primary framework of the cell wall, so its disappearance must clearly cause a large change in the cell wall structure. Our observations indicated that most of the cell wall remained intact after digestion, but as characterized by lighter staining than in fresh samples, an observation that suggests that the differences observed were unlikely to have been caused by reductions in cellulose. Previous studies concluded that hemicellulose had high utilization rate and digestibility by giant pandas (Schaller et al., 1985; Pan et al., 1988). As the main substance of the cell wall, hemicellulose is distributed evenly throughout the cell wall; therefore, its loss may not greatly affect the basic morphology of the cell wall. Our electronic microscopy work showed that following digestion, differences in stain levels seen in *Fargesia* shoot cells were mainly caused by the changes in the abundance of hemicellulose. That is, following digestion of hemicellulose by pandas, we observed a decrease in its constituent elements, which became combined with the high molecular weight atoms present in the coloring agent. This observation confirmed that pandas were able to digest at least some of the bamboo hemicellulose.

In addition to observing changes in color with digestion, we also noted differences in the thickness and morphology of the cell walls. Samples of panda feces (i.e., after digestion) displayed noticeably thinner parenchyma cell walls than fresh samples, and these cell walls were sometimes distorted and deformed. To confirm this analysis, we viewed semithin sections under a light microscope, recording each measurement (Table 11.1). These measurements confirmed that differences in cell wall thickness arising from digestion were significant: the thickness of *Fargesia* shoot parenchyma cell walls was 38% lower following digestion by pandas. This provided further evidence that hemicellulose, the main substance in the cell wall, is partially digested by giant pandas. Because of this decrease in the matrix, which reduced the stability of the cellulose framework, mechanical abrasion (which occurred during the movement of the bamboo through the digestive tract) had the effect of altering cellular shape.

Thus, various evidence came together to support the view that pandas are capable of using hemicellulose as part of their nutritional resource, a condition virtually unknown among carnivores. Other research has suggested that this hemicellulose absorption may occur in the process of the bamboo passing through the panda's stomach as it encounters an environment that destroys certain chemical structures and facilitates transformation of hemicellulose into simpler carbohydrate molecules. This transformation would facilitate absorption while the digesta pass through the lower portions of the digestive tract (Zhang, 1985).

We have already discussed the general structure and composition of the cell wall; that cellulose is the main structural component of the skeleton of the cell wall. The extensive network of cellulose molecules and their close connections ensure a sturdy basic configuration of the cell wall. Using electron microscopy, we observed fissures and losses in parts of the primary and secondary wall in various cells in panda feces (Figure 11.3). This suggested that, with digestion, cellulose also suffered from some destruction. We observed significant reductions in portions of primary cell walls in *Fargesia* shoots during digestion. This type of damage would seem unlikely simply from mastication or from the mechanical aspects of digestion because cell shapes generally remained intact (particularly noticeable from viewing the secondary cell wall). Nor does it seem plausible for this to arise from inherent structural characteristics of the cell because the pitted structure generally resulted in loss of the secondary cell wall, causing it to collapse inward and form a dent-like shape, whereas the primary wall was kept intact (Figure 11.3). Fecal samples did display a certain degree of loss in secondary cell walls, but it was much less than the loss seen in the primary walls. Analyses have shown that lignin is much more abundant in secondary than in primary cell walls and that there are also important differences in the molecular structure of the two types of cell walls. These changes likely account for most of the differences we observed as the consequence of digestion. These losses of the cell wall suggest the possibility that under certain conditions pandas can also digest some components of the cell wall itself, such as cellulose. We remain skeptical of this theory, but it does seem to deserve further experimentation. Solving this problem in panda nutrition would be a great contribution to our ultimate understanding of the mystery of the giant panda.

There are two possible pathways for giant pandas to digest cellulose. The first would involve the secretion of digestive enzymes capable of breaking down cellulose. The second would involve the use of microbiological symbionts to ferment cellulose. We have no evidence bearing on the first possibility, and because of the endangered status of pandas, we are not willing (nor are we permitted) to obtain digestive fluids directly from live animals for analysis. With regard to the second possibility, there have been studies on the structure of and microorganisms inhabiting the panda digestive system, and these indicate that the panda digestive tract is quite short and simple, only 6.1 times its body length (measured in the Qinling in February of 1987), and lacks a long and meandering caecum (Wang et al., 1983). This anatomical study of the giant panda's digestive tract suggests that the microflora required for fermentation of cellulose would not find a hospitable environment. The primary symbiotic microorganism found in the giant panda's digestive tract was Enterobacteriaceae family member *Escherichia coli*; minor symbionts included *Lactobacillus* and *Streptococcus faecalis*, none of which are known to be capable of digesting cellulose (Zhang and Zhang, 1995).

Verifying and Supplementing Morphological Results

CHEMICAL ANALYSIS OF PANDA FORAGE AND FECES

Morphological research results are capable of directly revealing the degree of use and digestion of various food items but do not provide quantitative results. For this, we must dig deeper. It is widely known that chemical analyses provide a quantitative, rapid, and accurate method of assessing nutrition. Using this method allows us to estimate the digestive efficiency of various foods and in this way calculate the daily energy acquired. This method also allows us to distinguish the nutritive elements of food and thereby understand relationships between various foods and feeding behavior. Results from chemical analysis thus not only form an important verification and supplementation of the morphological studies but also provide other useful information on panda use of foods. We have worked on these issues continually since 1985 and provide a brief overview of our results.

MATERIALS

Bamboo samples were all obtained either from the Huayang District on the southern slopes of the Qinling (now part of the Changqing Natural Reserve) or from Foping Nature Reserve. Samples were selected from species known to be favored by pandas and from bamboo parts known to be eaten. We obtained fresh fecal samples from the same season and area that we sampled fresh bamboo. After obtaining fresh weights of both types of samples, they were dried either by exposure to fresh air or in ovens and then crushed for lab analysis.

METHODS

We used liquid chromatography to discriminate amino acids. Following hydrolysis with hydrochloric acid, we used an automated amino acid analyzer (Beckman 121 MB) to separate 16 amino acids (using tryptophan in the acid decomposition process). Following this procedure, the amino acids were combined with NH_3 to estimate the approximate amount of crude protein in each sample. We used the Van Soest analysis method (Van Soest, 1975). All samples were thoroughly washed, resulting in the following products: neutral detergent fiber (NDF, which contained all the cell wall components), acid detergent fiber (ADF, all cell wall elements remaining after elimination of hemicellulose), a mixture of ash and lignin, and the remaining ash following burning (Table 11.2). Thus, we calculated the contents of each cell wall component as

Cell wall material = NDF,
Hemicellulose = NDF − ADF,
Cellulose = ADF − the mixture of ash and lignin,
Lignin = the mixture of ash and lignin − remaining ash.

Table 11.2.

CELL WALL CONTENT (G/100 G OF DRY MATTER) IN FRESH BAMBOO, PANDA FECES, AND BAMBOO RAT FECES. NDF = NEUTRAL DETERGENT FIBER; ADF = ACID DETERGENT FIBER. A DASH (—) INDICATES NO DATA ARE AVAILABLE.

Sample	NDF	ADF	Hemicellulose	Cellulose	Lignin	Ash
Bashania young leaf	63.70	33.72	29.97	20.22	9.99	3.52
Bashania leaf in panda feces	69.93	40.08	29.86	24.29	12.06	3.73
Bashania stem in panda feces	81.12	47.66	33.46	32.58	17.42	—
Bashania young stem	81.29	45.03	36.26	30.78	14.65	—
Fargesia stem	91.80	46.24	45.56	32.59	13.65	—
Fargesia stem in panda feces	91.95	47.34	44.23	34.65	15.63	—
Bamboo rat feces	77.41	50.25	27.17	32.09	18.24	—

We used radio spectroscopy to measure inorganic elements. Following either the process of transformation into ash or high-pressure decomposition with nitric acid, samples were subjected to absorption of nitric acid with spectrum pure carbon powder and then loaded for analysis of their pure spectral properties. We used a Jarrell-Ash 3.4m large-grating spectrometer, 1,180 g/mm, grating wave length of 3,000 Å, and direct current arc excitation at 12 A, and we used Kodak No. 1 photographic light-sensitive plates to document quantitative and quasi-quantitative results.

Experimental Results and Discussion

On the basis of our analysis of cell wall content (see Appendixes 12.1, 12.3, and 12.4 in the Chinese edition), we discuss two points: The first deals with the relationships between the nutritional value in various foods and the volume of these foods ingested by pandas. That is, we can use chemical analysis of panda forage to explain panda forage selection. The second is that we can make use of the relative amount of material in panda forage and panda feces to estimate daily digestibility and actual intake of various foods. On the basis of these estimates, we can then estimate daily caloric intake.

Chemical Analysis of Panda Forage Selection

CRUDE PROTEIN AND AMINO ACIDS IN BAMBOO

Although present in the cells of bamboo, protein is not abundant within the cell walls themselves. Protein is an indispensable daily requirement for pandas, and because they have no way to acquire it other than through forage, the amount of protein in forage is an important indicator of protein balance for pandas. There are 20 principal amino acids that constitute proteins, of which most can be synthesized by mammals internally. However, there are a few amino acids, called essential amino acids, that cannot be synthesized internally. For mammals these amino acids include isoleucine, leucine, lysine, phenylalanine, threonine, valine, methionine, and tryptophan (Shen et al., 1980), and these amino acids must come entirely from forage. Also, when the body metabolizes food, particularly protein, a certain proportion of various amino acids, known as the amino acid balance (also called the essential amino acid model), must be present for the body to properly use these amino acids for normal homeostasis. Therefore, the proportions of the various amino acids in forage constitute a criterion to assess the protein value of various forage items. As of now, amino acids required by giant pandas remain unknown. However, considering that humans and rats, two mammals that are taxonomically very distantly related, have similar amino acid requirements, it is reasonable to extrapolate these same requirements to giant pandas. On the basis of this extrapolation, we can use chicken eggs as a standard reference from which to estimate the ratio of amino acids in various panda forage because the amino acids in chicken eggs are almost 100% utilized. In relation to an egg, the difference in the amino acid score in forage limits the utilization of other amino acids by the giant panda.

We present the levels of crude protein, essential amino acids, and amino acid ratios in various bamboo samples, along with comparisons with beef and maize, in Table 11.3. On the basis of two criteria (crude protein content and amino acid ratio), we further ranked various food items (plant parts within each species). On the basis of the criteria of crude protein, the ranking for *Bashania* is young leaf of first-year shoot > young leaf of large bamboo > young stem > epidermis > old leaf > old stem. For *Fargesia*, the ranking is leaf > new shoot > young stem > old stem. On the basis of the criterion of amino acid ratios, the ranking for *Bashania* is young leaf of first-year shoot > young leaf of large bamboo > young stem > epidermis > old leaf > old stem. For *Fargesia* it is new shoot > leaf > young stem > old stem.

By comparing these ranks with the actual intake by pandas, we notice certain patterns. Pandas choose parts of *Bashania* for foraging on the basis of not only the amount of crude protein but also amino acid ratios. That is to say, pandas select forage items on the basis of the maximum assimilation of protein. However, in addition to stems and leaves, pandas consume copious amounts of *Bashania* shoots, especially during May, and we were unable to analyze the nutritional value of the shoots.

We found a similar pattern in *Fargesia*, with one puzzling exception: crude protein content (11.3%) and amino acid ratio (21.0%) in leaves far surpassed those in young stems (crude protein of 2.1% and an amino acid ratio of 4.1%), but pandas paid scant attention to *Fargesia* leaves. One possible explanation for this is the fact that *Fargesia* leaves are short and thin, making them unwise investments in terms of energetic expenditure and time, and thus, nutritional gains are possibly outweighed by these losses. This behavior suggests that panda select forage based on multiple considerations, and thus, our analysis should similarly consider factors beyond crude protein content and amino acid ratios.

Bamboo Cell Wall Contents

Cell walls constitute important components of plant cells, as well as comprising a large proportion of panda diets. However, from the standpoint of nutritional value, much of the cell wall content is not easily digested by pandas, including lignin, cellulose, and ash (Table 11.2). Thus, the relative abundances of these cell wall components in panda diets, particularly the indigestible lignin and cellulose, may also be useful criteria in assessing nutritive value.

We quantified the constituent elements of *Bashania* cell walls from young leaves (including leaves from first-year plants as well as from larger plants), as well as from *Bashania* stems and *Fargesia* stems (Table 11.2). As

Table 11.3.

CRUDE PROTEIN AMINO ACID CONTENT AND AMINO ACID RATIOS OF VARIOUS BAMBOO PARTS (IN G/100G OF DRY MATTER). A DASH (—) INDICATES NO DATA ARE AVAILABLE. ILE = ISOLEUCINE; LEU = LEUCINE; LYS = LYSINE; PHE = PHENYLALANINE; MET = METHIONINE; THR = THREONINE; VAL = VALINE.

Sample	Bamboo part	Crude protein	Amino Acid							Amino acid content (%)
			Ile	Leu	Lys	Phe	Met	Thr	Val	
Egg			0.43	0.57	0.40	0.37	0.34	0.31	0.46	100
Beef			0.33	0.52	0.54	0.25	0.24	0.28	0.35	69.3
Maize		9.77	0.35	0.84	0.18	0.42	0.21	0.22	0.38	44.9
Bashania	Young leaf	9.61	0.44	0.54	0.51	0.60	0.13	0.52	0.62	37.7
	Old leaf	10.20	0.11	0.34	0.47	0.62	0.81	0.67	0.02	4.3
	Young leaf of first-year shoot	15.24	0.64	1.37	0.73	0.87	0.19	0.79	0.94	56.1
	Young stem	1.58	0.06	0.11	0.15	0.07	—	0.06	0.13	13.3
	Old stem	0.98	0.04	0.01	0.09	0.05	—	0.05	0.08	1.9
	Epidermis of stem	1.494	0.05	0.10	0.18	0.06	—	0.06	0.11	11.4
Fargesia	Leaf	11.30	0.33	0.81	0.57	0.56	0.07	0.79	0.44	21.0
	New shoot	3.99	0.10	0.32	0.22	0.16	—	0.20	0.16	23.6
	Young stem	2.11	0.07	0.13	0.14	0.08	0.01	0.76	0.13	4.1
	Old stem	1.55	0.04	0.09	0.15	0.05	0.01	0.06	0.11	2.0

the site of photosynthesis in bamboo, leaves contain the most cell content (carbohydrates, lipids, etc.), whereas stems, which function primarily as structural and vascular support, contain more cell wall content. Schaller et al. (1985) used a nutritional quality ratio (protein/[lignin + cellulose]) to assess the nutritional level of bamboo in Wolong. Here we adopted the same approach in analyzing the nutritional value of bamboo in the Qinling and ranked each as follows:

young leaves from first-year *Bashania* shoots (ratio = 0.50) > young leaves of large bamboo (0.32) > *Fargesia* bamboo stem (0.05) > *Bashania* young stem (0.03).

Leaves from *Bashania* far surpassed the stems of either species in terms of this ratio, and those from first-year *Bashania* shoots were even higher than those from the most nutritious plant in Wolong, which was young bamboo *F. spathacea* shoots (0.48). This result illustrates that pandas derived maximum nutritional benefits from feeding on the young leaves of first-year *Bashania*. Pandas did select this forage for almost half the year, eating primarily leaves from young *Bashania* from October through the following April. From this, it follows that pandas avoided foods with a high lignin and cellulose content. Our earlier morphological analysis, as well as work conducted by other researchers, showed that pandas use one constituent element of bamboo cell walls (hemicelluloses), but we found no correlation with panda foraging. This may be because hemicellulose is common in every food item available to pandas. Moreover, utilization of hemicellulose is far lower than protein and may not function as a criterion used by pandas in forage selection.

INORGANIC ELEMENTS IN BAMBOO

Quasi-quantitative measurements indicate that inorganic elements in bamboo may be divided into three categories (see Appendix 12.1 in the Chinese edition): common elements (e.g., P, Ca) with percentages ranging from 1% to 5%; elements of moderate abundance (e.g., Si, K, Na, Mg, Fe, Al), with percentages ranging from 0.01% to 0.1%; and uncommon elements (e.g., B, Mn, Pb, Cu, Zn), with percentages below 0.01%. Inorganic elements in the leaves of both *Bashania* and *Fargesia* bamboo were more abundant than those in stems, as well as being more abundant in older than younger bamboo. Thus, inorganic elements were highest in old *Bashania* leaves and lowest in young stems. However, we noted no correlations between inorganic element content and panda forage selection. This may be because inorganic elements play only a minor role in panda nutrition, and thus, pandas displayed no obvious response to them. Although it is true that inorganic elements are crucial for survival and reproduction, it may be that they are sufficiently abundant in both forage and water that pandas have no need to select forage on the basis of their abundance. However, we have observed pandas seeking out unusual forage. For example, in May 1993, in Shuidonggou, we observed Jiaojiao excavate and consume the remains of a wild boar. At that time, she was lactating and caring for her 8-month-old cub. In another example, in Shanshuping in August 1997, we observed Xiwang excavate and consume some old pork bones that had been left as trash by people. According to her behavioral characteristics, we inferred that she was, at that time, late in gestation. Whether these unusual forage items were driven by the need to supplement inorganic elements remains unknown.

Panda Acquisition of Nutrients and Energy

We examined the absolute abundance and nutritive value of food items assimilated by pandas on a daily basis and how these provide the necessary caloric requirements. To this end, we first examined data on the volume of food ingested daily (see chapter 10). This analysis makes clear that daily intake varied greatly depending on panda development stage and physiological condition. Even among individuals of the same age class and in similar physical condition, daily food intake displayed considerable heterogeneity. From data presented in chapter 10, we know that subadults (e.g., Xiwang) consuming leaves and stems of *Bashania* consumed 18.01 ± 4.33 kg/day fresh weight ($n = 10$), whereas old individuals (e.g., Da Shun) eating the same type of forage consumed only 13.71 ± 2.06 kg/day fresh weight ($n = 16$), and that adults (e.g., Huzi) foraging on *Bashania* shoots consumed 41.7 ± 5.6 kg/day fresh weight ($n = 3$). On the basis of these data, we can extrapolate the basic daily intake of food by a wild giant panda. For this extrapolation, we used Xiwang's feeding rate as representative and used data from each healthy adult panda to generate a generic daily forage intake (see Table 11.6). We noted no statistically significant differences in daily food intake among these individuals; thus, we take their mean to represent the standard of normal daily forage intake. We can also use this mean to estimate the daily forage intake on a dry-weight basis (Table 11.6). During the September–April time period, the percentage of *Bashania* leaves eaten was 88.3%, whereas *Bashania* stems comprised 11.7% by weight. From this, we calculate daily dry-weight forage consumption for leaves (19.09 kg/day bamboo consumed × 88.3% eaten× 40% weight of bamboo parts) + stems (19.09 kg/day bamboo consumed × 11.7% eaten × 50% weight of bamboo parts) ($n = 10$). An adult giant panda consuming *Bashania* shoots would consume a dry weight of 8.34 kg/day ($n = 3$), and an adult giant panda consuming *Fargesia* shoots would consume a dry weight of 10.14 kg/day ($n = 3$).

Our analysis also showed that daily forage intake decreases in cases where a panda experienced particular physical conditions, such as old age, illness, gestation, and lactation. Mean daily forage intake for elderly pandas in dry weight would be 5.64 kg/day (n = 16), about 72.8% of the daily forage intake of a healthy adult.

Acquisition of Various Nutrients

PROTEIN REQUIREMENTS

We determined the protein content of various panda forage items and panda feces, which we use to calculate panda digestibility of protein. Protein levels in feces were quite high, resulting in our calculations of protein digestibility being relatively low (approximately 50%), whereas previous research using captive pandas has indicated that protein digestibility can be as high as 90% (Schaller et al., 1985). Some researchers have suggested that microorganisms (Dierenfeld, 1981) or large quantities of secondary metabolites (Schaller et al., 1985) deposited on the surface of feces when food passes through the digestive tract produce this upward bias in fecal protein. Thus, we have adopted the figure from captivity of 90% digestibility in our calculations of daily protein acquisition. To determine the minimum daily protein requirements for pandas, we relied on the equation for ruminants developed by Moen (1973). This equation defined an animal's minimum daily protein requirement as that which permits the animal to maintain its body mass. This calculation consists of three parts: (1) gross protein needed (g)

$$Q_{nb} = 0.1245W^{0.75},$$

where W is the animal's mass (kg), (2) endogenous protein excreted in urine (g)

$$Q_{eum} = 0.48 \times 70W^{0.75} \times 6.25/1000,$$

where 0.48 is the basal metabolic rate of nitrogen (mg/kJ) and $70W^{0.75}$ is the mean standard value for mammals, and (3) endogenous protein excreted in feces (g)

$$Q_{mfn} = 5F,$$

where 5 indicates the estimated protein excreted daily by a herbivore from each kilogram of dry matter ingested (g) and F is the daily intake of dry matter (kg).

Assuming a standard weight of 100 kg for a panda, we calculated the minimum daily protein required as

$$Q_{nb} + Q_{eun} + Q_{mfn} = 3.94 \text{ g} + 23.4 \text{ g} + 5F \text{ g} = 27.34 \text{ g} + 5F \text{ g}.$$

However, this minimum value may not actually satisfy the daily requirement for all the essential amino acids because the essential amino acids exist in low concentrations in forage. We therefore applied a multiplier of 1.5 to the above formula (Crampton and Harris, 1969), which we take as sufficient to allow for

Table 11.4.

MINIMUM REQUIRED PROTEIN CONSUMPTION AND ACTUAL PROTEIN CONSUMPTION OF GIANT PANDAS. PROTEIN CONTENT OF *BASHANIA* SHOOTS WAS NOT ESTIMATED.

Time period	Forage	Daily forage consumption (kg)	Crude protein requirement (g/day)	Crude protein consumption (g/day)
September–April	*Bashania* leaf and stem	7.86	100.0	613.5
June–August	*Fargesia* old shoot and young stem	4.70	75.4	168.9
May	*Bashania* shoot	8.34	103.6	

the lower values of some essential amino acids. This increased the minimum daily protein requirement for pandas to 41.01 g $+ 7.5F$ g (Table 11.4).

Except for *Bashania* shoots in May (for which we lacked data on intake), panda protein intake exceeded by a considerable margin the theoretical daily requirement (Table 11.4). This evidence indicates panda forage is able to provide sufficient nutrition to support pandas' year-round protein requirement. This result is particularly notable in the case of *Bashania* leaves and stems, which have very high protein levels and which, even in winter and spring, provide >6 times the required crude protein. This level of intake may be particularly meaningful for lactating females, who have increased protein demands, as well as for all pandas producing a denser fur coat in winter.

CARBOHYDRATE AND LIPID REQUIREMENTS

Our calculations have demonstrated that hemicellulose is greatly reduced during the process of digestion of giant pandas. We know from other research that lignin in plant cell walls is essentially indigestible, regardless of the animal species considered (Van Soest, 1980). This knowledge allowed us to use the amount of lignin present as a constant marker and thus calculate the digestibility of various forage items by dry weight.

The assimilation and utilization of hemicellulose by pandas are relatively high (Table 11.5). During winter and spring, the daily percentage intake of *Bashania* stems was 11.7% and that of leaves 88.3%, from which we calculated the digestibility of hemicellulose to be 18.1%. In summer, pandas largely subsisted on *Fargesia* stems, from which we calculated the digestibility of hemicellulose to be 15.2%. These results are similar to those obtained at both the National Zoo in Washington, DC, and Wolong. Captive pandas at the National Zoo feeding on a particularly rigid species of bamboo had a hemicellulose digestibility of 27% (Dierenfeld, 1981); pandas in Wolong had hemicellulose digestibilities of 21.5%, 26.0%, and 18.2% during spring, summer, and winter, respectively (Schaller et al., 1985). These data illustrate that digestion of hemicellulose provides an important source of nutrients for giant pandas and that their ability to make use of such a low-quality food

Table 11.5.

GIANT PANDA DIGESTIBILITY OF HEMICELLULOSE, CELLULOSE, AND DRY MATTER.

Sample	Hemicellulose		Cellulose		Dry matter digestibility (%)
	Content (g/100 g plant material)	Digestibility (%)	Content (g/100 g plant material)	Digestibility (%)	
		Winter/spring			
Bashania leaf					
Young leaf	29.972	17.5	20.22	0.48	17.2
Feces	29.856		24.29		
Bashania stem					
Young stem	36.262	22.4	30.78	10.9	15.9
Feces	33.460		32.58		
		Summer			
Fargesia stem					
Stem	45.561	15.2	32.59	7.20	12.7
Feces	44.232		34.65		

resource as bamboo reflects an important adaptation. Soluble carbohydrates are quite low in most bamboo species, with means generally only 2%–3% fresh weight (Schaller et al., 1985), which forces pandas to find alternative ways to acquire necessary energy, including the polysaccharide hemicellulose, a trait that has tremendous implications in the daily lives of pandas.

PANDA USE OF CELLULOSE

Because whether or not giant pandas can digest cellulose has been debated for so long, we decided to use the same approach as for hemicellulose (Table 11.5). This shows a much higher digestibility than we had expected. That said, it may be that the cellulose digestibility suggested here for first-year *Bashania* stems is biased high because pandas peel off and do not consume the epidermis layer, whereas our measured cellulose included the intact epidermis (and our analysis also indicated that cellulose in the epidermis is higher than in the stem interior). The rather large disparity in the evident cellulose digestibilities suggested for young *Bashania* leaves and young *Fargesia* stems is cause for some skepticism. However, further analysis and consideration provide us with a possible explanation based on differences in the nutrient contents of the two foods. The cellulose content in young *Bashania* leaves is low (20.2%), whereas the protein content is high (9.6%). In contrast, the cellulose content in young *Bashania* stems is much higher (32.5%), and its protein content is quite low (2.1%). When pandas feed on *Bashania* leaves, the high protein content provides them high levels of energy. In addition, *Bashania* leaves contain more cell content, which includes soluble carbohydrates and lipids. These, supplemented by hemicellulose (271

Acquisition of Nutrients   291

g/day), provide additional important energy sources. Thus, pandas acquire sufficient energy from *Bashania* leaves without recourse to the rather difficult to digest cellulose. This situation differs considerably from that when pandas forage on *Fargesia* stems. In that case, the low levels of protein and relatively small amount of cell contents force pandas to seek other sources of nutrients. Thus, in addition to using a large amount of hemicellulose (533.5 g/kg) from cell walls, pandas must make use of cellulose (179.1 g/day). Although this species-specific difference provides a possible explanation for the difference in cellulose digestibility among bamboo species, the questions of just how pandas manage to digest cellulose and if cellulose digestibility varies with forage type remain.

GIANT PANDA ACQUISITION OF ENERGY

The basal metabolic rate for the giant panda remains a topic for future research, but we can provide a first-order approximation using Moen's formula (Moen, 1973):

$$W^{0.75} \times 70 = 9254.5 \text{ kJ/day},$$

where W = 100 kg. According to Moen's equation and taking into account energy loss from normal daily activity, a giant panda weighing 100 kg would consume 13,092 kJ/day. If we also add in estimates of the cost of growth and reproduction, this figure might rise to approximately 14,000–17,000 kJ/day. Dierenfeld (1981) used a bomb calorimeter to estimate the specific heat of parts of the bamboo *Phyllostachys aureosulcata* as 20,064 kJ/kg for leaves and 19,228 kJ/kg for stems. In Wolong, Schaller et al. (1985) used these results in their analyses of energy acquisition. We did likewise in our Qinling research because our data indicated that Qinling bamboo was chemically very similar to bamboo in Wolong. Specifically, our calculations were daily forage mass × dry matter digestibility × forage specific heat ratio = actual energy supply = 26,762.9 kJ/day.

We estimated daily energy acquisition only for *Bashania* because we lacked data on *Fargesia* (eaten only during summer) and the pandas in the Qinling feed on *Bashania* bamboo for nearly 10 months of the year. Available energy far exceeds our estimated daily requirement of 14,000–17,000 kJ/day during both winter and spring (Table 11.6). This result further suggests that although the nutritional value of bamboo is quite low, pandas' behavioral and functional adaptations allowed them to acquire the necessary energy to maintain normal activity.

In most situations, carbohydrates and lipids are the nutrients providing the highest caloric content. However, soluble polysaccharides and lipids are not abundant in plant leaves or stems (Hladik, 1978; Milton, 1980). This limited availability provides us insight into why pandas must consume such large quantities of food daily; it is not only their requirement for a balance

Table 11.6.

GIANT PANDA ENERGY ACQUISITION DURING WINTER AND SPRING, SHOWING DAILY FORAGE CONSUMED, DRY MATTER DIGESTIBILITY, SPECIFIC HEAT, AND TOTAL ENERGY. DAILY FORAGE CONSUMPTION WAS 7.89 KG.

Forage	Ratio (%)	Dry matter digestibility (%)	Specific heat (kcal/kg)	Energy (kcal) each species	Forage total
Bashania leaf	88.3	17.2	4,800	5,730.0	6,402.6
Bashania stem	11.7	15.9	4,600	672.6	

of amino acids but, perhaps even more important, their need to acquire energy from carbohydrates and lipids. We have not measured the contents of the metabolically functional carbohydrates and lipids present in the panda's diet, but we provide an estimate in winter and spring based on consumption and specific heat values of protein (17.97 kJ/g) and hemicellulose and cellulose (16.72 kJ/g) of 7,702 kJ/day.

Thus, these calculations suggest that in winter and spring, energy provided by soluble carbohydrates and lipids makes up 28.8% of the total. This calculation may underestimate the true percentage because protein consumption does not function primarily as a source of energy, but rather is used primarily for bodily maintenance. Thus, protein likely contributes less to energy supply than we have estimated here, leading to greater importance of our estimates of carbohydrates and lipids. Perhaps panda intake of various food items arises from differences in carbohydrate and lipid content.

Pandas cannot maintain normal function if depending solely on energy provided from carbohydrates and lipids: there must be additional nutrients to supplement these resources. Indeed, these nutrients are provided by hemicellulose, which itself is a complex polysaccharide. Previous research has indicated that the panda's stomach acid and the alkaline environment of its small intestine may allow for the digestion of some sugar bonds in hemicellulose, allowing for part of it to be digested in the lower section of the intestine (Schaller et al., 1985). We estimate the specific heat of hemicelluloses (we use the specific heat of carbohydrates as 16.72 kJ/g) as 7,335.9 kJ/d. This estimate makes clear that the actual energy gained from hemicellulose makes up a large portion of the energy acquired by pandas: 27.4% of total energy during winter and spring. For pandas digestion and utilization of hemicellulose are therefore crucial. This valuable capability, so unusual for a carnivore, is among the most striking adaptations allowing pandas to subsist on bamboo (Table 11.7).

We point out that forage intake surrounding parturition was more complex. We noted the following patterns from our observations of Jiaojiao, Nüxia, and Ruixue during these time periods. Immediately following parturition, nursing panda females ceased feeding entirely, not resuming until

Table 11.7.

DAILY ACQUISITION OF NUTRITION AND ENERGY BY GIANT PANDAS BY PLANT CONSTITUENT.

Nutrient	Protein	Hemicellulose	Carbohydrate and lipid	Other
Acquisition of material (g/day)	613.5	479.0		
Acquisition energy (kcal/day)	2,638	1,755	1,682	328
Total energy (%)	41.2	27.4	28.8	5.1

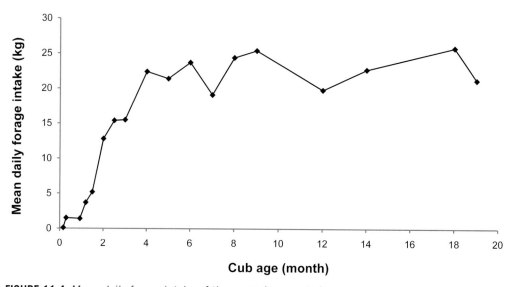

FIGURE 11.4. Mean daily forage intake of three nursing panda females (Jiaojiao, Nüxia, and Ruixue) during the 19 months following parturition.

some 10 days afterward. Thereafter, nursing females gradually increased their daily forage intake as the cub grew, recovering their normal intake rate when the cubs were about 4 months of age. For these three pandas, we estimated daily forage intake from 6–7 days postparturition until the cubs were 19 months old (n = 18; Figure 11.4). We note that this figure does display individual heterogeneity among these females, but the general trend is still clear. We see low daily forage intake until cubs reach approximately 4 months of age, suggesting that nursing females must have mobilized internal energy sources (assuming no dramatic change in daily energy consumption). Thus, possessing sufficient energy reserves that can be mobilized is seen as extremely important for the health and survival of both the female panda and her cub. This need may explain why females are smaller than males but do not differ from males in forage intake.

Comparisons in Panda Diets and Use with Wolong

Pandas in the Qinling and Wolong occupy distinct mountain systems and constitute discrete populations. However, they share some dietary characteristics: They both depend on a pair of bamboo species year-round (*Bashania* and *Fargesia* in the Qinling and *Bashania faberi* and *Fargesia* in Wolong). Our observations indicate that they prefer shoots in both cases, we suspect because shoots not only contain a large amount of moisture in primary tissue and are quite palatable but also contain relatively large amounts of primary products and are easily digestible. Pandas in these areas also display some differences. For example, diet selection by Qinling pandas follows a clearly marked seasonal pattern, in which they move to high elevations from June until September to feed on fresh and old shoots of *Fargesia*, whereas they remain at lower elevations the rest of the year, primarily feeding on leaves, stems, or shoots of *Bashania*. In Wolong, by contrast, pandas panda feed on stems and leaves of *Bashania faberi* year-round (its shoots are so small that pandas generally ignore them), except in May when they feed on *Fargesia* shoots. That Wolong pandas display such a reduced seasonal pattern probably results from differences in bamboo species and geographic pattern from that in the Qinling. As each bamboo species has a unique temporal and geographic pattern of growth and size, it is not surprising that pandas differ in strategies in using them.

Note

Chapter originally appeared as Chapter 12 in Chinese edition published by Peking University Press, Ltd.

References

Crampton, E. W., and L. E. Harris. 1969. *Applied Animal Nutrition: The Use of Feedstuffs in the Formulation of Livestock Rations*. San Francisco: W. H. Freeman.

Dierenfeld, E. S. 1981. "The Nutritional Composition of Bamboo and the Utilization by the Giant Panda." Master's thesis, Cornell University.

Hladik, C. M. 1978. "Adaptive Strategies of Primates in Relation to Leaf Eating." In *The Ecology of Arboreal Folivores*, edited by G. Montgomery, pp. 373–395. Washington, DC: Smithsonian Institution Press.

Li, R. A., M. X. Du, and Z. L. Li. 1984. 大熊猫竹食消化后的组织分析 [Histological analysis of bamboo residues after digestion by giant pandas]. *Acta Botanica Sinica* 26:112–114.

Li, X. B., and Q. Wu. 1993. 植物细胞壁 [Plant cell walls]. Beijing: Peking University Press.

Li, Z. L., and X. Y. Zhang. 1984. 植物解剖学 [Plant anatomy]. Beijing: Higher Education Press.

Milton, K. 1980. *The Foraging Strategy of Howler Monkeys: A Study in Primate Economics*. New York: Columbia University Press.

Moen, A. N. 1973. *Wildlife Ecology: An Analytical Approach*. San Francisco: W. H. Freeman.

Pan, W. S., Z. S. Gao, Z. Lü, Z. K. Xia, M. D. Zhang, L. L. Ma, G. L. Meng, X. Y. She, X. Z. Liu, H. T. Cui, and F. X. Chen. 1988. 秦岭大熊猫的自然庇护所 [The giant panda's natural refuge in the Qinling Mountains]. Beijing: Peking University Press.

Schaller, G. B., J. C. Hu, W. S. Pan, and J. Zhu. 1985. *The Giant Pandas of Wolong*. Chicago: Chicago University Press.

Shen, T., J. Y. Wang, and B. T. Zhao. 1980. 生物化学 [Biochemistry]. Beijing: People's Education Press.

Van Soest, P. J. 1975. "Physico-chemical Aspects of Fibre Digestion." In *Digestion and Metabolism in the Ruminant*, edited by I. W. McDonald and A. C. I. Warner, pp. 351–365. Sydney: University of New England Publishing Unit.

———. 1980. "Impact of Feeding Behavior and Digestive Capacity on Nutritional Response." In *Technical Consultation on Animal Genetic Resources Conservation and Management*. Rome: Food and Agriculture Organization of the United Nations.

Wang, P., C. Cao, M. S. Chen, and F. Li. 1983. 大熊猫 (*Ailuropoda melanoleuca*) 的组织学研究—I 消化道的显微结构 [Histological study of giant pandas—I. The microstructure of the digestive tract). *Acta Scientiarum Naturalium Universitatis Pekinensis* 5:67–73.

Zhang, Z. B. 1985. 熊猫对竹子的利用 [The usage of bamboo by giant pandas]. *Chinese Journal of Zoology* 20:44.

Zhang, Z. H., and A. J. Zhang. 1995. 大熊猫肠道正常菌群的研究 [The study of the giant panda's intestinal flora]. *Acta Theriologica Sinica* 15:170–175.

Zhu, L. X., N. Q. Cheng, and X. Z. Gao. 1983. 生物学中的电子显微镜技术 [The technique of electron microscope in biology]. Beijing: Peking University Press.

Chapter 12

Biodiversity on the Southern Slopes of the Central Qinling Mountains

The unique physiographic nature that characterizes the Qinling Mountains provides it a special biogeographic status. Particularly on its vast southern slopes, the diversity of mountain climate zones produces an abundance of biological resources, earning this portion of the Qinling recognition as one of 11 terrestrial key biodiversity regions within China (Editorial Committee of "China's Biodiversity: Country Study," 1998). The existence and structure of the area's species diversity are the result of the ecological and evolutionary processes that have taken place over the past 700,000-odd years. In this chapter, we review the biological diversity of plants, butterflies, and birds (for mammal diversity measure, see the Chinese edition).

Editor's note: We did not attempt to update plant and animal nomenclature for this chapter.

Diversity of Seed-Bearing Vegetation

SUMMARY

In this section, we summarize floral biodiversity in the main panda habitats within the Changqing National Nature Reserve. We established 20 plots (10×20 m each) at elevations of 1,400 to 2,400 m and for each plot documented the species richness and diversity of seed-bearing plants (primarily woody plants). We quantified the distribution of genera, species richness S, Shannon-Wiener diversity index H', and Jaccard similarity C_j at the tree, shrub, sapling, and seedling layers. We found that temperate genera occupied the largest area, and their proportional share increased at the expense of tropical species as elevation increased. A number of analyses illustrated that plant communities around 1,950 m elevation represent a boundary of types: above and below this elevation contour total species richness, tree species richness, number of lianas, and the diversity indices H' of trees, saplings, and seedling all differed. However, most indices did not differ if examined only on one or the other side of this elevational line. With time since disturbance, species diversity in logged forests tended to follow a pattern characterized by low to high to low, although confirming this pattern would require a longer time period of study than we had available.

EXISTING DIVERSITY

We began collecting and identifying plant specimens in the study area in 1996. During the summers of 1996 and 1997 we integrated line transect sampling by assessing the frequency of species at various levels to investigate vegetation plant diversity and regeneration. Our sample transects were primarily located in Xigou. We sampled in plots measuring 10 × 20 m (200 m²) located along an elevational transect at 100 m intervals. We supplemented these plots using the same methods for additional habitat types when required (e.g., naturally regenerating forests following timber harvest, replanted forests dominated by various species, and relatively undisturbed and uncut forests). In each case, we recorded plot location, elevation, slope, aspect, tree canopy density, bamboo status, and any indication of the effects on vegetation of wildlife or humans.

Within each plot, we randomly placed twenty-five 4-m² small plots (so that measurements were conducted on a total of 100 m², i.e., half the larger plot area), in which we made detailed measurements of trees, shrubs, and lianas. For mature trees (defined as diameter at breast height [DBH] ≥ 3 cm), we recorded species, height, and DBH. For young trees (defined as DBH <3 cm and height ≥ 1 m), we recorded species and number. For shrubs exceeding 1 m in height, we recorded species and number; for smaller shrubs, we recorded only species. We noted the presence of lianas within plots. From the 25 small plots, we further randomly selected 5 for detailed documentation of the species and relative abundance of vegetation in the herb layer.

We quantified species diversity using the S index and the Shannon-Wiener index H', as well as variation of these along the elevational gradient. We also analyzed community similarity and distribution by genera. Because of time constraints, our sampling was limited to the Huorenping area of Huayang Township at elevations of 1,100 to 3,070 m (with an emphasis on plots located at 1,500 to 2,400 m.). To supplement site visits, we also made use of satellite imagery.

DIVERSITY OF SEED-BEARING PLANTS

We quantified the α diversity of seed-bearing plants in each research plot as a measure of species richness.

α Diversity: Species Richness S

We provide summaries of the total number of species, trees, shrubs, herbs, and lianas in each plot, as well as the number of species in both the tree and shrub layers (Table 12.1, Figure 12.1). Above 1,930 m species richness

Table 12.1.

SPECIES RICHNESS S AND SHANNON-WEINER DIVERSITY INDEX H' OF QINLING VEGETATION PLOTS.

Plot	Elevation	Vegetation type	Total Species	Tree	Shrub	Herb	Liana	Tree[a]	Shrub[b]	Tree layer	Shrub layer
				Species richness S						Diversity index H'	
1	1,460	Replanted Chinese pine	59	15	13	25	6	5	11	1.41	1.78
2	1,480	Replanted Japanese larch and naturally regenerating Chinese pine	85	20	19	32	14	6	22	1.58	2.54
3	1,490	Replanted Japanese larch and naturally regenerating Chinese pine	68	15	19	23	11	4	18	1.35	2.59
4	1,550	Secondary mixed forest	70	19	17	27	7	12	15	2.15	2.63
5	1,640	Secondary mixed forest	52	15	15	13	9	8	15	1.99	2.35
6	1,650	Replanted northern China larch	39	13	10	9	7	5	12	0.47	2.22
7	1,650	Shrub	84	21	21	33	9	1	28	0	2.93
8	1,750	Secondary mixed forest	69	25	13	25	6	6	25	1.62	2.82
9	1,750	Replanted northern China larch and Chinese pine	54	14	15	22	3	2	9	0.32	1.89
10	1,850	Secondary mixed forest	56	18	14	17	7	4	20	1.39	2.50
11	1,860	Secondary mixed forest	42	12	13	13	4	4	8	1.39	1.32
12	1,860	Mixed pine-oak forest	41	12	17	10	2	4	16	1.01	2.1
13	1,910	Replanted spruce	66	17	20	24	5	0	24	0	2.44
14	1,930	Replanted spruce and northern Chinese larch	74	15	29	24	6	1	35	0	2.92
15	1,950	Secondary mixed forest	63	16	15	25	7	8	22	1.99	2.64
16	2,050	Secondary mixed forest	54	8	17	28	1	2	15	0.69	2.36
17	2,150	Secondary mixed forest	42	8	8	25	1	4	6	1.21	1.62
18	2,250	China paper birch, *Fargesia*	35	4	13	15	3	3	8	1.1	1.52
19	2,350	Fir, *Fargesia*	38	6	9	19	4	3	8	1.01	1.94
20	2,398	Primary China paper birch, *Fargesia*	34	3	11	20	0	1	5	0	1.36

[a] Number of species in tree layer.
[b] Number of species in shrub layer.

Biodiversity of the Central Qinling Mountains   299

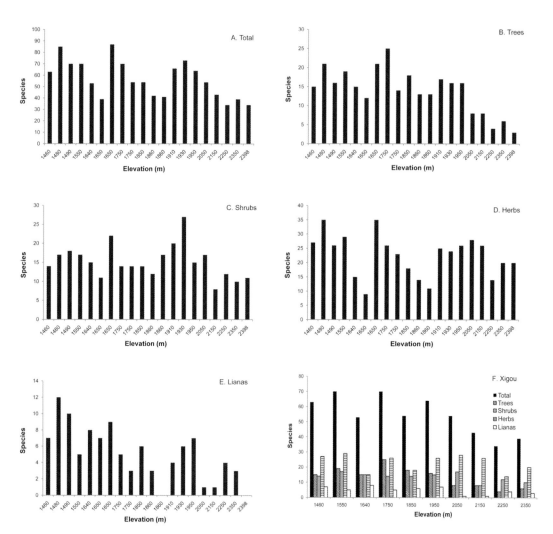

FIGURE 12.1. Plant species richness as a function of elevation for the entire study area: (a) total, (b) trees, (c) shrubs, (d) herbs, and (e) liana. (f) Plant species richness as a function of elevation for only the plots in the Xigou area.

declined with increasing elevation; below 1,930 m species richness fluctuated markedly. In Xigou, we noted a similar fluctuation of species richness with elevation below 1,930 m, but it was more muted (Figure 12.1). With 1,950 m elevation as a categorical distinction, analysis of variance indicated a significant difference ($P < 0.05$) between plots above and below this level among naturally regenerated stands. Similar results were found in analyses of just the tree and liana components.

Shannon-Wiener Index

We present Shannon-Wiener diversity indices H' among plots for the tree and shrub layers (Table 12.1), and these indices are displayed by plot

300   Chapter 12

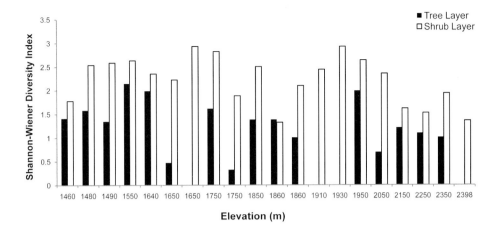

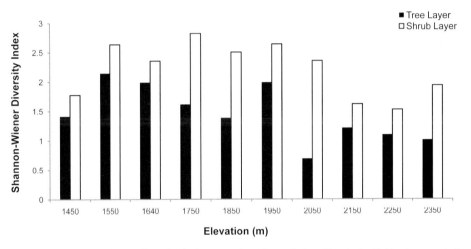

FIGURE 12.2. Shannon-Wiener diversity index as a function of elevation in both the tree and shrub layers (top) in the whole study area and (bottom) in only the Xigou area.

elevation (Figure 12.2, top) and for only the Xigou area (Figure 12.2, bottom). Again taking 1950 m as a dividing line, tree diversity among naturally regenerating stands was higher in lower-elevation plots than higher-elevation plots ($P < 0.05$). Diversity indices among shrubs did not differ with this elevational boundary. Diversity among trees was significantly higher among naturally regenerating plots than in artificially planted plots. No difference in diversity was evident among shrubs.

Characteristics of Plant Communities

PLANT SPECIES RICHNESS

We identified 440 species of seed-bearing plants, representing 261 genera and 97 families within the elevation range 1,100–3,070 m. These included 100 tree species (22.7%), 116 shrubs (26.4%), 180 herbs (40.9%), and 44 lianas (10.0%). These species were all commonly found in giant panda habitats of the Qinling and were representative and characteristic of these habitats.

According to Zhang (1992), the Qinling contains some 3,451 species of seed-bearing plants, which represent 1,005 genera and 175 families. This richness represents some 14.1%, 33.7%, and 58.1% of the corresponding number within all of China, despite the Qinling occupying only about 1% of China's area. Earlier estimates had put the number of species on the southern slope of the Qinling at as many as 2,500–3,000 (Department of Geography, Shaanxi Normal University, 1965; Nie, 1981). A preliminary survey of trees within just Yang County estimated some 321 species distributed in 152 genera from 72 families. These 152 genera constituted some 16% of the total of 959 trees species recorded from China. This is also 80% of the number of trees found in Sichuan's Emei Shan, an area well known for its high diversity.

Tropical plants constituted a substantial portion of those in the study area (46 genera, 17.6%), as well as those in the Qinling generally (274 species, 29.9%). However, some of these genera (e.g., *Ilex, Celastrus*) contained few species and played a minor role in forming plant communities. This indicates that the Qinling is not a center of tropical flora but is located at the northern periphery of the subtropical zone.

PLANT COMMUNITIES AND SUCCESSIONAL STAGES

Qinling floral community types are varied and diverse. According to the classification system in "Chinese Vegetation", natural vegetation includes conifer forest, broadleaf forest, shrublands, meadows, and aquatic vegetation. Coniferous forests include stands dominated by *Larix chinensis, Abies fargesii, A. chensiensis, Pinus armandii, P. massoniana, P. tabuliformis, Tsuga chinensis,* and *Cunninghamia lanceolata*, as well as the junipers *Platycladus orientalis, Sabina* spp., and *Juniperus* spp. Broadleaf forests include two broadleaf evergreen types (*Quercus acrofonta* and *Q. oxyphlila*) and deciduous forests of *Quercus, Betula,* and *Acer.* Overdevelopment has resulted in many foothills becoming dominated by shrubs. In addition to growing beneath the canopy in mature forests, bamboos are distributed continuously near mountain ridges. Alpine shrublands and meadows are located above the tree line (Figure 12.3; Fang and Gao, 1963; Pan et al., 1988).

Vegetation types near our research center included forests dominated by *Larix, Abies, Pinus armandii, P. massoniana, P. tabuliformis, Quercus,* and *Betula.* The southern slopes of the Qinling are home to 11 species of bamboo representing 5 genera. Of the 11, 5 species (*Bashania fargesii, Fargesia qinlingensis, F. dracocephala, F. decurvata,* and *Phyllostachys nigra* var. *henonis*) are eaten by giant pandas. The latter species has been cultivated and is found primarily around areas currently or formerly inhabited by people and thus has a

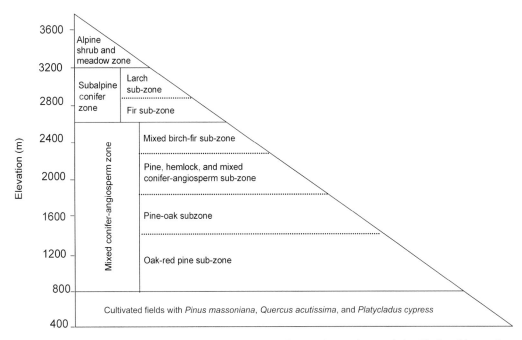

Elevation (m) — y-axis values: 3600, 3200, 2800, 2400, 2000, 1600, 1200, 800, 400

Alpine shrub and meadow zone

Subalpine conifer zone — Larch sub-zone; Fir sub-zone

Mixed conifer-angiosperm zone — Mixed birch-fir sub-zone; Pine, hemlock, and mixed conifer-angiosperm sub-zone; Pine-oak subzone; Oak-red pine sub-zone

Cultivated fields with *Pinus massoniana*, *Quercus acutissima*, and *Platycladus cypress*

FIGURE 12.3. Vertical zonation of vegetation types on the southern slopes of the Qinling Mountains.

restricted distribution (we know of only two such places within our study area). Recent taxonomic work has recognized three *Fargesia* species on the Qinling southern slopes; however, because these results were not available during most of our field work (and because the species are similar), we did not differentiate or map species-level distribution and refer to all simply as *Fargesia*. These bamboos form the primary summer diet for pandas at elevations of 1,700 to 3,000 m. *Bashania* is primarily located at elevations of 1,000–1,900 m. Within this elevational range, they form the main winter diet for pandas at 1,400–1,800 m.

Our preliminary analysis of vegetation conditions based on remotely sensed data was centered around Landianzi ridge. This study area of 900 km² (30 × 30 km) extended from a line connecting Huayang village and Jiuchiba in the south to a line connecting the Xushui River and the former county town in the north. We used bands 3, 4, and 5 from a Landsat Thematic Mapper (TM) image obtained on August 15, 1994, and produced a composite image using IDRISI. We conducted ground investigations on a portion of the study area and classified images on the basis of our training data. We used coalescence and stretching methods to clarify vegetation types and to minimize the influence of topographic shading.

However, the resulting composite image still contained some shadows, as well as overlap of spectral frequencies of vegetation types we knew were distinct. For example, the dry lands surrounding Huayang Village appeared identical in color to immature, reforested stands on midmontane slopes. Similarly, irrigated rice fields, subalpine meadows, and dry lands surrounding the old county town were indistinguishable. Thus, our final results were somewhat imprecise. However, we were still able to use the imagery, supplemented by ground surveys, to develop a preliminary classification of vegetation types in the study area, as well as to map their distribution (Plate 2). By integrating these three factors, we ultimately developed nine natural vegetation types as well as three reforested vegetation types, and two additional types accounted for developed areas (see Table 2.3): (1) subalpine shrublands (including *Fargesia*), distributed above 2,900 m to the tops of the mountains, (2) *Larix chinensis* stands (including *Fargesia*), distributed on midmontane slopes, primarily on northerly aspects, above 2,600 m, (3) *Abies* stands (including *Fargesia*), distributed in restricted areas below *Larix* stands, at 2,500–2,700 m, (4) *Betula* stands (including *Fargesia*), consisting of both *Betula albosinensis* and its variant, *B. a.* var. *septentrionalis,* distributed at 2,200 to 2,500 m, (5) mixed conifer-broadleaf stands, mostly consisting of mixed *Pinus-Betula* forest on midmontane slopes and mixed *Pinus-Quercus* forest on lower-elevation slopes, with understories including either *Fargesia* or *Bashania*, (6) *Quercus* stands (including *Bashania*), distributed on the lower portions of montane slopes, (7) low-elevation shrub savanna, distributed on southerly aspects, consisting primarily of the initiation stage of *Quercus* and associated shrubs, (8), riparian mixed stands, along creek and river courses, consisting primarily of broadleaf trees, (9) subalpine meadows, including both naturally formed meadows and secondary meadows created by timber harvest, (10) reforested conifer stands, which at higher elevations were primarily *Larix kaempferi, L. principis-rupprechti, Abies,* and *Picea* and at lower elevations were primarily *Pinus tabuliformis*, (11) immature reforested stands, planting subsequent to clear-cutting, (12) low-elevation grasslands, distributed within valleys, (13) two types of agricultural fields (dry land and rice paddies), and (14) towns and villages.

DISCUSSION

Species used in artificial reforestation programs in the study area included a native species (*Pinus tabuliformis*) and also three exotics (*Larix kaempferi, L. principis-rupprechti,* and *Picea asperata*). The two larch species (particularly *L. kaempferi*) were planted widely in clear-cut areas in the region, primarily because of their rapid growth, which allows a shorter rotation length and greater economic efficiency.

On the basis of our quantification of α and β diversity comparing artificial with natural regeneration, we conclude that the effects on biodiversity depend on the species planted, as well as the season, method, and postplanting management of reforested areas. The effects of artificial reforestation on biodiversity evidently vary with the time since planting. Over a short time scale, we characterize the pattern of diversity through time as low to high to low to high. Soon after replanting, diversity is low because of the effects of artificial cultivation. After artificial cultivation has stopped, however, species present in the form of seeds are free to develop, and seeds are also introduced through the agency of wildlife or other natural factors. Each species quickly takes up its niche within the developing system, but as none dominates, diversity remains high. As the canopy develops and closes, reducing solar radiation, competition among species below the canopy has the effect of decreasing their density, resulting in a tendency toward declining species richness. However, with encroachment of species from the periphery of the stand toward the stand interior, diversity within the forest again increases. Unfortunately, because we lacked sufficient replication of plots as well as plots within undisturbed forests that could serve as a control, our results are primarily descriptive.

Timber harvest can be classified as either clear-cut or selectively cut. As most areas subjected to timber harvest in our study area were clear-cut and then replanted, we lack samples allowed to regenerate naturally. Fortunately, we located an area in Shuidonggou (at 1,560 m elevation) where planned artificial regeneration never took place, which allowed us to qualitatively supplement our information. This area had been harvested (and shrubs removed) in 1989 and 1990; the aspect was westerly, the slope was 30°, and it was located near a small creek and about 100 m from a main road. We were unable to conduct a detailed, quantitative survey here, but from our look at the physiognomy and species composition of the stand, it was clearly different from selectively cut areas that had been allowed to regenerate naturally. When we visited it in 1997, the site was primarily occupied by shrubs and lianas, with just a few live *Anacardium occidentale* poking up among the stumps. The shrub layer had a canopy cover of 100%, primarily of *Pterocarya hupehensis, Meliosma cuneifolia*, and *Rubus corchorifolius.* Canopy cover of lianas was also high, primarily *Pueraria phaseoloides*, as well as *Clematoclethra* spp., *Paederia scandens*, and the grasslike lianas *Codonopsis pilosula* and *Cucubalus baccifer. Bashania* was present but growing poorly, covered in lianas. Although our sample size was limited, we provisionally conclude that diversity in naturally regenerating stands subjected to selective cutting was higher than in stands never subjected to harvest.

RELATIONSHIPS BETWEEN VEGETATION DIVERSITY IN HABITAT AND PANDA ACTIVITY

Our results, considering α and β diversity, among other factors, suggest that a zone of transition existed around 1,950 m elevation: above this line, species richness declined, whereas below it species richness fluctuated considerably. Diversity indices, whether viewed in terms of all species, trees species only, or liana species, differed on either side of this elevation line.

Among plots in Xigou, all plots around the 1,950 m elevation area were similar to one another, although similarity decreased markedly when comparing neighboring plots.

Coincidentally, our years of studying giant pandas revealed that it was just this elevation zone (about 1,900–2,000 m) that the animals often selected for denning and rearing cubs. This elevation zone was the one in which Jiaojiao denned and raised all five of her young; although she used different dens, all were located at similar elevations. Radio-collared females Momo and Ruixue also raised their cubs in similar areas. Was high plant diversity related to the presence of this transition zone? High diversity tends to reflect favorable habitat conditions. Habitat selection by pandas for denning and cub-rearing areas has evolved over a long time period and reflects their own assessment of habitat quality. This remains a topic for further research.

Diversity and Status of Butterflies

SUMMARY

We documented 191 species of butterflies within giant panda nature reserves on the southern slopes of the Qinling Mountains, Yang County, Shaanxi Province, and analyzed the elevational distribution and ecozone membership of the 179 species. Butterflies represented both Oriental and Palaearctic ecozones, and they were found intermingled; their distributions were related to specific aspects of Qinling geography. Species of Oriental origin tended to predominate our survey. Because the habitat conditions in the montane temperate conifer and mixed broadleaf-conifer (800–2,600 m) forests were conducive and human disturbance in these forests was minor, butterfly richness and diversity was high.

RATIONALE FOR SELECTING BUTTERFLIES AS A RESEARCH SUBJECT

Compared with vertebrates, insects are characterized by small body size, high population abundance, short life cycle, high reproductive potential, and intimate reliance on environmental conditions. The distribution of insects is also highly influenced by habitat conditions, and thus, they can serve as good representatives for all invertebrates. With the exception of a few strong fliers who migrate long distances, most insects have little ability to move far from their place of birth, particularly those lacking wings, who spend their entire lives within rather narrow ranges. At the same time, many herbivorous insects specialize on a single host, limiting themselves to just a few plant genera or even species. In turn, these plant hosts are themselves restricted to specific habitat conditions, and their range can be quite narrow. Thus, the geographic distribution of these herbivorous insects closely overlaps that of their plant hosts. Studying the distribution of butterflies can elucidate underlying environmental conditions. Moreover, insect abundance responds quite strongly and rapidly to environmental changes, further facilitating the usefulness of their study.

METHODS

We conducted surveys of the butterfly fauna of the southern slopes of the Qinling on four separate occasions over an 8-year period: July 23 to August 4, 1990; July 24 to August 5, 1993; August 4–17, 1996; and July 24 to August 9, 1998. We collected specimens along a 45-km transect line, which sampled an area of approximately 650 km^2 (of which approximately 100 km^2 was considered the core area). During the same time period, we also obtained approximately 2,000 butterfly specimens from other collectors working in the area.

BUTTERFLY DIVERSITY AND STATUS

The butterflies collected comprised 191 species from 96 genera and 9 families. Despite our relatively restricted collecting season, our documented species represented over one-seventh of the total number recorded within China (i.e., 1,223 species according to Zhou, 1994). We collected three species that had not earlier been documented as occurring in China: *Aphantopus arvensis, Ochlodes lanta,* and *Sovia lucovsii.*

Characteristic species in our study area included many in the genus *Lethe* (13 species plus 1 unnamed species), of which 9 were Oriental, 2 Palaearctic, and 2 cosmopolitan species. Of these 13 species, 7 had not previously been documented in Shaanxi. Although *Lethe* was well represented by species, the number of individuals was relatively low. Species in the genus *Lethe* depend on bamboo and are partial to shady areas. Bamboos are common in the study area and tend to grow in moist, shady areas within the forest. The genus *Ypthima*, whose members also depend on grass family plants, was also well represented.

Common species in the study area included *Papilio bianor, P. xuthus, Eurema hecabe, Pieris napi, Ypthima balda, Argynnis paphia, Limenitis homeyeri, Neptis sappho, N. rivubaris, N. pryeri, Polygonia caureum, Pseudozizeeria maha, Celastrina argiola,* and *Parnara guttata.*

VERTICAL ZONATION OF DIVERSITY

The Qinling is characterized by marked vertical zonation in both geography and vegetation types; the distribution of butterflies in the area follows this zonation. We have categorized the southern slope of the Qinling Mountains into five zones on the basis of vegetation elevation and cover type of the primary vegetation: (1) northern subtropical broadleaf evergreen forest zone (400–800 m), (2) temperate deciduous broadleaf forest zone (800–1,350 m), (3) montane mixed conifer-broadleaf zone (1,350–2,600 m),

Table 12.2.

BUTTERFLY SPECIES FOUND IN THE SOUTHERN SLOPES OF THE QINLING BY ELEVATION ZONE AND ECOZONE. ZONE 1, NORTHERN SUBTROPICAL MIXED EVERGREEN-DECIDUOUS FOREST (400–800 M); ZONE 2, TEMPERATE DECIDUOUS BROADLEAF FOREST (800–1,350 M); ZONE 3, MONTANE MIXED CONIFER-BROADLEAF FOREST (1,350–2,600 M); ZONE 4, SUBALPINE CONIFER FOREST (2,600–3,400 M).

Zone	Palaearctic		Oriental		Cosmopolitan		Total	Percentage of all zones
	Species	Percentage	Species	Percentage	Species	Percentage		
Zone 1	6	10.0	22	36.7	31	51.7	60	31.4
Zone 2	21	19.1	43	39.1	44	40.0	110	57.6
Zone 3	33	23.7	56	40.3	44	31.7	139	72.8
Zone 4	6	31.6	5	26.3	8	42.1	19	9.9
All zones	40	20.9	78	40.8	61	31.9	191	—

(4) subalpine coniferous forest zone (2,600–3,400 m), and (5) alpine shrub and meadow zone (>3,400 m). We collected specimens from as far south as the Changqing forestry area (at 500 m) to as far north as Huorenping (at 3,071 m), which included four of the above-listed five zones (Table 12.2). When we exclude the cosmopolitan species, Oriental species are more abundant in the two lower zones, and Palaearctic species are more abundant in the upper two zones.

Butterfly diversity was considerably higher in the montane zones (zones 2 and 3) than in lower (zone 1) and higher (zone 4) elevations. In the lower-elevation zone, human disturbance had dramatically altered the original vegetation, and urbanization and agricultural development had also led to a reduction in vegetation species richness, in places to a monoculture. The disappearance of specific host plants required by various species of butterflies led naturally to a marked reduction in butterfly species. Those butterflies that we found in this zone were chiefly associated with crops, such as *Pieris rapae, Papilio xuthus,* and *P. polytes.* The primary reason for the low number of species in the alpine zone was low temperature, making it difficult for any poikilotherm to survive. High winds in the alpine zone were an additional challenge for winged insects.

Dominant species in the northern subtropical broadleaf evergreen forest zone (400–800 m) were *Papilio protenor, P. bianor, P. xuthus, P. machaon, Eurema hecabe, Pieris rapae, Polygonia c-aureum,* and *Pseudozizeeria maha.* We documented no endemics in this zone; all species we found here were also present in the temperate deciduous broadleaf forest zone (zone 2). In theory, vegetation in this area would consist of evergreen-broadleaf forest. However, because the Qinling constitutes a north–south barrier, the climate here, unlike that in Guanzhong, is more similar to that in Sichuan, being moist and rainy. Crops grown in the Qinling are mostly of southern origin, such as tung oil trees, citrus trees, tea, and rice. The wild fauna is primarily of Oriental ecozone origin. Moreover, because this zone is low in elevation, it is more similar in climate to higher-elevation zones at lower latitudes and thus bears more similarity to subtropical regions such as the Sichuan Basin. Normally, one would expect the percentage of Oriental species to be higher in this zone than at higher-elevation zones, which we did not find. Cosmopolitan species constituted a larger portion in this zone than in others. We speculate that the reason so many cosmopolitan species were found in this low-elevation zone is human disturbance. From the perspective of zoogeographic classification, the study area is located close to the Hanjiang River Valley and is part of the western mountain plateau subregion of the central China region. Having been cultivated and used for fuel wood for many years, the primary vegetation and associated fauna are relegated to the foothill zone. In contrast, river and basin areas have been left with only small remnants of original vegetation, and fauna and flora existing there are primarily those created or allowed to exist by humans.

Butterflies living at low elevations mostly have close associations with agriculture. Farming has a very long history in the Hanzhong Basin, perhaps going back as far as about 5000 BCE. Eventually, these agriculturally adapted species came to occupy a large proportion of the local butterfly fauna. Examples of such species we collected in the study area include *Papilio polytes*, which uses pepper plants for its host, *Papilio xuthus,* whose hosts are citrus trees, and *Papilio machaon*, considered a pest of vegetables and medicinal herbs in the Umbelliferae family, as well as some members of the genus *Pieris*, which eat crops of the mustard family such as cabbage.

We also noted that cosmopolitan species were particularly well represented among collections we made near roads where cultivated fields were common but native vegetation was less common. This further supports the notion that human activity was capable of influencing the type of insect fauna present. Finally, because our study area was located on the southern fringe of the Qinling, some Oriental species adapted to low elevations had evidently been able to find ways to persist within these man-altered environments. Examples include two species in the genus *Mycalesis* as well as *Troides aeacus.*

The temperate deciduous broadleaf forest and montane mixed conifer-broadleaf zones (800–2,600 m) are the most distinctive and characteristic regions of the southern slopes of the Qinling. Dominant species include *Papilio bianor, Eurema hecabe, Pieris napi, Ypthima balda, Argynnis paphia, Limenitis homeyeri, Neptis sappho, Neptis rivularis, Celastrina argiola,* and *Parnara guttata*. We documented 72 endemic species in this zone (of which 13 were from the temperate deciduous broadleaf forest zone), including *Ninguta schrenkii, Euthalia thibetana,* and *Vanessa indica.* Particularly notable was that species in the genus *Lethe* were mainly found in this zone. Species of Oriental origin tended to dominate, comprising 39.1% and 40.3% of these two zones (79%

and 72% excluding cosmopolitan species). These two zones occupy most of the Qinling and are primarily covered with northern Chinese temperate vegetation. The main deciduous broadleaf trees are willows, oaks, maples, and birches, and the main conifers are pines, firs, and spruces. Broadleaf evergreen trees are also present. Because the study area was located close to the southern boundary of Shaanxi's deciduous broadleaf forest zone, most plant species represented transitional types; vegetation showed broad similarity to that of the southern portion of the northern China flora. However, because the area is largely unaffected by the cold winter winds blowing from the north and also unaffected by winter and spring droughts, the flora also exhibits differences from that of northern China. Some secondary trees, shrubs, and herbs are more typically found in southern or western China, and lianas, found in valleys of the study area, are not northern Chinese species at all. Both of these zones also share the characteristic of being relatively uninfluenced by human activity.

Because of constraints on our time and ability to collect in the subalpine coniferous forest zone (2,600–3,200 m), we collected relatively few specimens, representing only 19 species. Although we suspect that our sampling did not fully represent the community, we are confident that the paucity of species found reflects reality. We found no obvious dominant species in this zone, although characteristic species included the high-elevation specialists *Erebia alcmena* and *Lethe helle*. According to a report by researcher Liu Huaxun, vegetation displays a marked response to latitude because it is influenced by the wet Pacific and Indian monsoons as well as the expected gradual decrease in temperature from south to north. The basis of vertical zonation is identical to that of latitudinal zonation, with higher elevations very much like traveling north in latitude (Wang and Jiang, 1988).

THE POSITION OF THE QINLING BUTTERFLY FAUNA AMONG CHINA'S BUTTERFLIES

Considering the similarity of the butterfly fauna in the study area to that in other areas within China (Table 12.3), we see that the areas with the highest similarity were Wudang Shan (49.8%) and Shennongjia (45.6%). Both areas are located quite close to the southern slopes of the Qinling, and both are included within the central China region according to Chinese entomological classification. The Qinling is considered to represent the boundary between this region and the adjacent northern China region (Zheng et al., 1959). The butterfly fauna of Taibai Shan, located on the northern slope of the Qinling, had a similarity of only 39.9% with that in the study area. That these two faunas, located geographically so close to one another, were more divergent than the study area's butterfly fauna were with the somewhat more distant areas of Wudang Shan and Shennongjia further confirms the close relationship of the study area butterflies with those from the central

Table 12.3.

SIMILARITY (SPECIES IN COMMON) OF BUTTERFLIES IN THE SOUTHERN SLOPE OF QINLING TO OTHER REGIONS WITHIN CHINA.

	Palaearctic		Oriental		Cosmopolitan		Overall	
Location	Species	Percentage	Species	Percentage	Species	Percentage	Total	Percentage[a]
Western Tian Shan	5	45.5	0	0	6	54.5	11	9.6
Mao'er Shan	21	38.8	3	5.5	30	55.6	54	38.3
Taihang Shan	9	32.1	0	0	19	67.9	28	23.6
Tai Shan	7	20.6	1	2.9	26	76.5	34	28.3
Taibai Shan	23	39.7	5	8.6	30	51.7	58	39.9
Wudang Shan	15	20.8	14	19.4	43	59.7	72	49.8
Shennongjia	15	21.4	21	30.0	34	48.6	70	45.6
Cang Shan	4	8.9	15	33.3	26	57.8	45	24.5

[a] Twice the number of species in common divided by the sum of species from the southern slope of Qinling and species in the region being compared.

China region, although the study area's butterfly fauna also retains some similarities to that of the north China region.

What Do Butterflies Teach Us about the Value of Protecting Panda Habitat?

Butterflies represent the richness of all insect species and have a great value worth protecting for their own sake: (1) The existence of multiple subspecies and ecotypes reflects the area's habitat diversity and complexity, which facilitates providing resources for a large array of species. (2) The discontinuous distribution of some butterfly species reflects the fact that the area contains some subtropical characteristics as well as heterogeneity of habitats. (3) The facts that the area's butterfly fauna is transitional among various ecozones and that it features marked verticality make this area an ideal location for ecological research. (4) Wildlife living in the Qinling's southern slopes has already been greatly affected by mankind's activity, and their distributions have been greatly altered from the original ones, making the area a true "refuge." As such, it provides an important place to study wildlife conservation. (5) The study of butterfly distribution has shown that the midmontane region (at about 800 to 2,600 m) has the best habitats, the most favorable climate, and the least human disturbance. Unsurprisingly, this zone features the highest species richness and diversity, and it is in part because of this that it also functions as the last redoubt of the giant panda. Although climatic conditions in lower-elevation zones (below 800 m) are also ideal, the magnitude of human disturbance has simply been too high to allow large-scale flourishing of wildlife. Conversely, climatic conditions above 2,600 m are harsh. Thus, despite the fact that human influence here is low, biodiversity is also low, and habitat conditions for wildlife are less than ideal.

Avian Diversity

WITH QIN DAGONG

SUMMARY

From 1984 to 1999, we recorded 242 bird species, representing 16 orders, 44 families, and 140 genera, on the southern slope of the Qinling Mountains, from as low as the Han River Valley (450 m elevation) to the top of Xinglongling Peak (3,071 m). On the basis of genera, we noted characteristic patterns formed by elevational zonation, with relatively few avian species found within cold temperate alpine zones >2,700 m. The number of cosmopolitan species was negatively associated with elevation. Palaearctic species seemed to prefer the montane temperate zone between 1,350 and 2,600 m. According to taxonomic composition, the study area's birds represent 60% of the world's bird orders and 76% of China's bird orders. The total number of families represented is over one-fourth of the world's bird families and one-half of China's. The number of species identified comprises 2.6% of the world's bird species and 19.5% of China's bird species. During our study, we recorded only 18 fewer bird species than recorded in the 1950s and 1960s. This suggests that the southern slopes of the Qinling still retain suitable habitats for the survival of bird species. However, obvious changes have been observed in summer migrants, winter migrants, and passing migrants. Deforestation and pesticide overuse are the main threats at present to the diversity of bird species on the southern slopes of the Qinling Mountains.

Because birds occupy a special place within natural ecosystems and because they are generally diurnal and can therefore be relatively easily observed, we made observations and documented avian diversity continuously on the southern slopes of the Qinling beginning in November 1984 from our base at the research station. During uninterrupted fieldwork from 1984 to 1999, we took advantage of intervals in our research of the giant panda to accumulate data on the diversity of bird species. This work has important reference value for our understanding of the biodiversity conditions on the southern slopes of the middle Qinling Mountains in the latter half of the 20th century.

METHODS OF DOCUMENTING SPECIES

During the period November 1984 through the end of 1999, we conducted long-term field work based at our research camp in the southern slopes of the Qinling on 39 separate occasions. In collecting information on the distribution of bird species, we used a variety of field methods, which can generally be summarized as follows:

- From 1992 to 1995, during clear mornings in winter, spring, and summer, we used the 1-hour period from approximately 0700 to 0800 hours to observe and record migrant and resident birds along a forest road at elevations of 1,430 to 1,650 m along the east bank of the Cha'eryan River, a tributary of the upper Youshui River.
- We kept long-term records of bird species observed around our field stations at the following time periods and locations: November 1984 to May 1985 at Sanguanmiao, Foping Nature Reserve (1,500 m), and March 1986 to April 1999 at Huatanwan (500 m), followed by Huayangzhen (1,100 m), Shanshuping (1,500 m), Baioyangping (1,750 m), and, finally, Daping (2,100 m), all in the Changqing National Nature Reserve.
- From March 1986 to March 1995, we observed and recorded winter migrants on approximately 20 occasions along the Han River, the north bank of the Han River, the Huatanwan Reservoir, and the surrounding hills from a base at the former Changqing Forestry Bureau, using binoculars, cameras equipped with telescopic lenses, and video recorders.
- At least twice each year from November 1984 to April 1999, we drove a jeep along the Zhou-Cheng Highway from the northern slopes of the Qinling Mountains over the main ridge at 2,100 m to the southern slopes, descending along the banks of the Jiaoxi River, crossing over Tudiling, and entering the Jinshui River Valley and continuing our descent. We then crossed the watershed between the Jinshui and Youshui Rivers and entered the central Han River Basin from its northeast corner, followed the foot of the mountain to the west 15 km, and then turned onto the road from Yangxian to Huayang. We drove along the Youshui River Valley from its mouth at 500 m in elevation to our research station at Shanshuping at 1,500 m elevation, a total distance of 72 km. During these trips, we made efforts to observe all the bird species along the route.
- During the summers of 1987 and 1990 as well as from winter 1994 through spring 1995, we captured and marked birds using mist nets at Shanshuping and Daping.

The data we obtained were clearly insufficiently systematic and precise to obtain a complete understanding and analysis of the distribution of birds within the giant panda's habitat on the southern slopes of the Qinling Mountains. However, given the existence of work conducted prior to our surveys (Zheng et al., 1973) and the fact that following our panda research, it would likely be difficult for others to conduct such long-term and continuous research on the bird species in this area, we present our findings. We

Table 12.4.

SUMMARY STATISTICS OF BIRD SPECIES ON THE SOUTHERN SLOPE OF THE QINLING BY ELEVATIONAL ZONE AND ECOZONE.

Elevational zone	Climate zone and landscape	Palaearctic		Oriental		Cosmopolitan		Total number of species	Overall percentage
		Number of species	Percentage	Number of species	Percentage	Number of species	Percentage		
Foothills (450–800 m)	Northern extent of northern subtropical zone, valley bottoms, foothills, cultivated fields, sparse forests	19	41.30	62	51.67	54	78.26	135	57.45
Lower montane (800–1,350 m)	Temperate band and agricultural zone of montane zone, mixed conifer-angiosperm forests increase with elevation	29	63.04	93	77.50	48	69.56	170	72.34
Midmontane (1,350–2,700 m)	Montane temperate zone, forest production zone, secondary mixed forests	33	71.73	90	75.00	43	62.31	166	70.63
Subalpine (2,700–3,071)	Subalpine cool temperate zone, steep cliffs with pine forests on ridges and peaks	20	43.47	38	31.67	17	24.63	75	31.91
All zones		46	100	120	100	69	100	235	100

314   Chapter 12

provide these bird species lists as a reference in appraising the biodiversity conditions of the southern slopes of the Qinling Mountains over the last 20 years of the 20th century.

BIRD DIVERSITY

We documented 242 species of birds, representing 16 orders, 44 families, and 140 genera, in the area from the Han River Valley at 450 m to Huorenping, the main peak of Xinglong Mountain at 3,071 m. We classified species as either winter migrants, summer migrants, or resident/temporary visitors on the basis of their migratory status in the study area. Because the Qinling functions as a dividing line between China's northern and southern climates and because the study area encompasses a considerable elevational range, geographic characteristics influence the current distribution pattern of birds. To illustrate relationships between the current distribution pattern of each bird species and its evolutionary history, we also included the ecozone association of each bird species in all of our lists (e.g., Appendixes 12.1 through 12.5 in the Chinese edition). We also categorized vertical distributions of birds into four elevational zones: foothills (450–800 m), lower montane (800–1,350 m), midmontane (1,350–2,700 m), and subalpine (2,700–3,071 m). Because of the large geographic and climatic differences between the northern and southern slopes of the Qinling Mountains, the elevation range for each of our ecozones differs from those of Zheng et al. (1973).

ELEVATIONAL DISTRIBUTION OF AVIAN SPECIES

We observed an elevational distribution of birds on the southern slopes of the Qinling (Table 12.4). We infer that alpine areas >2,700 m are part of the cold temperate montane zone as bird species documented there were relatively few; cosmopolitan species declined with increasing elevation. The upper elevations of the midmontane zone (1,350 to 2,600 m), where a temperate montane climate prevailed, provided the most ideal habitats for Palaearctic species. In general, species richness declined with increasing elevation.

The lower elevations of the midmontane zone and the upper elevations of the foothill zones (800 to 1,350 m), with a warm temperate climate, provided the most ideal habitats for Oriental species. A transition zone between Palaearctic and Oriental bird faunas existed in the broad midmontane zones, located at roughly the boundary between the warm temperate montane and temperate montane zones at 1,350 m.

HIGH DIVERSITY OF BIRD SPECIES

The following bird species merit particular attention: Species endemic to China are Temminck's tragopan (*Tragopan temminckii*) and Reeves's pheasant (*Syrmaticus reevesii*). Species included on China's list of protected species are as follows: crested ibis (*Nipponia nippon*), scaly-sided merganser (*Mergus squamatus*), golden eagle (*Aquila chrysaetos*), black kite (*Milvus migrans*), Chinese sparrow hawk (*Accipiter soloensis*), goshawk (*Accipiter gentilis*), Eurasian sparrow hawk (*Accipiter nisus*), besra (*Accipiter virgatus*), upland buzzard (*Buteo hemilasius*), rough-legged buzzard (*Buteo lagopus*), cinereous vulture (*Aegypius monachus*), hen harrier (*Circus cyaneus*), saker falcon (*Falco cherrug*), Eurasian hobby (*Falco subbuteo*), red-footed falcon (*Falco vespertinus*), common kestrel (*Falco tinnunculus*), Temminck's tragopan (*Tragopan temminckii*), and Koklass pheasant (*Pucrasia macrolopha*). Species with unique evolutionary characteristics include the forest wagtail (*Dendronanthus indicus*), water cock (*Gallicrex cinerea*), and common hoopoe (*Upupa epops*). Several genera are of special interest regarding species divergence: 4 of 9 redstart species and 7 of 33 laughing thrush species found in China can be found in the Youshui River Valley. Considering species with particular economic value, 9 species of pheasant and 11 species of waterfowl were earlier valued for economic reasons but are now protected species. The following species elicit interest among bird watchers: Eurasian skylark (*Alauda arvensis*), red-billed leiothrix (*Leiothrix lutea*), and mynas (*Acridotheres* spp.). Birds used in traditional Chinese medicine (Collaborating Group of "Medical Zoography of China," 1979, 1983) include the Eurasian cuckoo (*Cuculus canorus*), common magpie (*Pica pica*), and various species of robins and owls.

Changes in Avian Composition

Earlier work on bird diversity on the southern slopes of the Qinling and a broad area of the Daba Shan was conducted during 1956–1965 by Shaanxi Northwestern University, the Chinese Academy of Science's Institute of Zoology, and the former Sichuan Agricultural Institute (Zheng et al., 1973). Although our methods and survey intensity varied from the earlier work, we find it useful to compare our results with theirs in order to provide a comparison over the 30-year time frame. Although our surveys during the 1980s and 1990s were limited to only a single tributary of the upper Han River (the Youshui River Basin), whereas the earlier work covered 10 tributaries of the upper Han River (Figure 12.4), we found only 26 fewer species (10% of total) than during the 1950s and 1960s. Specifically, we found 18 fewer residents (13.5% of the originally documented species), 9 fewer summer migrants (14.5%), and 13 fewer temporary visitors (25.5%). However, we documented 20 more winter migrants than during the earlier period, an increase of 2.2 times.

These data suggest the following: (1) Habitats on the southern slopes of the central Qinling remain generally suitable for birds; thus, the total number of species (particularly residents) has changed only marginally. (2) Changes in the number of summer and winter residents over the past 30 years have been somewhat larger. Considering both the decrease in summer

Legend

▲ Peak • Township — River ☐ Province

0 25 50 100 km

FIGURE 12.4. Survey regions, Qinling Mountains, 1956–1965.

migrants and the increase in winter migrants, a total of 30 species (42% of all migrants) displayed a difference. (3) The number of temporary visitors declined by 27.4%, suggesting declines in bird communities beyond the Qinling during the latter half of the 20th century.

Biodiversity Values

Biodiversity is often distinguished as having either instrumental (i.e., utilitarian) value or existence (i.e., inherent) value. Instrumental value involves appraising the advantages of biological resources by considering their benefits to people, e.g., the nutritional value of paddy rice. Existence value refers to the concept that living things have their own intrinsic value and do not exist to satisfy the needs and wants of man. In this section, to avoid immediately linking the concepts of biodiversity value to such materials aspects as wood, fiber, and medicine, we have deliberately placed our discussion of intrinsic value first and only later discuss biodiversity's instrumental value.

EXISTENCE VALUE

This issue involves considering the ethics of the relationships between mankind and other living beings. Until the 1980s, when considering the rationale for conserving plant and animal species lacking tangible economic value, most authors felt constrained to link their discussion to the economic benefits these species could produce for mankind. For example, in 1958 when Mao Zedong mobilized a mass campaign to exterminate sparrows because of their purported damage to agricultural crops, many species almost disappeared within just a few short days. When ornithologists later responded to this near disaster in the late 1970s, however, they defended protection of birds based on the argument that in feeding their young, sparrows actually benefitted agricultural protection by consuming insect pests.

This situation improved considerably beginning in the 1980s. Emerging ethics would hold that all wild fauna and flora, regardless of their economic value, or lack thereof, possess an inherent right to continued existence. The World Charter for Nature ratified by the General Assembly of the United Nations in October 1982 clearly pointed out that "mankind is a part of nature" and that "every form of life is unique. . .warranting respect regardless of its worth to man." This brief history illustrates how advances in an ethical framework toward nature have followed the development of human society.

These simple moral precepts existed even in ancient China. Over 2,000 years ago, ancient Chinese philosophers wrote that "it is forbidden to develop the mountain forests and river marshes at this time" and emphasized the concept of the "harmony between man and nature." The Bible records that in the age of Ezekiel and Isaiah, it had already been pointed out that plundering land was not only inadvisable but also simply wrong (Leopold, 1996). However, these progressive ideas were not actually adopted by societies during these time periods. It was not until the 1980s, with the earth's biodiversity facing a sixth major extinction wave in its evolutionary history (this one being caused almost entirely by humans), that people began suddenly awakening to the realization that they must respect the other species with which they share the earth and acknowledging that biodiversity, just as mankind, has a right to continued existence. With this, people have increasingly recognized that they must become responsible members of the ecological community rather than conquerors who desire the power to exploit the earth.

What, then, are the intrinsic biodiversity values of the southern slopes of the Qinling?

1. Biodiversity is a legacy bequeathed to us and our descendants by nature. Beginning with the Quaternary, the faunal system south of the Qinling became inhabited by both the giant panda and the stegodon. With the advent of the Holocene, these communities quickly declined, and their geographic distributions contracted markedly. Many members of this faunal system became extirpated, leaving only species in isolated areas to persist to the present time, including the southern slopes of the Qinling. Examples of species that have persisted here from early Pleistocene times include the Sichuan golden monkey, Rhesus macaque, Malayan porcupine, hog badger, masked palm civet, dhole, and muntjac; examples of species that persist today and represent the earlier giant panda–stegodon fauna of the mid-Pleistocene are tufted deer, forest musk deer, serow, larger Indian civet, bamboo rat , yellow-throated marten, and Edward's long-tailed giant rat (Pan et al., 1988).

318   Chapter 12

Monotypic genera have particular importance, are systematically unique, and display evolutionarily primitive characteristics. Globally, the presence or absence of such plants is an important criterion of whether a botanical area is ancient or not. The southern slopes of the central Qinling contain many monotypic genera, including *Decaisnea fargesii, Hemiptelea davidii, Kalopanax septemlobus, Pteroceltis tatarinowii, Poliothyrsis sinensis, Ceridiphyllum japonicum* var. *sinense, Tetrapanax papyrifera, Sinowilsonia henryi, Tetracentron sinenseis, Gingko biloba, Eucommia*, and other trees. Monotypic lianas include *Sinofranchetia chinensis* and herbaceous plants species *Kingdonia uniflora, Platycodon grandiflorum,* and *Tussilaga farafara* (Pan et al., 1988).

The Qinling also contains endemic plants, including *Larix chinensis, Paulownia shensiensis, Acer miotaiense,* and *Lonicera shensiensis*. Among birds and mammals, the southern slopes of the Qinling also harbor monotypic families, genera, and species, as well as species found only in China. In addition to the giant pandas, these unique mammals include the Chinese jumping mouse, Sichuan golden monkey, takin, and tufted deer. Unique birds include the golden pheasant, Elliot's pheasant, forest wagtail, water cock, and hoopoe.

We documented 440 seed-bearing plant species, 242 bird species, 93 species of mammal, and 200 butterfly species on the Qinling's southern slopes. These reflect species that were relatively easily documented, but there are no doubt others we were unable to collect. Each of these known or unknown species is an integral part of the area's life history; each species has a magnificent and irreplaceable evolutionary history. Today, they form part of our global natural heritage. Their continued existence is a rich heritage that nature has bequeathed to us and our descendants.

2. Biodiversity is a resource mankind can use to obtain new knowledge. Every species embodies a unique genetic system. The unique structure, physiological system, and behavior patterns of every species are capable of providing for mankind information about new food sources and processes of producing medicines. Lacking the graces bestowed upon us by these potential contributions of biodiversity, mankind faces a future trapped in a vicious cycle of increasing environmental deterioration and poverty as human populations increase.

Bamboo is a notable natural resource characterizing the southern slopes of the Qinling, and both the giant panda and the bamboo rat represent species that could not persist without it. Bamboo rats have digestive systems that are more fully developed, both in structure and function, than those of pandas. These digestive abilities suggest that bamboo rats may play host to symbiotic microorganisms. Given the magnitude of practical lessons regarding the process and mechanisms of digestion we have to learn from species such as giant pandas, bamboo rats, and herbivorous ungulates, who can say what new and practical knowledge may come in the future from studying

Biodiversity of the Central Qinling Mountains 319

hedgehogs, Asiatic black bears, tragopans, larches trees, or *Byasa impediens* butterflies?

3. Biodiversity maintains the stability of communities and ecosystems. The shrubs and forbs that form part of the Qinling's natural vegetation, if considered merely from the perspective of economic benefits to mankind, would seem to lack value, but they too are a part of the intact biological community. If they were arbitrarily eliminated or replaced, the stability of the Qinling's biological community would be put at risk.

Prior to 1970, a diverse array of aquatic plants found habitats within the clear waters of the Youshui River. Boulders in rivers were covered with mosses, and there was abundant water year-round. The river's currents were gentle, and banks on either side harbored a rich array of species. The selective forestry practiced by forestry operations in the early 1980s had little effect on the area's biodiversity; in fact, the moderate and spatially limited disturbances caused by these operations even increased species richness in some areas. However, with the advent of large-scale clear-cutting beginning in the late 1980s, reforestation was increasingly focused on producing rapidly growing monocultures of larch and similar species. Shrubs and forbs were manually cleared for the initial 5 to 7 years following replanting. Although this expedited rapidly growing, high-yield forests, it also had the effect of diminishing the natural biological community. This led to a serious loss of biodiversity as well as destruction of soil structure on mountain slopes, which resulted in erosion, reduced water flow, and even occasional instances of rivers running dry. With this change, the formerly clear waters became increasingly turbid.

This example illustrates the unintended results that can occur when people fail to adopt an ecological perspective in considering their problems and instead pursue merely their immediate economic gain. In this case, clearing forbs and shrubs that appeared to have no economic benefit (but in fact had important ecological function) had disastrous consequences. Empirical data showed that these plants, which appeared economically unimportant, in fact functioned to ensure the health of the area's soils and constituted an indispensable part of its biological community. The presence of these seemingly unimportant elements turned out to be crucial to the survival of and, indeed, to the continued ability to extract economic value from those elements that had more obvious and direct economic value.

Instrumental Value

In discussing the value of biodiversity, we must not, of course, forget that human welfare is an important point of departure. In a sense, the reason for people to protect biodiversity today is to be able to sustainably use it. We categorize instrumental values of biodiversity on the southern slopes of the Qinling Mountains into four aspects.

1. Ecosystem function can ensure safe living conditions and production for residents. From the point of view of ecological function, the midmontane and alpine zones of the southern slopes of the Qinling Mountains (i.e., >1,350 m) contain relatively well preserved forest and bamboo forest ecosystems, forming an important protective catchment area for the upper reaches of the Han River. Tree leaves (including the layer of fallen leaves deposited each fall) intercept rainfall, thus reducing the erosive power of precipitation. The root systems of innumerable species combine to form a network that serves to loosen and increase permeability of the soil, increasing pore space and enhancing the ability of montane and alpine zones to hold moisture.

The vegetation of the southern slopes of the middle Qinling Mountains also performs important functions in protecting the soil and water of the upper Han River, purifying the air, and moderating the climate. Pollination conducted by the many species of insect functions to ensures the production of fruit trees and many other crops. The existence of many scavengers and microscopic decomposers purifies the environment of the Qinling Mountains and guarantees the cycling of materials within the ecosystem.

2. Biodiversity can also provide mankind with new pharmaceutical resources. The rich variety of animal and plant species on the southern slopes of the central Qinling can provide society with important medicinal resources. For example, the Pacific yew tree (*Taxus brevifolia*) is found at 1,200–1,450 m. Because its branches and trunk are weak, it is worthless as lumber. However, in recent years, scientists have extracted the compound taxol from the needles and bark of the Pacific yew and found it to be an effective treatment for many kinds of cancer (Joyce, 1993).

One can find a few large *Ginkgo biloba* trees scattered about at elevations of 700–800 m on the southern slopes of the Qinling. From the diameter of their trunks, we estimate that they have been growing there for several hundred years. Each autumn, these trees are filled with fruit, indicating that this region is favorable for the growth of ginkgo trees. We now know that a substance can be extracted from the leaves of gingko trees that is useful in treating stroke patients (Del Tredici, 1991; Del Tredici et al., 1992), and this medicine has already been applied clinically in Asia and Europe.

Two other medicinal plants known to grow in the area include the rare tree *Eucommia ulmoides*, which grows at 900–1,000 m, and the Chinese dogwood, *Cornus chinensis,* which grows at 1,200 to 1,450 m. Among useful herbs that can provide important resources, we find *Gastrodia elata*, *Codonopsis pilosula*, and *Fritillaria* spp. growing throughout the entire southern slopes of the Qinling Mountains. Some can grow as high as 3,000 m in the alpine zone. With the continuing development of new biotechnology, we expect that new and effective pharmaceuticals will continually be recognized and developed from these resources.

3. The diversity of species, ecosystems, and landscapes can function as resources for ecotourism. During the preceding 30 years, the timber industry has flourished on the southern slopes of the middle Qinling Mountains, playing an important role in the economic life of Shaanxi. However, beginning several years ago, the adverse effects of the timber industry gradually became apparent. On the one hand, it damaged the native ecosystem of the southern slopes of the Qinling Mountains and its biodiversity; on the other, it threatened the natural environment needed by local residents and, in so doing, functioned to obstruct sustainable development of the economy. For the southern slopes of the central Qinling, a place described in ancient Chinese poetry by the line "The road to Sichuan is harder than the road to Heaven," developing ecotourism is the best way to develop the local economy. The southern slopes of the central Qinling harbor unique landforms and precious and unique life forms, providing outstanding resources for the development of ecotourism. Among these beautiful green mountains and clear rivers, people can search for signs not only of the giant panda but also of the golden takin, golden monkeys, and other rarely seen species. Trekkers and birdwatchers can search for the rare crested ibis, flocks of parrotbills, laughing thrushes, and the rare pheasants, such as Temminck's tragopan, the Satyr tragopan, the Koklass pheasant, the blood pheasant, and Reeves's pheasant. They can climb and explore among the high-altitude ridges and peaks and examine the remains of ancient Quaternary glaciers. These experiences can allow tourists to experience the joy of life in the wilderness while also providing economic and moral support for the protection of the area's biodiversity. Under sound scientific management, ecotourism can develop into an important local revenue source, similar to ecotourism developed in Rwanda based on the opportunity to view mountain gorillas (Vedder, 1989) and lion and elephant ecotourism in Amboseli National Park in Kenya (Western and Henry, 1979). Ecotourism can offer the area the largest and most sustainable income at the smallest cost in terms of ecosystem function.

4. Biodiversity can provide the resources to support people's lives and welfare. This is the aspect of biodiversity that is most generally appreciated and into which society is most willing to invest research effort. In reality, however, this is only the most elementary value of biodiversity. The Qinling's native flora and fauna provide local residents a rich variety of wild food resources, including honey, kiwi fruits, bamboo shoots, mushrooms (including wood ear mushrooms), chestnuts, and wild game, among others. Historically, these wild foods have provided a nutritional buffer in times of war and chaos. At the same time, they have functioned as a storehouse of wild seeds that can be used as a resource to improve the quality of cultivated foods. The raw lacquer extracted from the pines on the Qinling's

southern slopes is of very high quality, can be used without additional processing, is durable, and is heat resistant. The many species of excellent timber, such as Manchurian ash (*Fraxinus mandshurica*), can meet local requirements for housing and construction. Additionally, the area produces numerous species of precious medicinal plants, such as *Magnolia officinalis,* safflower (*Carthamus tinctorius*), *Ophiopogon bodinieri*, the vine *Schisandra chinensis*, the saprophytic herb *Gastrodia elata*, the fungus *Polyporus umbellatus*, *Fritillaria thunbergii*, *Ginkgo biloba*, the ginseng-like herb *Codonopsis pilosula*, and medicinal dogwood (*Cornus officinalis*), all of which are famous for their high quality and quantity.

Important oil-bearing plants include the Chinese lacquer tree (*Toxicodendron vernicifluum*), *Sapium japonicum* and *S. sebiferum*, and horse mulberry (*Coriaria sinica*). Of the 30-odd plants producing aromatic oils, *Cinnamomum camphora*, *Lindera obtusiloba*, *Litsea sericea*, and *Cymbopogon jwarancusa* stand out. In addition to bamboo, there are over 200 species of plants that can provide fiber, including wild ramie (*Boehmeria nivea*), wild hemp (*B. platanifolia*), Chinese alpine rush (*Eulaliopsis binata*), and cogon grass (*Imperata cylindricaor*). The area also has many starch- and sugar-producing plants (such as many species of oak) and plants useful in industrial chemistry, such as rubbers, paints, and dyes.

From the perspective of cultivated plants, the southern slope of the central Qinling Mountains can be planted with both warm-loving and cold-resistant crops. Among these are more than 20 species of grain crops and more than 10 species of cash crops. The latter include tung oil trees (*Vernicia fordii*), Chinese prickly ash (*Zanthoxylum bungeanum*), Chinese tallow tree (*Sapium sebiferum*), and the Chinese windmill palm (*Trachycarpus fortunei*). Fruit trees include mandarin oranges (*Citrus reticulata, C. madurensis*), loquats (*Eriobotrya japonica*), figs (*Ficus carica*), plantains (*Musa basjoo*), pears (*Pyrus* spp.), peaches (*Prunus persica*), apricots (*P. armeniaca*), plums (*P. salicina*), chestnuts (*Castanea bungeana*), and walnuts (*Juglans regia*).

Note

Chapter originally appeared as Chapter 13 in Chinese edition published by Peking University Press, Ltd.

Appendix

We documented 191 species of butterflies within giant panda nature reserves on the southern slopes of the Qinling Mountains, Yang County, Shaanxi Province, and analyzed the elevational distribution and ecozone membership of 179 species. Butterflies represented both Oriental and

Palaearctic ecozones, and they were found intermingled; their distributions were related to specific aspects of Qinling geography. Species of Oriental origin tended to predominate our survey.

References

Collaborating Group of Medical Zoography of China. 1979. 中国药用动物志 (第1册) [Medical zoography of China, volume 1]. Tianjin, China: Tianjin Science and Technology Press.

Collaborating Group of Medical Zoography of China. 1983. 中国药用动物志 (第2册) [Medical zoography of China, volume 2]. Tianjin, China: Tianjin Science and Technology Press.

Del Tredici, P. 1991. "Ginkgos and People: A Thousand Years of Interaction." *Arnoldia* 51:2–15.

Del Tredici, P., H. Ling, and G. Yang. 1992. "The Ginkgos of Tian Mu Shan." *Conservation Biology* 6:202–209.

Department of Geography, Shaanxi Normal University. 1965. 陕西省汉中专区地理志 [Geography of Hanzhong, Shaanxi Province]. Xi'an, China: Shaanxi People's Publishing House.

Editorial Committee of China's Biodiversity: Country Study. 1998. 中国生物多样性国情研究报告 [China's biodiversity: Country study]. Beijing: China Environmental Science Press.

Fang, Z., and S. Z. Gao. 1963. 秦岭太白山南北坡的植被垂直带谱 [Elevational distribution of vegetation on the south and north slopes of Taibai in Qinling Mountain]. *Acta Phytoecologica et Geobotanica Sinica* 1:162–163.

Joyce, C. 1993. "Taxol: Search for a Cancer Drug." *Bioscience* 43:133–136.

Leopold, A. 1996. *A Sand County Almanac*. New York: Oxford University Press.

Nie, S.R. 1981. 陕西自然地理 [Physical geography of Shaanxi]. Xi'an, China: Shaanxi People's Publishing House.

Pan, W. S., Z. S. Gao, Z. Lü, Z. K. Xia, M. D. Zhang, L. L. Ma, G. L. Meng, X. Y. She, X. Z. Liu, H. T. Cui, and F. X. Chen. 1988. 秦岭大熊猫的自然庇护所 [The giant panda's natural refuge in the Qinling Mountains]. Beijing: Peking University Press.

Vedder, A. 1989. "In the Hall of the Mountain King." *Animal Kingdom* 92:31–43.

Wang, Q., and S. N. Jiang. 1988. 四川卧龙自然保护区天牛区系及其起源与演化的研究 [Studies on the cerambycid fauna of Wolong Nature Reserve, Sichuan, and its origin and evolution]. *Entomotaxonomia* 10:131–146.

Western, D., and W. Henry. 1979. "Economics and Conservation in Third World National Parks." *BioScience* 29:414–418.

Zhang, Q. W. 1992. 秦岭南坡植被的区系地理成分分析—以湑水流域为例 [An analysis on the floristic component of the vegetation on the southern side of Qinling Mountains—Taking Xushui Valley as an example]. *Geographical Research* 11:83–92.

Zheng, Z. X., Y. W. Qian, and Y. K. Tan. 1973. 秦岭鸟类志 [Avifauna of Qinling]. Beijing: Science Press.

Zheng, Z. X., R. Z. Zhang, and S. J. Ma. 1959. 中国动物地理区划与中国昆虫地理区划 [Zoogeographical region of China and insect ecogeography]. Beijing: Science Press.

Zhou, Y. 1994. 中国蝶类志 [Atlas of Chinese butterflies]. Zhengzhou, China: Henan Science and Technology Press.

Postscript

The People of the Qinling Study

Wang Dajun

In the summer of 1992 when I was an undergraduate student at Peking University, I had opportunity to take part in a 1-week field biology course in the mountains just outside of Beijing. Professor Pan Wenshi was one of the instructors in that course. Professor Pan had previously been one of the lead researchers on a wild panda study in Wolong Nature Reserve. I was very surprised to learn from him that giant pandas had not become endangered from the mass flowering and die-off of bamboo. This was counter to what I thought I knew about giant pandas; indeed, in the mid-1980s, when I was a junior high school student, I had donated my pocket money to buy bamboo for starving pandas. Pan also told us many stories about wild giant pandas, stories of the interactions between pandas and humans, and stories about researchers working and living in nature. I was mesmerized by these stories and applied to join his new research project as soon as we returned to Beijing. I had never given a thought to being a wildlife biologist before meeting Professor Pan. This kind of experience happened not only to me but to many others as well, including all seven coauthors of this volume. The research project started by Professor Pan Wenshi 28 years ago dramatically changed the lives of many young people in a way that is hard to express.

Pan gained his first experience as a wildlife researcher in 1980 as part of the Wolong panda research project; there he had the opportunity to work with internationally recognized field biologist Dr. George Schaller and also Professor Hu Jinchu, who had previously done work on pandas. The Wolong study was the first systematic scientific field research on wildlife in China and was a radical departure from Pan's earlier research on duck hepatitis virus. The opportunity to work with international scientists and his childhood interest in nature and wildlife resonated with him unlike any previous work. After the field work in Wolong finished in 1983, Pan returned to Peking University and assembled a group of young people to look for a new research site to continue studies of wild pandas. After several months of searching, they decided to set up a research station in the Changqing portion of the Qinling Mountains. The land was owned and managed by the Changqing Timber Company, a commercial forestry company operated by the provincial government.

Pan Wenshi and Lü Zhi, his first graduate student, came to the Qinling in 1984. The project did not start smoothly: tragedy struck when a 21-year-old student, Zeng Zhou, fell off a cliff and died while searching for signs of giant pandas. Zhou was memorialized by his tombstone being set in the place where he lost his life, Sanguanmiao (in what is now Foping Nature Reserve). It is difficult to imagine how Pan overcame the sadness of this event and still managed to do all that was necessary to initiate this study.

Pan's dream was to embark on a long-term research project. For that, the research team needed a base camp in the forest. At first they borrowed rooms or beds from the logging teams. Later the timber company generously provided them some vacant rooms for free. From 1984 until 1997 this field station was constantly bustling as graduate students came to do their research projects and undergraduate students came to attend field biology and ecology courses. A dozen graduate students conducted doctoral or master's degree research, and more than 100 undergraduate students came for field training. All the research and instruction was centered around two related questions: why is the giant panda endangered, and how can we protect it from going extinct? Although adopting modern research methods, we retained the Chinese cultural rules of cooperation and teamwork. All researchers and students focused on the whole of the project, not a separate piece, and we all cooperated to collect data. When it came time to create a thesis or dissertation, each student used a distinct subset of data, according to their interest and capacity.

We interacted more like a family than a research group during the fieldwork. Professor Pan was the "parent," all the students were brothers and sisters, and some eventually became husband and wife (such as Zhu Xiaojian and me). We worked together and shared all the housework, including carrying water, collecting and splitting fuel wood, making fires, cooking, and purchasing supplies. Our camp was located in a place named Shanshuping. We adapted the timber company facility (which had been abandoned in late 1989); it was 13 km from Huayang Town. We used motorcycles for transportation and carrying materials. For at least 10 months each year, the camp was occupied by students immersed in data collection.

Our daily research activities included capturing and radio-collaring pandas, radio tracking, behavioral observations, habitat surveys, population surveys and biodiversity surveys. Because of our many hours in the field with wild pandas, we established a sort of "friendship," or at least a trusting relationship, with some individuals. We could observe these select pandas very closely and even played with some of them. Although there may be

328 Postscript

some danger in researchers becoming too intimate with their study animals, I cannot forget the joy we had living with the pandas.

After 4 years of work, the first monograph about giant pandas in the Qinling was published in 1988: *The Giant Panda's Natural Refuge in the Qinling Mountains.* The coauthors included researchers from a variety of fields, such as biology, ecology, geography, geology, forestry, sociology, and history. The aim was to describe the state of the Qinling Mountains and panda habitat from multiple points of view. A very important conclusion was that, in the past, there was a balanced relationship between human activities and giant panda habitat: sustainable human development could coexist with giant pandas.

However, by the 1980s this balance between humans and pandas was rapidly changing. The rapid expansion of the forest industry was no longer in a sustainable balance and but was tilting away from pandas. By the early 1990s, giant panda habitat had decreased dramatically, and the landscape and its natural refuges no longer seemed able to ensure the survival of the world's most unique bear. In the face of these changes, Professor Pan Wenshi and his team shifted their efforts from research to conservation. We wrote letters appealing for the cessation of logging in Changqing and for the establishment of a nature reserve there. One letter to the state government written by Pan and his students reached the state deputy prime minister , Zhu Rongji, via Professor Pan's channel in the Federation of Returned Overseas Chinese (Pan himself was born in Thailand). Thankfully, a positive reply was received from the highest level of the Chinese government, and logging in this area was terminated in 1994, leading to the establishment of the Changqing National Nature Reserve in 1996. About ¥100 million (about $12 million) was provided by the government and the World Bank to establish this nature reserve and to set up other companies for the placement of the 2,000 workers displaced from the closing of the timber company.

Protection of the Changqing part of the Qinling occurred 4 years before a broader logging ban in the upper and middle Yangtze River watershed area. That logging ban was prompted by a massive flood that destroyed vast areas in southern China in 1998; this natural disaster made the government seriously realize the importance of forest protection for the benefit of people. An indirect consequence of this large-scale logging ban was that it covered the entire range of giant pandas. More nationwide projects followed to speed the process of forest recovery, including the National Nature Forest Protection Project and the Grain for Green Project, which used financial incentives to encourage farmers to abandon crop fields on steep mountain slopes. All these programs aided in the restoration of giant panda habitat. However, the timber companies affected by the 1998 logging ban were not as lucky as the one in Changqing to receive compensation for their workers.

The entire research team was proud of this achievement. Pan then felt content to leave politics behind and begin looking toward assembling all of the acquired information for a new monograph on giant pandas in Qinling. This long-term data set required a rigorous statistical analysis, which was quite a challenge for us. We could not ensure that all of our analyses were in line with the most currently acceptable procedures, but we thought the results, in combination, told a compelling story that answered our initial questions about the status and conservation of wild pandas. The 2001 monograph, *A Chance for Lasting Survival*, was the culmination of this effort.

A 1995 National Geographic Society documentary on our work, featuring Professor Pan, put it in the international spotlight. Later, Pan received several international prizes for his efforts in giant panda conservation and research. By the time these awards were given, however, he had left the Qinling Mountains, as the administration of the newly established nature reserve changed its policy on research and prohibited further radio-collaring of pandas.

In 1996, Pan shifted his research to white-headed langurs (*Presbytis leucocephalus*) in the Guangxi Province of southern China. Pan would later comment that this species shares numerous similarities with giant pandas, as both are black and white in color and both are endangered (considering population numbers, the langur is more endangered than giant pandas). Like the panda study, Pan established a research base and brought in a number of students who investigated the langur's behavior and social structure. As usual, however, he focused more on conservation than on research. He once said, "If we watch species going extinct in front of us, how useful is that we publish 100 or even 1,000 papers by studying them?" In addition to research, he has also helped the local people, improving their tap water system and the local health clinic and establishing biogas pools. In exchange for helping the local residents he has asked them not to damage the vegetation and to protect langurs. The population of langurs in his study area has increased remarkably, from less than 100 in 1996 to about 700 in 2012. It is a pity that white-headed langur populations elsewhere do not have a champion like Pan to help them recover.

In 2002 Professor Pan took on the issue of marine protection, perhaps because he grew up on the coast in southern China. As a result of his efforts, Qinzhou Prefecture in Guangxi abandoned plans for industrial development in favor of protecting the Chinese white dolphin (*Sousa chinensis*). Pan is still actively working for conservation today even past his seventy-fifth birthday.

Pan provided a life-changing experience for all those who worked under him. The authorship of the 2001 volume was complex because we shared in each other's lives as we shared in our research and writing. Professor Pan was the inspiration and editor of the volume, wrote chapter 1, and contributed heavily to chapters 2, 3, and 13 (chapter numbers in this section refer to the Chinese edition). His students took the lead on various chapters and helped each other with the analysis and writing. Many of the coauthors progressed to become researchers and conservationists in their own right, following their experiences in the Qinling Mountains. Among the research family, **Lü Zhi** (author of chapter 10 in the original volume) was the eldest sister. Together with Pan, she initiated the research project and established the research base in Qinling. She did most the field work in the period from 1985 to 1992 and finished her PhD on the primary study of giant panda behavior and social structure. Her work established the basis for all the research in the subsequent studies. During 1992–1995, Lü Zhi left the mountains to work on her postdoctoral project in Dr. Steven O'Brien's genetics lab in the United States. She focused on the genetic diversity of giant pandas using the blood samples collected from wild pandas between 1991 and 1994. She returned to Peking University in 1995 and worked part time for the World Wildlife Fund (WWF): she followed Pan's belief that to change the world, we needed more conservation practice. After 5 years at WWF she went to recharge herself again, with 2 years at Harvard and Yale universities, taking courses and giving lectures. She came back to Peking University in 2003, was made a professor, and began to establish her own team focused on conservation activities in China. Her studies and experiences led her to believe that the root of the conservation crisis was overuse of natural resources for the developing economy and the key to conservation of wildlife and natural systems was to change development to be more in harmony with the natural world. She established a local conservation nongovernmental organization (Shan Shui) in 2007 after 4 years of working with Conservation International, and in 2009 she became the head of her own research center, the Center for Nature and Society, at Peking University.

Zhu Xiaojian (author of chapters 8 and 9) joined the project as an undergraduate in 1992 and finished with a PhD in 1999. Her research focused on behavior, especially the relationship between mother and cub, and the social structure of giant pandas. Her studies, and the increasing international contacts of Professor Pan, allowed her to finish her analysis at the US National Zoo and San Diego Zoo. She returned from the United States in 1998 and remained within Peking University as an instructor. Presently, Xiaojian is in charge of the general ecology curriculum for undergraduate students in Peking University. She also supervises graduate students on their research, including a study of Asiatic black bears in southwest China.

Wang Hao (author of chapters 6 and 7) joined the project in 1994 as an undergraduate. He was primarily interested in population estimation and dynamics. He received his PhD in 1999 and stayed at Peking University as a lecturer. Hao's experience in Qinling led to his role as a technical expert in the Third National Giant Panda Survey (1999), and he has retained that role ever since. Two important contributions he has made have been the establishment of the monitoring system in panda reserves based on transect sign surveys (now adopted by most reserves) and the first biodiversity rapid assessment projects in China.

Long Yu (author of chapters 11 and 12) joined the research team in 1995, focusing on panda nutrition. She devoted much of her time to organizing the 2001 monograph. She joined Professor Pan on the white-headed langur project and white dolphin research and also teaches courses with Professor Pan.

Fu Dali (coauthor of chapter 13) joined the team in 1995. She studied vegetation ecology of the study area. She received her master's degree in 1998 and went to Dartmouth University in 1998 for her doctoral research on plant community ecology. Lü Zhi, Wang Hao, and I had a reunion with her in her home in New Hampshire before Christmas of 2000 to celebrate the birth of her first baby. She now lives in Houston.

Zhou Xin (coauthor of chapter 13) had loved insects since childhood. In 1990, while still in high school, he made his first visit to the Qinling Mountains to collect insects. He made several other visits as an undergraduate and then became a graduate student of the research team in 1997. After receiving his master's degree in 2000, he attended Rutgers University for a PhD project on the genetics of insects and then worked on a postdoctoral project on DNA barcoding of insects in Canada. He came back to China and joined the Beijing Genomics Institution in Shenzhen, becoming a department chief of scientific research, continuing his work on insect genomics and barcoding.

The coauthors of this volume were part of a larger research community working and living together. We all shared good times together in the field and learned from each other's thoughts and experiences. Some left the project before the monograph was compiled, but much of the data and information are attributable to their efforts. They included, in particular, Zhou Zhihua, Lajiacairen, Mao Xiaorong, and Zhang Yingyi. Zhou Zhihua graduated from Peking University with a master's degree on population viability analysis of wild pandas in 1995 and then began to work for the State Forestry Administration. Lajia moved to the United States with his family after working 2 years for the Chinese Wildlife Conservation Association; his master's thesis was on takin ecology. Mao Xiaorong also moved to United States with his family after working several years for an agriculture product company. Lajia and Xiaorong are no longer working on wildlife conservation. Zhang Yingyi made great contributions to the feeding behavior analyses detailed in chapter 11, and then she followed Professor Pan to assist with the white-headed langur project.

Besides the researchers, additional people and institutions played an important role in the project. The leaders of Changqing Timber Company warmly welcomed us and provided much support to the research team during the early years. Even before we started, they had established a patrolling team to protect giant pandas. This patrolling team was essential to the researchers, helping with data collection and logistics. Gao Zhengsheng, the chief manager of the timber company, was a well-educated person, who fully supported our research efforts. Such support was not always appreciated by his colleagues in the timber company. After the patrolling team was dissolved in 1988, some workers continued to help the research team, such as Xiang Bangfa and He Mingxing. Xiang Bangfa was an army veteran who had grown up in the Qinling Mountains, so he was very familiar with the wildlife and had good survival skills that were helpful to the students. Most of my colleagues and I were city folk who became good field researchers because of his help and training. After 13 years of work with researchers, he retired back to his home village in 1997, a 2-hour drive from the reserve.

Xiang Bangfa also brought his then 19-year-old son, Xiang Dingqian, to the project. Like his father, he knew about the forest and wildlife as if it were his backyard, and he became an indispensable partner for the young researchers trying to find their way in the field. He later became an employee of the timber company and then for the nature reserve. He is nationally famous among nature reserve staff for his fieldwork skills, including patrolling and monitoring, handling animals, and wildlife photography. We have recently started working together in research projects in Changqing.

Another man from the timber company, He Mingxing, provided excellent logistical support to the field project for many years, including his work as an accountant. I still miss the delicious noodles he made for the field team. After his retirement in 1997, he went back to his hometown north of the Qinling Mountains. Like Xiang's family, He Mingxing also had a son, He Baisuo, who is still working for the nature reserve, as a department director of conservation and research.

We also hired assistants from the local community to supplement the workers from the timber company. Wang Wenshu was one of these assistants. Wenshu's home was 7 km from our field base; he worked with us from 1994 to 1997 as a "family member." He was the most capable person in the village and good at almost everything related to the forest, including logging, firewood, charcoal, herb collection, and collection of other forest products. He was also skilled at farming, medicinal plants, carpentry, masonry, and cooking. His involvement not only provided much support to the research team but also helped us communicate better with the local people. Wenshu is currently making a good living with his abilities in the village, including hosting visitors by providing accommodations and food; his homemade food is the best in that area.

I (coauthor of chapters 2 and 3 and author of chapters 4 and 5) came to the Qinling Mountains to join Professor Pan's team in 1993. After several years of research and experience there, I came to believe that the giant panda is endangered only because of habitat degradation and fragmentation. I expanded my research to other giant panda range mountains after 1999 and combined all the data into an analysis of panda habitat fragmentation for my 2002 doctoral dissertation. I later began working with Bill McShea from the Smithsonian Institution on a biodiversity project in Sichuan Province. I have used the same model taught to me by Professor Pan of creating a research community with students from Peking University. Beyond research, we have provided training to nature reserve staff and encouraged them to do their own surveys and to build long-term data sets similar to what we created in the Qinling Mountains. Recently, I have expanded my research to the cold and dry Tibetan Plateau, where I am trying to understand ecosystem management issues within a unique Tibetan culture based on Buddhism. I have worked on endemic species of the plateau, including the snow leopard (*Panthera uncia*) and Przewalski's gazelle (*Procapra przewalskii*). During this work it has become clear to me that one of the primary barriers to wildlife conservation in China is the lack of good scientific information. To overcome this barrier, we need long-term field ecology research to answer detailed questions, much as we did in Changqing area. I have started to build a new field project and base camp in Sichuan, hoping to generate good science through years of hard work and international collaboration.

After 9 years of studying the giant panda, I left Changqing with the rest of the panda research team. I returned there with students in 2008 to work with the reserve staff on biodiversity monitoring. We found a high diversity of mammals and birds after 15 years of habitat restoration. An especially remarkable event was a leopard photographed by one of our remote trip cameras; this was evidence to me of a very healthy ecosystem in Changqing. A second event, however, made us feel both happy and worried: the development of government-sponsored tourism for economic growth. We were happy that local people could improve their living conditions following the economic decline after the 1994 logging ban, but we are worried about controlling the development such that it will avoid damaging wildlife habitat. On July 28, 2010, a sudden flood, greater than any in the past 50 years, swept through the area and destroyed recently constructed roads and buildings for the tourism trade, and some people lost their lives. It reminded us that the power of nature sometimes adjusts the balance between wildlife and humans in a sudden and aggressive way. This may be how the natural refuge of giant pandas (and other wildlife) has historically been "reset" from human disturbance. More studies on the effect of natural disturbances in the recovery process are warranted.

Professor Pan once mentioned to me that he was unsure if we did the right thing by helping stop the logging in Changqing because Huayang town was in economic decline. He visited Changqing again in April 2010 after an 11-year absence. A farmer recognized him and came to shake his hand; he

334   Postscript

told Pan that local people have to come to appreciate our work, which led to restoration of the forest. This, of course, made Pan very happy—he could now rest assured that he had done the right thing 20 years before, as both nature and people clearly benefitted from our research and conservation work in the Qinling Mountains.

Index

Page numbers followed by f and t indicate figures and tables.

340   Index